THE QUAILS, PARTRIDGES, AND FRANCOLINS OF THE WORLD

The Quails Partridges, and Francolins of the World

PAUL A. JOHNSGARD

University of Nebraska–Lincoln

Colour plates of paintings by
MAJOR HENRY JONES
from the collection owned by
THE ZOOLOGICAL SOCIETY OF LONDON

Oxford   New York   Tokyo
OXFORD UNIVERSITY PRESS
1988

Oxford University Press, Walton Street, Oxford OX2 6DP
Oxford New York Toronto
Delhi Bombay Calcutta Madras Karachi
Petaling Jaya Singapore Hong Kong Tokyo
Nairobi Dar es Salaam Cape Town
Melbourne Auckland
and associated companies in
Beirut Berlin Ibadan Nicosia

Oxford is a trade mark of Oxford University Press

*Published in the United States
by Oxford University Press, New York*

© (text) Paul A. Johnsgard, 1988
© (Plates 1–21, 23–42, 44–62, 64–98, 100–105, 107–27) Zoological Society of London, 1988
© (Plates 22, 43, and 99) Mark Marcuson, 1988
© (Plates 63 and 106) Timothy Greenwood, 1988

*All rights reserved. No part of this publication may be reproduced,
stored in a retrieval system, or transmitted, in any form or by any means,
electronic, mechanical, photocopying, recording, or otherwise, without
the prior permission of Oxford University Press*

British Library Cataloguing in Publication Data
Johnsgard, Paul A.
*The quails, partridges and francolins
of the world.*
1. Partridges   2. Quails
I. Title
598'.617   QL696.G27
ISBN 0–19–857193–3

Library of Congress Cataloging in Publication Data
Johnsgard, Paul A.
The quails, partridges, and francolins of the world.
1. Quails.   2. Partridges.   3. Francolins.   I. Jones,
Henry, 1838–1921.   II. Zoological Society of London.
III. Title.
QL696.G27J65   1988   598'.617   87–31366
ISBN 0–19–857193–3

*Set by
Wyvern Typesetting Ltd, Bristol
Printed in Great Britain by
The Alden Press, Oxford*

Preface and acknowledgements

During the spring of 1985, as I was completing the proof-reading on my *Pheasants of the world*, I began considering the desirability of writing a companion volume for this work, to include all of the relatively closely related quail- and partridge-like members of the galliform group. These generally small and often somewhat inconspicuous game birds have heretofore been almost totally neglected as to comprehensive treatment, in part no doubt because of the very large number of known taxonomic forms. However, the group merges almost imperceptibly with the typical pheasants, and as such provides a necessary foundation for any attempt to understand the evolution of the larger, more diversified, and more elaborately plumaged pheasant-like birds. I had already extensively summarized the biologies of most of the New World quails (Johnsgard 1973), and it seemed that perhaps a comparable treatise on all of the world's partridge- and quail-like birds would be a useful reference for ornithologists, aviculturists, sportsmen, and other bird enthusiasts.

With this in mind, while visiting England early in 1985, I reviewed and indexed all of the water colours of the galliform birds that were done by Major Henry Jones at the turn of the century (now part of the Zoological Society of London's library collections), and which had provided the primary illustrative basis for my pheasant volume. This survey revealed that Major Jones had made paintings of about 90 per cent of the 134 currently recognized species of partridges, quails, and francolins, and that only 12 additional species (most of which were described after Jones's death) would have to be painted to provide a complete illustrative treatment of this large and important component of the gallinaceous birds, which have not been monographed in this century. I asked the Zoological Society of London to consider undertaking a collaborative project involving their permission for the use of these plates and my preparation of an accompanying text, and suggested that they approach Oxford University Press as to their possible interest in publishing such a volume. Both organizations greeted the idea with enthusiasm. Thus I immediately began serious work on the manuscript, which continued until late 1986. My literature search was essentially terminated with 1986 titles; only a few additional changes could be incorporated into the text thereafter.

Inasmuch as there are nearly three times as many species of partridge- and quail-like birds as there are of pheasants, it was necessary that some substantial condensations be made so as to keep the two volumes reasonably comparable in size. First, a reduced section on comparative biology was deemed appropriate, inasmuch as at least some of the comparative chapters of my recent pheasant book might apply equally well to the present group. Secondly, less attention in the individual species accounts has been paid to subspecies (e.g. identification keys have not been extended to include subspecies, nor have subspecies been mapped individually). Furthermore, all of the descriptive species accounts have been kept to a substantially shorter average length, which was achieved in part by a late decision to excise from the manuscript detailed plumage descriptions for each species. This was done reluctantly, but in the belief that species identification should be possible using the keys, head profile drawings, comments on in-hand identification, and the comprehensive set of colour plates. It is hoped that the pages thereby saved have been used for information of more general contemporary interest, and that those people requiring such detailed descriptions will in general know where in the literature to seek them out. The excellent although now totally antiquated catalogue by W. R. Ogilvie-Grant (1893) of the gallinaceous birds then present in the British Museum (Natural History) has particularly useful plumage descriptions, and additionally is invaluable for determining synonymies. Descriptions of North and Central American forms may be found in Ridgway and Freidmann (1946), and shorter descriptions of Neotropical species are to be found in Blake (1977). Very late in the preparation of the manuscript Volume II of *The Birds of Africa* (Urban et al. 1986) appeared, which also provides brief plumage descriptions of the African species. Its authors suggested the taxonomic mergers of many previously accepted races of *Francolinus*, and altered the taxonomic species sequence (as well as making some suggested vernacular name changes) that had been previously recommended by Hall (1963) for that large genus. I have noted these recommendations in the text wherever appropriate, but the manuscript was too close to completion upon receipt of this volume to justify altering my sequence of species to conform fully with it.

To provide complete illustrative coverage in colour

for all of the currently recognized species of quails, partridges, and francolins it was necessary to have five paintings done specifically for this book, and I am especially indebted to Timothy Greenwood, of Orpington, Kent (Plates 63 and 106), and Mark Marcuson, of Lincoln, Nebraska (Plates 22, 43, and 99), whose splendid efforts have allowed me to reach this goal. All of the line illustrations, including the distribution maps, are my own.

Museum work for this book was performed in, or specimens were borrowed from, the British Museum (Natural History), Tring; the US National Museum of Natural History, Washington, DC; the American Museum of Natural History, New York; and the Field Museum of Natural History, Chicago; to all these institutions and their staffs I offer my sincere appreciation.

Several other persons and institutions helped me in various ways, particularly in facilitating my library research. I am especially indebted to Scott Johnsgard, Drs Gordon and Marion Sauer, the University of Kansas libraries, the Van Tyne Memorial Library of the University of Michigan, and the Edward Grey Institute Library of Oxford University for their valuable help toward this end. Reginald Fish, Librarian of the Zoological Society of London, provided great assistance in repeatedly making the paintings by Major Jones available to me, and in arranging for the necessary photographic work associated with colour plate preparation. Geoffrey Welch provided photographs of *Francolinus ochropectus* for use in illustration preparation, and David Rimlinger also loaned me photographs of live birds and made available unpublished avicultural data from the files of the San Diego Zoo. Mr Rimlinger and Kenneth Fink also provided enthusiastic assistance during various trips to the San Diego Zoo to photograph and make observations on live birds. I also greatly appreciate access to the private avicultural collections of Keith Howman and Mickey Ollsen, and the help of Dr G. W. H. Davison in sending me an advance copy of an in-press manuscript as well as encouraging me to write this book. Mr Kenneth Fink kindly read an early draft of the manuscript.

Lincoln, Nebraska   P. A. J.
1987

Contents

List of colour plates	xi
List of figures	xiii
List of distribution maps	xiv
Introduction	xvii

I COMPARATIVE BIOLOGY

1 Taxonomy, phylogeny, and zoogeography	3
2 Reproductive biology	10
3 Ecology and population dynamics	16
4 Ontogenetic growth and development	25
5 Adult vocalizations and non-vocal behaviour	31

II SPECIES ACCOUNTS

1 **Subfamily Odontophorinae** (New World Quails)	41
Key to the genera of Odontophorinae	41
2 Genus *Dendrortyx* Gould 1844	42
Key to the species of *Dendrortyx*	42
Long-tailed tree-quail	42
Bearded tree-quail	43
Buffy-crowned tree-quail	45
3 Genus *Philortyx* Gould 1846	47
Barred quail	47
4 Genus *Oreortyx* Baird 1858	49
Mountain quail	49
5 Genus *Callipepla* Wagler 1832	51
Key to the species of *Callipepla*	51
Scaled quail	51
Elegant quail	53
Gambel's quail	54
California quail	55
6 Genus *Colinus* Gouldfuss 1820	60
Key to the species of *Colinus*	60
Northern bobwhite	61
Black-throated bobwhite	64
Spot-bellied bobwhite	66
Crested bobwhite	68
7 Genus *Odontophorus* Vieillot 1816	71
Key to the species of *Odontophorus*	71
Marbled wood-quail	72
Spot-winged wood-quail	73
Rufous-fronted wood-quail	74
Chestnut wood-quail	76
Dark-backed wood-quail	77
Rufous-breasted wood-quail	78
Black-breasted wood-quail	78

	Gorgeted wood-quail	79
	Tacarcuna wood-quail	80
	Venezuelan wood-quail	82
	Black-fronted wood-quail	83
	Stripe-faced wood-quail	83
	Starred wood-quail	84
	Spotted wood-quail	85
8	Genus *Dactylortyx* Ogilvie-Grant 1893	87
	Singing quail	87
9	Genus *Cyrtonyx* Gould 1844	90
	Key to the species of *Cyrtonyx*	90
	Montezuma quail	90
	Ocellated quail	92
10	Genus *Rhynchortyx* Ogilvie-Grant 1893	94
	Tawny-faced quail	94
11	**Tribe Perdicini** (Old World Quails, Partridges, and Francolins)	96
	Key to the genera of Perdicini	96
12	Genus *Lerwa* Hodgson 1837	98
	Snow partridge	98
13	Genus *Tetraophasis* Elliot 1871	100
	Key to the species of *Tetraophasis*	100
	Verreaux's monal-partridge	100
	Szechenyi's monal-partridge	101
14	Genus *Tetraogallus* J. E. Gray 1832	103
	Key to the species of *Tetraogallus*	103
	Caucasian snowcock	103
	Caspian snowcock	106
	Himalayan snowcock	107
	Tibetan snowcock	108
	Altai snowcock	109
15	Genus *Alectoris* Kaup 1829	111
	Key to the species of *Alectoris*	111
	Arabian red-legged partridge	111
	Przhevalski's rock partridge	114
	Rock partridge	114
	Chukar partridge	116
	Philby's rock partridge	118
	Barbary partridge	119
	Red-legged partridge	120
16	Genus *Ammoperdix* Gould 1851	122
	Key to the species of *Ammoperdix*	122
	See-see partridge	122
	Sand partridge	124
17	Genus *Francolinus* Stephens 1819	126
	Key to the species of *Francolinus*	126
	Black francolin	129
	Painted francolin	132
	Chinese francolin	133
	Red-necked francolin	134
	Swainson's francolin	137
	Grey-breasted francolin	138
	Yellow-necked francolin	139
	Erckel's francolin	141

Pale-bellied (Djibouti) francolin	142
Chestnut-naped francolin	143
Jackson's francolin	144
Handsome francolin	145
Cameroon Mountain francolin	146
Swierstra's francolin	147
Ahanta francolin	147
Scaly francolin	148
Grey-striped francolin	150
Double-spurred francolin	150
Yellow-billed (Heuglin) francolin	152
Clapperton's francolin	153
Hildebrandt's francolin	154
Natal francolin	155
Hartlaub's francolin	156
Harwood's francolin	157
Red-billed francolin	158
Cape francolin	159
Crested francolin	160
Ring-necked francolin	161
Montane red-winged (moorland) francolin	162
Shelley's francolin	164
Grey-winged francolin	166
Acacia (Orange River) francolin	167
Levaillant's (Red-winged) francolin	169
Finsch's francolin	170
Coqui francolin	171
Schlegel's banded francolin	172
White-throated francolin	173
Latham's forest francolin	174
Nahan's francolin	175
Grey francolin	176
Swamp francolin	178
18 Genus *Perdix* Brisson 1760	180
Key to the species of *Perdix*	180
Grey partridge	180
Daurian partridge	183
Tibetan partridge	184
19 Genus *Rhizothera* G. R. Gray 1841	186
Long-billed wood-partridge	186
20 Genus *Margaroperdix* Reichenbach 1853	188
Madagascar partridge	188
21 Genus *Melanoperdix* Jerdon 1864	190
Black wood-partridge	190
22 Genus *Coturnix* Bonnaterre 1791	192
Key to the species of *Coturnix*	192
European migratory quail	193
Asian migratory quail	195
African harlequin quail	197
Black-breasted quail	199
Stubble quail	200
Brown quail	201
Asian blue quail	202
African blue quail	202
23 Genus *Anurophasis* van Oort 1910	206

Contents

Snow Mountain quail	206
24 Genus *Perdicula* Hodgson 1837	208
Key to the special of *Perdicula*	208
Jungle bush-quail	208
Rock bush-quail	210
Painted bush-quail	211
Manipur bush-quail	212
25 Genus *Ophrysia* Bonaparte 1856	214
Indian mountain-quail	214
26 Genus *Arborophila* Hodgson 1837	215
Key to the species of *Arborophila*	215
Necklaced hill-partridge	216
Sichuan hill-partridge	218
Red-breasted hill-partridge	219
Collared hill-partridge	221
Rufous-throated hill-partridge	221
White-cheeked hill-partridge	223
White-throated hill-partridge	224
White-eared hill-partridge	225
Javan hill-partridge	225
Bare-throated hill-partridge	226
Brown-breasted hill-partridge	228
Orange-necked hill-partridge	229
Chestnut-headed hill-partridge	230
Borneo hill-partridge	231
Red-billed hill-partridge	231
Scaly-breasted hill-partridge	232
27 Genus *Caloperdix* Blyth 1861	235
Ferruginous wood-partridge	235
28 Genus *Haematortyx* Sharpe 1879	237
Crimson-headed wood-partridge	237
29 Genus *Rollulus* Bonnaterre 1791	239
Crested wood-partridge	239
30 Genus *Ptilopachus* Swainson 1837	241
Stone partridge	241
31 Genus *Bambusicola* Gould 1862	244
Key to the species of *Bambusicola*	244
Mountain bamboo-partridge	244
Chinese bamboo-partridge	245
32 Genus *Galloperdix* Blyth 1844	247
Key to the species of *Galloperdix*	247
Red spurfowl	247
Painted spurfowl	248
Ceylon spurfowl	249
Bibliography	251
Index	261

Colour plates

Plates 1–32 appear between pp. 44 and 45; Plates 33–63 between pp. 108 and 109; Plates 64–95 between pp. 156 and 157; Plates 96–127 between pp. 220 and 221.

1. Long-tailed tree-quail
2. Bearded tree-quail
3. Buffy-crowned tree-quail
4. Barred quail
5. Mountain quail
6. Scaled quail
7. Elegant quail
8. Gambel's quail
9. California quail
10. Northern bobwhite
11. Black-throated bobwhite
12. Spot-bellied bobwhite
13. Crested bobwhite
14. Marbled wood-quail
15. Spot-winged wood-quail
16. Rufous-fronted wood-quail
17. Chestnut wood-quail
18. Dark-backed wood-quail
19. Rufous-breasted wood-quail
20. Black-breasted wood-quail
21. Gorgeted wood-quail
22. Black-fronted wood-quail and Tacarcuna wood-quail
23. Venezuelan wood-quail
24. Stripe-faced wood-quail
25. Starred wood-quail
26. Spotted wood-quail
27. Singing quail
28. Montezuma quail
29. Ocellated quail
30. Tawny-faced quail
31. Verreaux's monal-partridge
32. Szechenyi's monal-partridge
33. Snow partridge
34. Caucasian snowcock
35. Caspian snowcock
36. Himalayan snowcock
37. Tibetan snowcock
38. Altai snowcock
39. Arabian red-legged partridge
40. Przhevalski's rock partridge
41. Rock partridge
42. Chukar partridge
43. Philby's rock partridge
44. Barbary partridge
45. Red-legged partridge
46. See-see partridge
47. Sand partridge
48. Black francolin
49. Painted francolin
50. Chinese francolin
51. Red-necked francolin
52. Swainson's francolin
53. Grey-breasted francolin
54. Yellow-necked francolin
55. Erckel's francolin
56. Chestnut-naped francolin
57. Jackson's francolin
58. Cameroon Mountain francolin
59. Ahanta francolin
60. Scaly francolin
61. Grey-striped francolin
62. Double-spurred francolin
63. Pale-bellied, handsome, Hartlaub's, Swierstra's, and Nahan's francolins
64. Yellow-billed francolin
65. Clapperton's francolin
66. Hildebrandt's francolin
67. Natal francolin
68. Harwood's francolin
69. Red-billed francolin
70. Cape francolin
71. Crested francolin
72. Ring-necked francolin
73. Montane red-winged francolin
74. Shelley's francolin
75. Grey-winged francolin
76. Acacia francolin
77. Levaillant's red-winged francolin
78. Finsch's francolin
79. Coqui francolin
80. Schlegel's banded francolin

81. White-throated francolin
82. Latham's forest francolin
83. Grey francolin
84. Swamp francolin
85. Grey partridge
86. Daurian partridge
87. Tibetan partridge
88. Long-billed wood-partridge
89. Madagascar partridge
90. Black wood-partridge
91. European migratory quail
92. Asian migratory quail
93. African harlequin quail
94. Black-breasted quail
95. Stubble quail
96. Brown quail
97. Asian blue quail
98. African blue quail
99. Snow Mountain quail
100. Jungle bush-quail
101. Rock bush-quail
102. Painted bush-quail
103. Manipur bush-quail
104. Indian mountain-quail
105. Necklaced hill-partridge
106. Sichuan, orange-necked, and chestnut-headed hill-partridges
107. Red-breasted hill-partridge
108. Collared hill-partridge
109. Rufous-throated hill-partridge
110. White-cheeked hill-partridge
111. White-throated hill-partridge
112. White-eared hill-partridge
113. Javan hill-partridge
114. Bare-throated hill-partridge
115. Brown-breasted hill-partridge
116. Borneo hill-partridge
117. Red-billed hill-partridge
118. Scaly-breasted hill-partridge
119. Ferruginous wood-partridge
120. Crimson-headed wood-partridge
121. Crested wood-partridge
122. Stone partridge
123. Mountain bamboo-partridge
124. Chinese bamboo-partridge
125. Red spurfowl
126. Painted spurfowl
127. Ceylon spurfowl

Figures

1. Dendrogram of postulated generic phylogeny of the Odontophorinae — 7
2. Taxon-density map of the Odontophorinae — 7
3. Dendrogram of postulated generic phylogeny of the Perdicini — 8
4. Taxon-density map of the Perdicini — 9
5. Diagram identifying major morphological features of galliforms, and method of numbering remiges — 26
6. Postures associated with social behaviour in *Alectoris* — 33
7. Postures associated with social behaviour in the grey partridge — 34
8. Postures associated with egocentric behaviour in the Asian blue quail — 35
9. Postures associated with egocentric and social behaviour in the Asian blue quail — 36
10. Postures associated with high-intensity agonistic and sexual display in the northern bobwhite and Asian blue quail — 37
11. Adult male plumage variation in the northern bobwhite — 63
12. Adult male plumage variation in the spot-bellied and crested bobwhites — 67
13. Adult plumage variation in the rufous-fronted, dark-backed, chestnut, and rufous-breasted wood-quails — 76
14. Adult plumage variation in the black-breasted, black-fronted, Tacarcuna, gorgeted, and Venezuelan wood-quails — 81
15. Adult plumage variation among species of *Tetraogallus* — 105
16. Adult plumage variation among species of *Alectoris* — 113
17. Adult plumage variation in the red-necked francolin — 136
18. Adult plumage variation in the montane red-winged, grey-winged, and Shelley's francolins — 164
19. Adult plumage variation in the Levaillant's red-winged and acacia francolins — 168
20. Adult plumage variation in the Sichuan, necklaced, and rufous-throated hill-partridges — 218
21. Adult plumage variation in the white-cheeked, white-eared, white-throated, collared, red-breasted, and Javan hill-partridges — 220
22. Adult plumage variation in the brown-breasted and bare-throated hill-partridges — 227
23. Adult plumage variation in the scaly-breasted hill-partridge — 233

Distribution maps

1. Distribution of the tree-quails (*Dendrortyx*) — 42
2. Distribution of the mountain and barred quails — 47
3. Distribution of the scaled and elegant quails — 52
4. Distribution of the Gambel's quail — 55
5. Distribution of the California quail — 57
6. Distribution of the northern bobwhite — 61
7. Distribution of the black-throated, spot-bellied, and crested bobwhites — 65
8. Distribution of the marbled and spot-winged wood-quails — 72
9. Distribution of the rufous-fronted, chestnut, dark-backed, and rufous-breasted wood-quails — 75
10. Distribution of the black-breasted, gorgeted, Tacarcuna, Venezuelan, and black-fronted wood-quails, the singing quail, and North American introduced distribution of the grey partridge — 80
11. Distribution of the stripe-faced, starred, and spotted wood-quails — 84
12. Distribution of the Montezuma, ocellated, and tawny-faced quail, and North American introduced distribution of the chukar partridge — 90
13. Distribution of the snow partridge — 99
14. Distribution of the Verreaux's and Szechenyi's monal-partridges — 100
15. Distribution of the snowcocks (*Tetraogallus*) — 104
16. Distribution of the red-legged partridges (*Alectoris*) — 112
17. Distribution of the see-see and sand partridges — 123
18. Distribution of the spotted group of francolins — 130
19. Distribution of the bare-throated group of francolins — 135
20. Distribution of the montane group of francolins — 141
21. Distribution of the scaly group of francolins — 148
22. Distribution of the vermiculated group of francolins — 151
23. Distribution of the striated group of francolins — 160
24. Distribution of the red-winged group of francolins — 163
25. Distribution of the red-tailed group of francolins — 171
26. Distribution of the forest francolins — 174
27. Distribution of the grey and swamp francolins — 176
28. Distribution of the typical partridges (*Perdix*) — 181
29. Distribution of the long-billed and black wood-partridges — 186
30. Distribution of the European and Asian migratory quails — 193
31. Distribution of the African harlequin, black-breasted, and brown quails — 197
32. Distribution of the stubble and Snow Mountain quails — 200
33. Distribution of the Asian and African blue quails, and of the Madagascar partridge — 203
34. Distribution of the jungle and rock bush-quails — 209
35. Distribution of the painted and Manipur bush-quails — 212
36. Distribution of the necklaced, Sichuan, red-breasted, and collared hill-partridges — 216
37. Distribution of the rufous-throated, white-cheeked, white-throated, and white-eared hill-partridges — 222

38. Distribution of the Javan, bare-throated, and brown-breasted hill-partridges   226
39. Distribution of the orange-necked, chestnut-headed, Borneo, red-billed, and hill-partridges   229
40. Distribution of the scaly-breasted hill-partridge and ferruginous wood-partridge   232
41. Distribution of the crimson-headed and crested wood-partridges   237
42. Distribution of the stone partridge, bamboo-partridges (*Bambusicola*), and spurfowl (*Galloperdix*)   242

Introduction

The partridges, quails, and francolins comprise a nearly cosmopolitan group of 134 species of small to relatively large gallinaceous birds that are mostly non-migratory and terrestrially adapted. A short hind toe is always present, and at least all the forward-pointed toes have sturdy, blunt claws that are highly suitable for scratching and digging, as well as a short and blunt bill that serves for crushing hard food items. The wings are relatively short and rounded, making take-offs and flying energetically expensive, and protracted flights unusual except for a few migratory or nomadic species. Attempted escape from danger by crouching or running rather than by flight is frequent in many species. Most species are diurnal and all are primarily vegetarians, with well-developed crops for temporary food storage as well as muscular gizzards adapted to grinding hard food materials with the aid of grit. Grit consumption and dust-bathing are characteristic of all.

Most species are monogamous, with pair-bonds sometimes persisting between breeding seasons. Males of perhaps all species utter advertisement calls during the breeding season, proclaiming their own sexual availability, the holding of a breeding territory, or both. Most species are single-brooded, although renesting following loss of a clutch is frequent, and incubation of separate clutches by each of the pair members has sometimes been observed. The carrying of materials for nest construction is apparently lacking in all. The nests vary from simple, unlined scrapes on the ground to well-concealed and sometimes completely domed-over nests hidden within heavy vegetation. The average clutch size is highly variable but often fairly large (normal range about 2–16), and the eggs are sometimes spotted but not concealingly patterned. Except in unusual cases of double-brooding or death of the female, males typically do not assist in incubation, which begins with the laying of the last egg, resulting in synchronous hatching. The young are sometimes fed directly by the parents, but more often the parents direct their young to food items with specific behaviour and 'tidbitting' calls. The young are precocial, are initially largely insectivorous, and are often able to fly within two or three weeks of hatching. Probably all species become sexually mature within a year, and some may mature in less than three months. Parental distraction behaviour by injury-feigning is typical and may be performed by either sex. Participation by the male in brood care and defence is frequent but not universal.

Families of adults and their grown young form the basis of post-breeding social units, and in some tropical species these units are often small coveys, consisting of single pairs and their immature offspring. However, larger aggregations of less closely related individuals occur in some situations, especially where flocking is promoted by the distribution of limited resources such as safe roosting sites or water supplies. In a few temperate-latitude species the group size may also be influenced by the optimum number of individuals required to form outwardly facing circular clusters of roosting birds that maintain maximum individual visibility while simultaneously sharing body heat. Relatively high annual mortality rates appear to be typical of the entire group, which are counterbalanced by a rapid attainment of sexual maturity and a moderately high reproductive potential.

These birds collectively differ from the pheasants in being generally smaller, having less adult sexual dimorphism in weight and plumage, lacking facial combs and wattles in males, and having little or no iridescence in their adult plumages. Additionally, their tails are typically shorter and less ornately patterned than those of pheasants, and usually are flat and rounded rather than vaulted and/or elongated. All of these differences appear to be related to the monogamous mating systems typical of the quail- and partridge-like birds, as compared with the usual non-monogamous mating patterns of pheasants, and the associated differential evolutionary influences of sexual selection on male signalling devices in the two groups. However, some partridge genera such as *Galloperdix* approach the typical pheasants in their tail shape and the extent of their bare facial skin or male plumage iridescence. Similarly, some genera of pheasants have short and partridge-like tails as well as little or no male plumage iridescence or adult sexual dimorphism, making it difficult if not impossible to draw an entirely satisfactory morphological or behavioural distinction between these two groups.

Although it may thus be convenient to characterize collectively the partridges, quails, and francolins for purposes of such comparisons, it is equally important to realize that the New World quails (subfamily Odontophorinae) are probably much less closely related to the Old World partridges, quails, and francolins (tribe Perdicini) than the latter group is to the pheasants (tribe Phasianini), in spite of the very great similarities in the morphology and biology of the two former groups. Indeed, these two taxonomic groups of gallinaceous birds offer a significant opportunity to appreciate and investigate the influences of parallel and convergent evolution occurring independently in the eastern and western hemispheres, which is one of the primary advantages of including both of them in a single comprehensive volume.

A second reason for providing this comprehensive volume lies in the fact that no collective survey of all the partridge, quail, and francolin species of the world has been undertaken in this century, and none has ever attempted to illustrate them all in colour or to map their distributions. The very large number of forms to be considered is a probable reason for this, and dealing equitably with all of them represented a daunting if not impossible challenge. A few of the American and European quails and partridges are major game birds, and indeed the technical literature of these species is at least as large as that of various North American and European grouse and the ring-necked pheasant. However, many of the species of quails and partridges are native to tropical forests of Africa, Asia, and Latin America, where they are of little economic importance and usually have effectively eluded the few biologists who have attempted to investigate their natural histories. At this point in history we are losing many of these tropical woodland and forest habitats, together with their associated biota, sometimes even before we have been able to learn much about these organisms beyond the mere evidence of their existence. Indeed, one species (the Indian mountain-quail) is already certainly extinct, and several other species or races have become rare or endangered as a probable result of habitat loss and degradation in recent years. At minimum this book provides a means of readily identifying all of the extant species of the world's quails, partridges, and francolins, and attempts to document the sometimes rudimentary state of our knowledge concerning their distributions and their biological attributes. As such, it should also be useful in promoting future research on the group and in pointing out species or areas of special interest and concern to conservationists and ecologists.

A few comments on the appropriate English application of the vernacular names 'quail', 'partridge', and 'francolin' might be in order. The English noun 'quail' has traditionally been used for the very small, short-tailed, and migratory species of *Coturnix*, which in Old French and Middle English were known as *quaille*. (The verb 'quail', to cower or shrink with fear, was probably more recently derived from the huddling or crouching behaviour of these birds when threatened.) The English term 'partridge', however, has traditionally been applied to the somewhat larger and longer-tailed perdicine species such as the grey partridge, which derives from the Middle English *pertriche*. Both the vernacular and scientific names of the grey partridge are derived from the Latin word *perdix*. In Greek mythology Perdix was the nephew of Daedalus, and whose soul was carried off by (or who was transformed into) a partridge-like bird after being pushed off a cliff by his jealous uncle. The term 'francolin', which serves not only as an English vernacular name but also is the basis for the generic name *Francolinus*, is derived from a diminuative of the Portuguese *franqo*, a hen. It also derives directly from the Italian *francolino*, the vernacular name for the black francolin.

In North America the term 'quail' has unfortunately been applied to virtually all of the smaller native odontophorine species, although in both size and relative tail length they might better have been called partridges. An even more suitable term would have been 'colins', which is derived from *zolin*, the Aztecan-language (Nahuatl) name for these birds and is also the basis for the generic name *Colinus*. Perhaps using a contrived term such as 'toothed-quail', in reference to the somewhat serrated lower mandible condition of all the Odontophorini, might provide a taxonomically more acceptable if less elegant alternative vernacular name for these New World forms.

I have tried to follow both tradition and common sense as much as possible in dealing with these problems, applying the vernacular name 'francolin' to all species of *Francolinus* and 'partridge' to those of *Perdix* and a few other similar genera that traditionally have been called partridges. The term 'quail' has been applied in this book to all species of *Coturnix*, the apparently closely related genus Margaroperdix, and to various smaller odontophorine species that lack suitable alternative English names. However, whenever possible distinctive compound names, such as 'tree-quail' for *Dendrortyx* and 'wood-quail' for *Odontophorus*, have been adopted for the larger odontophorine forms. Similarly, distinctive hyphenated names have been employed whenever appropriate for distinguishing various perdicine genera, such as 'bush-quail' for *Perdicula* and 'mountain-quail' for *Ophrysia*, always avoiding exact semantic overlap with any New World counterparts. Likewise, all species of the genus *Arborophila* have been consistently referred to as 'hill-partridges', and several similar and apparently closely related monotypic genera of tropical woodland partridges (*Caloperdix, Haematortyx, Rollulus, Rhizothera,* and *Melanoperdix*) have been collectively designated as 'wood-partridges'. Finally, the name 'harlequin quail' has been preferentially bestowed on the African *Coturnix delegorquei*, where it seems to have priority of use, rather than applying it to the New World *Cyrtonyx montezumae* for which the eminently suitable alternative name 'Montezuma quail' is available.

I have drawn range maps for all but one species (the extinct *Ophrysia*), and in this regard I have generally relied upon reliable published sources, especially *An atlas of speciation in African non-passerine birds* (Snow 1978) in the case of African species and the *Manual of Neotropical birds* (Blake 1977) for Neotropical forms. For those species that seemingly lack available weight data altogether I have provided estimated adult weights, using the cube of the average wing length measurement and establishing regression lines of estimated weights based on separate known-weight samples from the Odontophorinae and Perdicini. In general I believe that these estimated average weights are accurate to about 10–15 per cent, and thus may prove useful for ecological or other modelling purposes in the absence of actual weights. Calculated egg weights are from those of Schönwetter (1967) or, if necessary, were calculated by me according to methods described earlier (Johnsgard 1973), which have proven to give results virtually identical to those of Schönwetter.

I · COMPARATIVE BIOLOGY

1 · Taxonomy, phylogeny, and zoogeography

A brief review of the systematic classifications of the quails, partridges, and francolins might begin with the landmark synopsis of Ogilvie-Grant (1893), who provided the first essentially complete and modern descriptive taxonomic overview of the group. He recognized 59 genera of Phasianidae, of which 11 were New World forms here included in the Odontophorinae. Another 23 were in the Old World group here considered to comprise the Perdicinae, while the rest consisted of typical pheasants, guineafowl, and turkeys. In a later (1896–7) less technical summary of his work, he proposed the following classification of the Phasianidae:

Family Phasianidae
  Subfamily Perdicinae (21 genera, 115 species)
  Subfamily Phasianinae (29 genera, 111 species)
  Subfamily Odontophorinae (11 genera, 47 species)

In this classification not only were the turkeys and guineafowl included in the pheasant subfamily Phasianinae, but additionally four genera that are now generally considered to be part of the perdicine assemblage (*Ptilopachys*, *Bambusicola*, *Galloperdix*, and *Ophrysia*) were regarded as part of the pheasant group. The seemingly excessive number of species recognized by Ogilvie-Grant resulted from his species-level recognition of numerous forms that in more modern revisions have generally been considered as comprising geographic races.

Not long after Ogilvie-Grant's work, Beebe (1914) proposed a classification of the pheasant-like galliforms that was based on their patterns of tail moult. He observed that the Old World partridges and their relatives might be separated from the pheasants proper by their centrifugal (medial to lateral) tail moult, while the typical pheasants exhibit other kinds of moulting patterns, most commonly centripetal (lateral to medial). Although he was unable to examine specimens of all the Old World partridge, quail, and francolin genera, he became satisfied that this character prevails in the entire perdicine group. He thereby defined his subfamily Perdicinae as including all the just-mentioned Old World forms plus two additional pheasant-like genera (*Tragopan* and *Ithaginis*) that he found to conform to the perdicine pattern of tail moult. A complete survey of the Perdicinae to test the validity of this trait has seemingly never been done, although Marien (1951) noted that at least in the post-juvenal moult of *Perdix perdix* and *Ammoperdix griseogularis* the moulting pattern diverges from a strictly centrifugal one.

In his monograph of the pheasants, Delacour (1951) considered all of the quails (including the New World quails), partridges, and pheasants to be part of a single subfamily Phasianinae, but he later (1961) indicated that the New World quails might be given tribal (Odontophorini) recognition within this subfamily. Contrary to Beebe, Delacour included both *Ithaginis* and *Tragopan* with the more typical pheasants in his tribe Phasianini.

Primarily on the basis of mensural osteological evidence, Verheyen (1956) suggested the following classification of the quail- and partridge-like phasianids:

Subfamily Perdicinae
  Tribe Odontophorini: all New World forms
  Tribe Coturnicini: including *Coturnix* (*sensu lato*) and probably *Perdicula* and *Ophrysia* (*Margaroperdix* not studied)
  Tribe Perdicini: remaining Old World perdicines

Verheyen based his tribal distinction between the Old World *Coturnix*-like forms and the more typical partridge-like assemblage of perdicines largely on proportional differences that would appear to be of questionable phyletic significance, and such separation of these groups is not supported by recently available DNA hybridization data (Sibley and Ahlquist 1985). Interestingly, however, *Coturnix* has not been recorded as hybridizing with any of the pheasants or partridge-like genera, nor with any of the New World quails (Gray 1958).

Potential taxonomic distinction of the New World odontophorines from the Old World quails has always been unequivocal; serrations on the lower mandible's cutting edges typify all the extant species of the former. A thorough osteological study of the New World quails was performed by Holman (1961), who concluded that this group is sufficiently unique anatomically as to warrant full family status ('Odontophoridae'). This degree of taxonomic separation of the odontophorines from the outwardly very similar Old World galliforms might at first seem questionable, but recently Sibley and Ahlquist (1985, 1986) have concluded on the basis of preliminary DNA hybridization studies that the New World quails may be more isolated from the Old World forms than are, for example, the turkeys and the guineafowl. These authors estimated that the New World quails may have diverged from phasianoid progenitors as long as about 70 million years ago, while the perdicines may have separated from the typical phasianids about 50 million years ago. However, Helm-Bychowski and Wilson (1986) have estimated divergence times for various galliform genera that are substantially more recent than these estimates, including a separation of *Alectoris* from *Gallus* of at least eight but probably less than 20 million years ago. Their restriction mapping data for nuclear DNA placed *Alectoris* within the general cluster of typical pheasant genera (and *Meleagris*), but unfortunately they did not analyse any representative New World quails. However, they believe that phyletic splitting within the Phasianidae probably occurred less than 31 million years ago, and that the Phasianoidea originated some time earlier, but no more than about 41 million years ago. Finally, Stock and Bunch (1982) concluded from comparative karyotype studies that the New World quails, grouse, turkeys, and Old World quails and pheasant-like birds are all closely related, and might therefore be allied in a single subfamily Phasianinae.

Probably the most comprehensive comparative analysis of the phasianids for taxonomic purposes was that of Hudson *et al.* (1966). They examined the wing and leg musculature of 35 genera of galliform birds, including six genera of odontophorines and five genera of perdicines. On the basis of their analysis they proposed the following classification of the Phasianidae:

Subfamily Odontophorinae: New World quails
Subfamily Phasianinae: pheasants (excepting peafowl) plus Old World perdicines
Subfamily Meleagridinae: turkeys
Subfamily Numidinae: guineafowl
Subfamily Pavoninae: peafowl (*Pavo* only genus studies)

Only a few recent taxonomic synopses have been sufficiently complete as to provide a basis for comprehensive taxonomic comparisons of all the extant genera and species. In Table 1 abbreviated versions of representative classifications presented by Peters (1934) and by Howard and Moore (1980) are compared with that which I have here adopted. My arrangement is based in part on reviews and assessments of all the above-mentioned studies, but at the same time attempts to retain as many of the traditional generic limits and sequences as seemed feasible. It thus takes into account the accumulating evidence that the New World quails are well isolated phyletically from the Old World perdicine forms, while reasons for comparable taxonomic separation of the Old World pheasants from the perdicine assemblage are substantially weaker, warranting in my opinion no more than tribal distinction. I also believe that no confirmatory evidence is presently at hand to warrant tribal distinction of the *Coturnix*-like forms from the rest of the perdicines. The table also reflects the adoption of a slightly broader concept of the genus than that of Peters, and indicates the considerable number of allospecies (well-differentiated taxa that owing to their essentially allopatric distributions cannot be certainly determined as to whether they represent full species or only well-defined subspecies) that exist in these groups and that following tradition are here considered full species. As suggested by Amadon (1966), the probable superspecies affiliation of forms considered here to comprise allospecies is indicated by parentheses in the listings of taxa following the generic diagnoses.

Finally, as a means of comparing some of the taxonomic characteristics of the New World quails with the Old World perdicine and phasianine groups, a comparison of various taxonomic statistics among these three groups is presented in Table 2. This table uses the taxonomic categories I established earlier (Johnsgard 1986) for the Phasianini, and those used here for the Odontophorinae and Perdicini. (A similar tabular comparison of biological traits among these groups was presented in the earlier book.) It may be seen that the percentage of monotypic genera in the pheasants is substantially higher than for the other two groups. This situation is a probable reflection of the higher incidence of non-monogamous mating systems in pheasants, and the resultant pressures of sexual selection for accelerated evolution of effective male secondary sexual characteristics. Subjective tendencies often promote the taxonomic 'inflation' of such conspicuous characteristics to the generic diagnosis level, even when they possibly merely serve as species-specific reproductive signals. There is a corresponding lower number of species per genus in the pheasants, but a similar array of subspecies per species exists among the three groups. Another structural reflection of the high incidence of non-monogamous mating in pheasants is the nearly universal occurrence of male tarsal spurs, a trait that is found in only about half of the perdicine species and in none of the Odontophorinae. Even the seemingly monogamous species of pheasants typically have tarsal spurs, as do many apparently monogamous (but possibly strongly territorial) species of perdicines. Spurs among the Galliformes in general tend to be better developed in larger and tropical species, and although more widespread among polygamous forms they may

Table 1. Taxonomic characteristics of the Odontophorinae, Perdicini, and Phasianini

	Odontophorinae	Perdicini	Phasianini
Total subspecies	145	302	150
Total species	31	103	49
Total genera	9	21	17
Average subspecies/species	4.7	2.9	3.1
Average species/genus	3.4	4.9	2.9
Monotypic species (%)	39	36	57
Monotypic genera (%)	44	48	35
Species with male spurs (%)	0	58	96
Non-monogamous species (%)	0	0 (or few)	72[2] 78[3]
Male/female weight ratio	1.05[1] ($n=11$)	1.07[2] ($n=34$)	1.33[2] ($n=30$)
Ecology of species (%)			
Open-country	13	54	10
Forest edge	23	9	36
Forest or woodland	64	37	53
Ecology of monotypic genera			
Open-country	*Philortyx*	*Lerwa* *Margaroperdix* *Anurophasis* *Ptilopachus* *Ophrysia*	*Ithaginis* *Catreus*
Forest edge	*Oreortyx*		*Pucrasia*
Forest or woodland	*Dactylortyx* *Rhynchortyx*	*Rhizothera* *Melanoperdix* *Caloperdix* *Haematortyx* *Rollulus*	*Afropavo* *Rheinartia* *Argusianus*

[1] Johnsgard (1973).
[2] Davison (1985).
[3] Johnsgard (1986).

well have originally evolved in monogamous forms in conjunction with intraspecific fighting behaviour (Davison 1985).

The ecological analysis in Table 1 suggests that in the Old World the grassland and savannah-like habitats are occupied primarily by perdicine forms, with the forest and forest-edge habitats being more commonly exploited by the phasianines. The majority of New World quail species are associated with forests or woodlands, with only the apparently more specialized types having colonized forest-edge, grassland, and savannah habitats. All three groups have a moderate number of monotypic genera, and these forms are ecologically distributed in an apparently unpredictable manner.

Phylogeny and zoogeography

As has been suggested earlier (Johnsgard 1973), it seems highly probable that the origin of the Odontophorinae occurred in the forests of Central America. The genera *Odontophorus* and *Dendrortyx* provide convenient examples of seemingly generalized New World quails that approximate the presumed ancestral odontophorine types. Radiation into forest-edge and non-forested environments of North America was evidently a later development. A phyletic dendrogram of the probable diversification of the New World quails has been already published; a redrawn depiction of this evolutionary scenario is here presented (Fig. 1). A taxon-density map of the Odontophorinae at the generic level (Fig. 2) provides a useful zoogeographic overview of the basic distributional patterns illustrated by this group, and a reflection of its predominantly Central American orientation. The highest geographic incidence of taxonomic diversity occurs in the general vicinity of southern Mexico and Guatemala, where eight species representing five genera (*Dendrortyx*, *Colinus*, *Odontophorus*, *Cyrtonyx*, and *Dactylortyx*) occur in close proximity. Not far to the south of this area the monotypic genus *Rhynchortyx* also may be found,

Table 2. Comparative taxonomies of the quails, partridges, and francolins

Peters (1934)	Howard and Moore (1980)	Present study
Subfamily Odontophorinae	Subfamily Odontophorinae	Subfamily Odontophorinae
Dendrortyx (4)	*Dendrortyx* (3)	*Dendrortyx* (3)
Oreortyx (1)	*Oreortyx* (1)	*Philortyx* (1)
Callipepla (1)	*Callipepla* (1)	*Oreortyx* (1)
Lophortyx (5)	*Lophortyx* (3)	*Callipepla* (4)
Colinus (4)	*Philortyx* (1)	*Colinus* (4/3[1])
Odontophorus (16)	*Colinus* (4)	*Odontophorus* (14/10)
Dactylortyx (1)	*Odontophorus* (14)	*Dactylortyx* (1)
Cyrtonyx (3)	*Dactylortyx* (1)	*Cyrtonyx* (2/1)
Rhynchortyx (1)	*Cyrtonyx* (3)	*Rhynchortyx* (1)
	Rhynchortyx (1)	
Subfamily Phasianinae	Subfamily Phasianinae	Subfamily Phasianinae
		Tribe Perdicini
Lerwa (1)	*Lerwa* (1)	*Lerwa* (1)
Ammoperdix (2)	*Ammoperdix* (2)	*Tetraophasis* (2/1)
Tetraogallus (5)	*Tetraogallus* (5)	*Tetraogallus* (5)
Tetraophasis (2)	*Tetraophasis* (2)	*Alectoris* (7/6)
Alectoris (4)	*Alectoris* (7)	*Ammoperdix* (2/1)
Anurophasis (1)	*Anurophasis* (1)	*Francolinus* (41/40)
Francolinus (34)	*Francolinus* (41)	*Perdix* (3)
Pternistis (4)	*Perdix* (3)	*Rhizothera* (1)
Perdix (3)	*Rhizothera* (1)	*Margaroperdix* (1)
Rhizothera (1)	*Margaroperdix* (1)	*Melanoperdix* (1)
Margaroperdix (1)	*Melanoperdix* (1)	*Coturnix* (8/5)
Melanoperdix (1)	*Coturnix* (5)	*Anurophasis* (1)
Coturnix (5)	*Synoicus* (1)	*Perdicula* (4)
Synoicus (1)	*Excalfactoria* (2)	*Ophrysia* (1)
Excalfactoria (2)	*Perdicula* (2)	*Arborophila* (16/13)
Perdicula (1)	*Arborophila* (15)	*Caloperdix* (1)
Cryptoplectron (2)	*Tropicoperdix* (3)	*Haematortyx* (1)
Arborophila (15)	*Caloperdix* (1)	*Rollulus* (1)
Tropicoperdix (3)	*Haematortyx* (1)	*Ptilopachus* (1)
Caloperdix (1)	*Rollulus* (1)	*Bambusicola* (2)
Haematortyx (1)	*Ptilopachus* (1)	*Galloperdix* (3)
Rollulus (1)	*Bambusicola* (2)	
Ptilopachus (1)	*Galloperdix* (3)	
Bambusicola (2)	*Ophrysia* (1)	
Galloperdix (3)		
Ophrysia (1)		
Total genera:		
36	34	30
Total species:		
134	135	134/118[1]
Monotypic genera:		
17	15	14
Monotypic species:		
54	54	50/35[1]

[1] Second number applies if forms designated in text as allospecies are reduced to subspecies rank.

while the remaining genera (*Philortyx*, *Callipepla*, and *Oreortyx*) occur variably farther to the north, and mainly are associated with edge or open-country habitats.

Gutierrez (1980) suggested with respect to the North American mountain quail and California quail that *Oreortyx* is much older and probably evolved in a brush or forest-like environment, while the California quail probably differentiated from its nearest living relative (the Gambel's quail) much more recently and in a more open environment. The four open-country *Callipepla* species that now occur in western North America must be regarded as among the most specialized of the New World quails. They

Taxonomy, phylogeny, and zoogeography 7

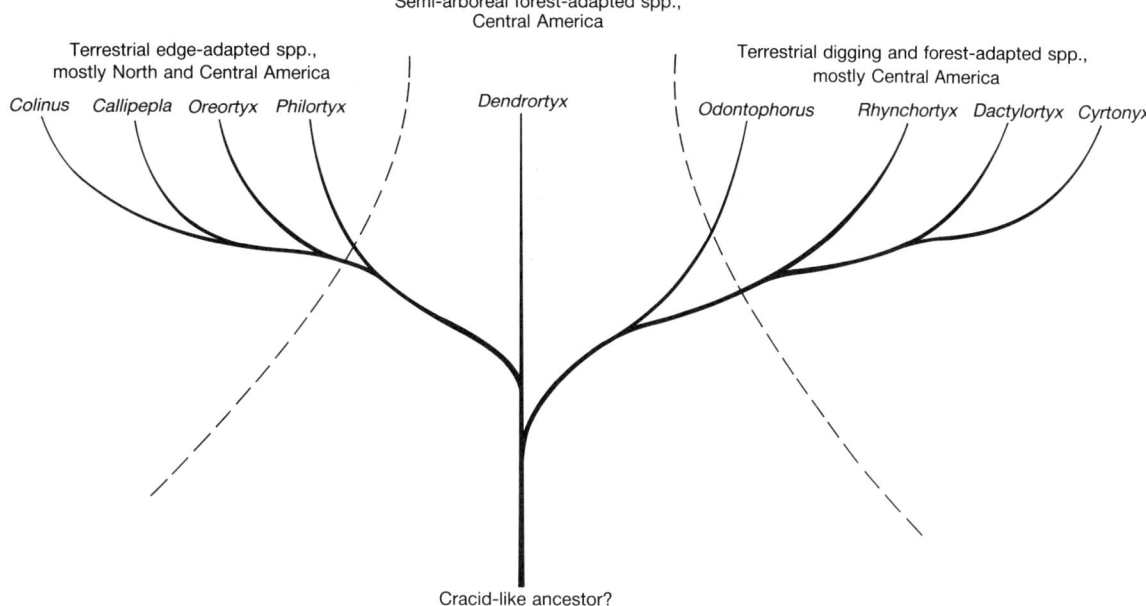

Fig. 1. Dendrogram of postulated generic phylogeny of the Odontophorinae. After Johnsgard (1973).

Fig. 2. Taxon-density map of the Odontophorinae, showing number of indigenous genera present regionally.

presumably initially differentiated during late Tertiary times as arid habitats in the south-western region expanded, and perhaps developed their current level of speciation during two cycles of glaciation (Hubbard 1974). Similarly, the high degree of geographic variability and speciation that is apparent in the *Colinus* populations of South, Central, and North America must reflect a fairly recent pattern of geographic isolation and diversification associated with Pleistocene effects on these forest-edge or savannah-adapted forms.

The early evolution of the Old World phasianids is somewhat more difficult to visualize, but has been discussed in my recent volume on pheasants (Johnsgard 1986). As suggested there, it seems most probable that the perdicine assemblage emerged from a root cluster of primitive galliforms of which the megapodes are the nearest modern Old World representatives. This early perdicine radiation most probably occurred in the tropical and forested parts of the Old World, perhaps in the general area of what now comprises the Malay Peninsula and Greater Sundas. The genus *Arborophila*, like *Odontophorus* in the New World, provides what appears to be a nuclear assemblage of rather unspecialized and forest-adapted forms whose ranges are now centred in south-east Asia, and from which many of the other extant Oriental and Palearctic partridge genera seem to represent possible morphologically derived types. No early phasianid fossils are available to support a postulated tropical Asian origin of the Phasianidae; the earliest reputed phasianid fossils would seem to be *Argillipes*, *Coturnipes*, and *Percolinus* from the

Lower Eocene of England (Harrison and Walker 1977). There are also some Eocene and Lower Oligocene galliform fossils from France, but these are not phasianoid (C. Mourer-Chauviré, cited in Helm-Bychowski and Wilson 1986). However, the present Austral-Asian distribution of the megapodes tallies well with an imagined tropical Asian origin of the phasianoid group.

A large assemblage of closely related perdicine forms is provided by the genus *Francolinus*, which is now distributionally centred in Africa. Hall (1963) postulated an Oligocene origin of *Francolinus*, but did not suggest any possible ancestral stock or speculate as to the group's possible geographic origin. However, the francolin species now occurring in Asia are shown in Hall's phyletic dendrogram as resulting from very early branching points, which might support a possible Asian origin for the genus. Interestingly, Olson (1974) described a fossil guineafowl (*Telecrex*) from the Upper Eocene of Mongolia, supporting a possible Asian origin of that distinctive phasianoid group, and Crowe (1978) suggested that the African guineafowl may have been derived from an Asian savannah-dwelling francolin-like ancestor. More recently, Crowe and Crowe (1985) proposed that *Francolinus* evolved from a quail-like ancestor colonizing Asia during Pliocene times.

On this basis and from current zoogeographic evidence one might postulate that the earliest (possibly Eocene) ancestral phasianids were tropical forest-adapted, partridge-like birds of south-east Asia that gradually moved both westward and northward into Europe, Africa, and central Asia, progressively colonizing seasonally dry or temperate forests, forest edges, savannahs, grasslands, and ultimately even alpine habitats. These ancestral types eventually radiated into such basic structural and ecological subgroups as are now represented by the specialized alpine- or montane-adapted forms (*Lerwa*, *Tetraogallus*, and *Tetraophasis*), the primarily grassland, semi-desert, and savannah or forest-edge forms (*Alectoris*, *Francolinus*, *Perdix*, *Coturnix*, and their relatives), and those forms that evidently remained largely forest- or scrub-adapted and primarily still occur in the apparently ancestral areas of south-east Asia (*Arborophila*, *Perdicula*, and several related monotypic genera). A few contemporary perdicine genera (especially *Galloperdix*) seem to provide direct morphological links with various groups of extant pheasants, particularly the junglefowl and gallopheasants, suggesting that the typical pheasants also radiated, perhaps polyphyletically, from various partridge-like ancestral stocks. However, although the alpine-adapted partridges would morphologically appear to have much in common with some of the rather atypical montane pheasants (such as *Ithaginis* and *Tragopan*), such outward similarities must certainly result at least in part from convergence.

Primarily on the basis of available morphological and zoogeographic criteria, a highly speculative

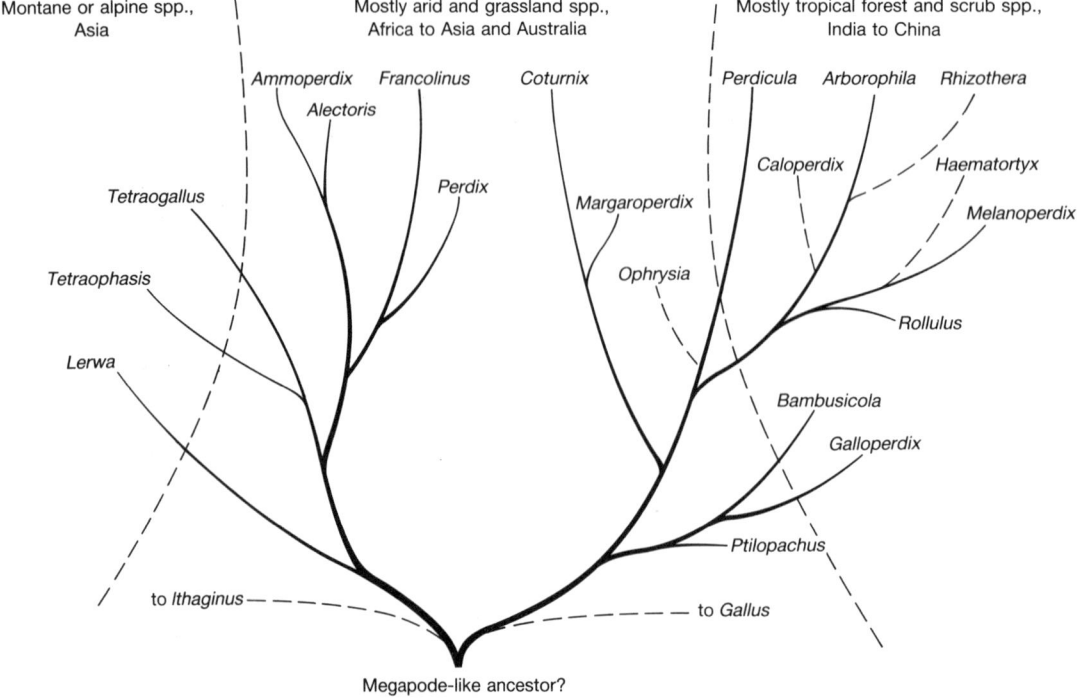

Fig. 3. Dendrogram of postulated generic phylogeny of the Perdicini.

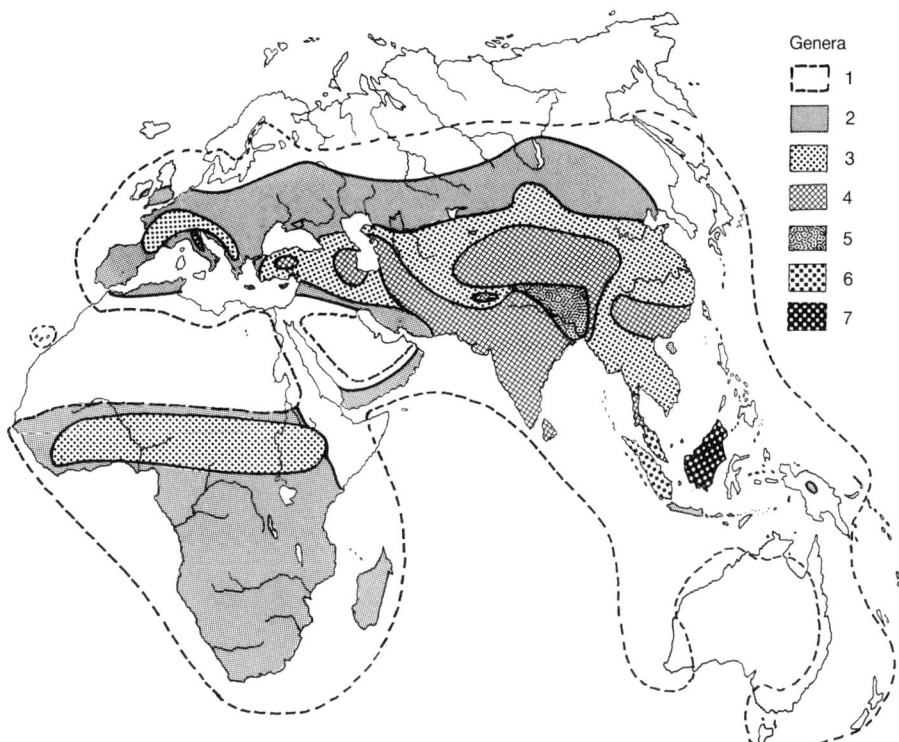

Fig. 4. Taxon-density map of the Perdicini, showing number of indigenous genera present regionally.

phyletic dendrogram of the Perdicini genera has been constructed (Fig. 3). It is obvious that, given the limited comparative anatomical, biochemical, and behavioural evidence currently available, any such proposed perdicine phylogeny is fraught with uncertainties at this point. It may be hoped that such promising new techniques as DNA hybridization and sequencing will help to reduce the number of apparent potential alternative phyletic pathways that now seemingly exist in this very large and morphologically relatively homogeneous group, and additionally provide an objective test of their level of phyletic separation from the pheasants.

A species-density map prepared for the Phasianini (Johnsgard 1986) indicates that the richest contemporary diversity of typical pheasants occurs along the southern edge of the central Himalayas and in the vicinity of the Sunda Platform (Greater Sundas, Borneo, and Malay Peninsula). A comparable map of species-density in the Perdicini would show a somewhat similar centre of zoogeographic distribution in south-east Asia, as well as a very strong African representation that is primarily a reflection of extremely rich adaptive radiation of francolins on that continent. Species-level diversity within the genera *Francolinus* and *Coturnix* is particularly great in east-central and southern Africa, with Kenya, Tanzania, and Zambia, for example, all exhibiting very high levels (12 or more species) of such diversity.

A taxon-density map for the Perdicini, which like that of Fig. 3 is drawn at the generic level, is presented here (Fig. 4). This approach provides a seemingly clearer index to fundamental zoogeographic tendencies in the Perdicini than would a comparable map based on species-level taxonomy. In particular it tends to emphasize the role of south-east Asia and especially the Sunda Platform as a probable primary centre for early perdicine evolution. For example, Borneo has more (seven) genera than are found in any other area of comparable size; indeed five of the seven Bornean genera (*Caloperdix, Rhizothera, Haematortyx, Melanoperdix,* and *Rollulus*) are monotypic, and all of these are endemic to the general vicinity fo the Sunda Platform. By comparison, continental Africa exhibits a much lower level of generic diversity, and has only a single monotypic and endemic genus (*Ptilopachus*).

2 · Reproductive biology

I have earlier (Johnsgard 1973) provided a fairly complete discussion of the reproductive biology of all the native North American quails and two representative Old World partridge genera (*Perdix* and *Alectoris*), to which the interested reader may be directed. In this brief summary I intend only to make such additional comments as seem desirable to bring that discussion up to date and to incorporate data from some additional species appropriate to a world coverage.

Pair-forming and pair-maintaining behaviour

Detailed studies on pair-formation activities have been performed on grey and rock partridges as well as on several species of New World quails. The study of Jenkins (1961a) on the grey partridge is particularly illuminating, and may be used as an example of what is probably the usual partridge pair-formation pattern. He found that pairing among birds of the same covey occurred only when a pair of the previous season was re-established, or when the bird taking the active role in male selection was the female. In the latter case the male chosen was usually a bird that had joined the covey the previous autumn, and was never the father or brother of the female. This apparent anti-incest adaptation may be fairly widespread in quails and partridges, but its behavioural mechanism is still uncertain. Observations on several species that a degree of initial hostility during early social interactions may be important in facilitating pair-bonding (Goodwin 1953; Raitt 1960; Ellis and Stokes 1966) may perhaps be a significant factor in this mechanism. Jenkins (1961) found that males breeding for the first time typically left their coveys and began to associate with other groups, performing aggressive or courtship behaviour toward members of those coveys. When a male displayed to a hen she would sometimes join him, and the incipient pair would then depart the covey together. Pairs that were thus formed by January or early February were usually maintained, while birds pairing relatively late in the mating season often changed their mates several times before forming a permanent union. Occasionally males deserted their mates during incubation, but rarely did so after the eggs had hatched. Excess unmated males often would associate with established pairs, attempted to display to the paired females, and occasionally successfully displaced the original males. When both members of a grey partridge pair survived to a second year they usually re-established their pair-bond with little apparent courtship.

Extensive studies on the pairing behaviour of the rock partridge have been performed by Menzdorf (1975a, 1976b; and in Cramp and Simmons 1980), indicating that pair-bonding in that species is usually monogamous and is often long-term. Some observed instances of successive bigamy by males as well as re-mating by widowed females has been observed, as well as some partner-switching (Pepin 1984). In this species pairing begins within the flocks during winter, which begin breaking down into smaller units about February. A similar pairing pattern has been reported in the red-legged partridge (Jenkins 1957), where both males and females have sometimes been found associating with two apparent mates. In these species of *Alectoris*, which essentially lack sexual dimorphism, initial male displays toward females are clearly aggressive, and only when the female fails to perform hostile responses in return are sexually oriented displays such as food-calling and mock-feeding ('tidbitting' behaviour) performed. At least in captivity, dominant males are prone to complete their pairing much later than more subordinate ones, apparently because of the former's overly aggressive tendencies toward females.

Although less is known of pair-bonding tendencies in the European migratory quail, it is evidently not exclusively monogamous, and pair-bonding may vary (in part depending on the local sex ratio) to include simultaneous bigamy by the male, successive polygamy, and promiscuity (Cramp and Simmons 1980). In captivity this species is promiscuously active throughout the entire year (Wetherbee 1961). In the Japanese quail it has been found that females prefer to mate with first cousins over siblings, thus balancing the genetic disadvantages of extreme inbreeding with limited outbreeding (Bateson 1982). The fairly closely related Australian stubble quail has been reported to form strong, apparently lifelong monogamous pair-bonds under captive conditions (Cruise 1966). Similarly, strong monogamous pair-bonding tendencies have been attributed to such other congeners as the harlequin quail, the black-breasted quail, and the Asian blue quail, all of which are more strongly sexually dimorphic in plumages than are stubble quails or migratory quails. Harrison

(1965) suggested that the conspicuous and apparently complex repertoire of postures and calls in the Asian blue quail might be related to its need for a complex communication system associated with a relatively long pair-bond, at least by comparison with that of the migratory quails. A similarly highly complex system of vocalizations appear to be typical of such tropical New World forms as *Dendrortyx*, *Dactylortyx*, and *Odontophorus*, all of which likewise presumably have extended pair-bonds but lack the marked plumage sexual dimorphism typical of most *Coturnix* species.

It is also possible that the considerable plumage dimorphism [as occurs in these tiny *Coturnix* (and *Perdicula*) forms, as well as in most of the smaller New World quails and in some small francolins, but which seems to be less prevalent in larger odontophorines and perdicines] may be related to maximizing visual signals associated with achieving effective male social dominance (and thus obtaining mates and holding territories) in those species where tarsal spurs are lacking and weight or strength differences among individual males may be insignificant. In larger and longer-lived species the higher probability of individual male variations in tarsal spur lengths and fighting abilities associated with age, experience, and strength differences may alone be sufficient to facilitate pair-bonding by the most fit individuals regardless of their possible degree of plumage dimorphism.

Field studies on individually marked California quail have been performed by Raitt (1960). From autumn until March these birds occur in coveys or flocks, and their behaviour then is largely directed toward maintenance of the social group. However, during March pair-forming behaviour began within the flock that Raitt had under observation, which numbered about 40 birds. Much hostile behaviour occurred among the males, and nearly complete segregation into pairs developed by early May. Pairing was associated with increased overall levels of within-flock hostility, and with chases by males toward other males, by females toward females, and by males toward females, as well as male-to-male and female-to-female fights. Copulation behaviour apparently began after preliminary pair-bonding had occurred, which appeared to be a very gradual process and apparently was not accompanied by any specific 'pairing' displays.

As in the grey partridge, a substantial number of unmated excess male California quail typically remain available after the available females have completed pairing; these birds persistently announce their location and sexual availability by uttering loud advertisement calls. This loud 'cow' call, which has close equivalents in the other species of *Callipepla*, is functionally identical to the 'bobwhite' call of unmated *Colinus* males. In both of these genera mated males rarely if ever utter loud advertisement calls. An equivalent unmated male advertisement call also occurs in *Oreortyx* and probably also in *Philortyx* and *Cyrtonyx* among the New World quails; possibly comparable calls by unmated males have been reported in other genera as well. On the other hand, loud duetting by paired birds has been observed in such forest-adapted genera as *Dendrortyx*, *Odontophorus*, and *Dactylortyx*, and this behaviour probably serves an important role in keeping pairs or families in contact with one another under the conditions of reduced visibility typical for such forest-dwelling birds.

The New World quails evidently lack calls that function purely for territorial advertisement and population dispersal of paired birds; in this way the odontophorines might differ from at least such Old World perdicines as *Alectoris*, where such a possible spacing function has been attributed to the 'rally call' of the chukar partridge (Williams and Stokes 1965). The corresponding call of the rock partridge is also evidently primarily used for territorial demarcation (Menzdorf 1977), and calls of probably comparable function also occur in the Barbary and red-legged partridges. Breeding-season territoriality has been attributed to all four of these *Alectoris* species, and in the rock partridge the territories of individual pairs reportedly are typically close together but lack contiguous boundaries (Menzdorf 1976c). Probable territoriality (at least, the existence of non-overlapping home ranges by paired males) has likewise been reported in the grey-winged and the Levaillant's red-winged francolin (Mentis and Bigalke 1980). The European migratory quail has also been described as being highly territorial (with males having individual calling territories of about 1.7–3.7 acres) during the breeding season, the territories being regularly patrolled and fiercely defended. The male's advertising call or 'song' is only infrequently uttered after pair-formation has occurred. Thus, territorial advertisement in this highly migratory species may well serve mainly as an efficient means of attracting mobile females, rather than primarily promoting population dispersal of breeding pairs.

Age of sexual maturity and incidence of non-breeding

Evidently virtually all species of quails, partridges, and francolins mature during their first year, based on such evidence as the early attainment of adult plumages, the apparent absence of significant numbers of non-breeding females in wild populations, and the regular breeding under captive conditions of hand-raised birds less than a year old. I have obtained fertile eggs from captive scaled quail within 160 days of hatching, and from only slightly older individuals of all the six species of *Colinus* and *Callipepla* that I have maintained in captivity. Highly precocial breeding is also known to occur in the genus *Coturnix*; Disney (1978) reported that some individuals of Australian stubble quail breed at the age of four months under natural conditions, and it is possible that some individuals of the European migratory quail may even breed the summer of hatching (Glutz von Blotzheim 1973). Indeed, hand-reared females of the European migratory quail have been known to lay fertile eggs when as young as 31 days, and males mature sexually at 28 days. One hand-reared and released female was known to have established a nest at only 67 days of age (Wetherbee 1961). Similar precocial sexual development has been commonly observed in hand-reared Asian blue quail. Possible breeding by free-living francolins no more than five months old has recently been reported for two African francolins (Mentis and Bigalke 1980).

If any examples of deferred sexual maturity were to be present anywhere in the entire group they might be expected to occur in the alpine-adapted genus *Tetraogallus*. In this group the period to sexual maturity is generally believed to be only one year (Cramp and Simmons 1980), but it has recently been observed that hand-raised females of the Himalayan snowcock usually do not breed until their second year (Stiver 1984).

During highly unfavourable breeding seasons extensive non-breeding by part of the adult quail or partridge population may occur, at least in some temperate-zone forms. In North America this has been reported for mountain, Gambel's, California, and scaled quail, and may also occur in northern bobwhites. Probable non-breeding during unusually dry years has also been reported in chukar and Barbary partridges among Eurasian species.

Number of nesting or renesting attempts per year

Multiple nesting, or at least renesting, has been observed with considerable frequency in both the New World quails and the Old World perdicines. Indeed, renesting following the failure of the initial nesting effort is probably the rule rather than the exception for most species of both groups. In the European migratory quail as many as two renesting attempts may be made following initial egg loss (Cramp and Simmons 1980). In the grey partridge second nesting efforts are common, but the clutch size of such repeat nestings averages about 30 per cent smaller than that of initial clutches (Lack 1947; Blank and Ash 1960). Leopold (1977) reported that as many as three nesting attempts per season may be made by the California quail, while Roseberry and Klimstra (1983) estimated that on their study area most female northern bobwhites must likewise have laid two or three clutches per season in order to have achieved their estimated productivity levels relative to observed nesting success rates.

Actual double-brooding (the successful hatching of two broods during a single breeding season) is much less well documented, and the available evidence on this subject is summarized in Table 3. As may be seen, some examples of apparent double-brooding exist for both the New World quails and the Old World partridges, but in many cases the evidence is only circumstantial. Possible double-brooding has been reported for at least two species of *Coturnix* (Cramp and Simmons 1980; Frith and Carpenter 1980). Circumstantial evidence for double-brooding has also been reported for two African francolins (Mentis and Bigalke 1980). Leopold (1977) believed

Table 3. Proven or presumptive cases of double-brooding in quails and partridges

Gambel's quail	Indirect evidence of double-brooding by wild birds during one year (Gullion 1956)
California quail	Circumstantial evidence of double-brooding in wild birds (McMillan 1964); two cases of double-brooding in semi-captive birds (Francis 1965)
Northern bobwhite	Nineteen cases of double-brooding in free-living and penned wild birds observed during 25 years (Stanford 1972a)
Rock partridge	Double-brooding observed by R. Meinertzhagen in a captive pair (Goodwin 1953)
Chukar partridge	One report of double-brooding in captivity (*British Birds* **17**, 315); double-brooding suspected in wild birds (Mackie and Buechner 1963)
Barbary partridge	A captive male incubated eggs and reared young (*Avicultural Magazine* **11**(4), 228)
Red-legged partridge	Double-brooding reported from captive birds (Goodwin 1953). Double-brooding also frequent in wild (Ricci 1984; Jenkins 1957); from 60 to 80 per cent of older females may double-brood (Green 1984)
See-see partridge	Double-brooding possibly occurs in wild birds (Cramp and Simmons 1980)
European migratory quail	A few records exist of males incubating or leading broods (Bannerman 1963)

that double-brooding in wild California quail perhaps occurs only one or two years per decade, during those summers that for various reasons are highly favourable for reproduction. However, in at least the red-legged partridge the incidence of double-brooding is certainly much higher, and in some areas may be virtually a yearly phenomenon (Jenkins 1957; Green 1984). In this species it is probable that the males regularly participate in incubation but may not brood as effectively as do females (Spano and Csermely 1980). The other three European species of *Alectoris* evidently do not regularly exhibit such double-brooding behaviour (Menzdorf 1975b).

Double-brooding in the quails and partridges may be facilitated by the fact that in a large number of species the male regularly assists in the rearing of the young. In a few species the male also regularly participates in incubation, and occasionally may even assume the entire incubation duties, especially if the female should die during the incubation period. It is but a short progression from this situation to one in which the female simply turns over incubation responsibility of her first clutch to her mate, and immediately begins a second clutch that she alone incubates. Examples of such behaviour have been observed or suggested for California and Gambel's quails, northern bobwhite, and several species of *Alectoris*.

Clutch sizes and egg-laying rates

The egg-laying rate of both the New World quails and the Old World perdicines approximates slightly less than one egg per day, judging from available data. Thus, northern bobwhites and California quails average 1.1–1.4 days per egg, while grey and chukar partridges average 1.3–1.5 days per egg (Johnsgard 1973; Cramp and Simmons 1980). As a result, the completion of a clutch usually requires about 30–40 per cent more days than there are eggs laid by the female. At least in captivity the Asian migratory quail typically lays an egg per day, and under such conditions from 200 to 365 eggs may be laid in a single year (Wetherbee 1961), suggesting that at least when adequate food supplies are present there are few if any physiological limitations on potential clutch size for this species.

Average clutch sizes are probably less easily and precisely estimated than are egg-laying rates, for although the tendency to lay a clutch of a particular mean size is certainly an adaptive heritable and species-typical trait, proximal effects such as the dietary history of the individual female, her age and health, the possible physiological drain resulting from earlier nesting efforts, and similar influences are likely to be important determinants of clutch size. Lack (1947) and Blank and Ash (1960) provided detailed analyses of such clutch-size variations in the grey partridge. From these and other studies it is apparent that although clutch size in various quail and partridge species is largely independent of the age of the female, it usually diminishes with repeat nestings during the same season. In the grey partridge the clutch tends to be slightly (about 8 per cent) higher in a female's first laying season than subsequently. Additionally, a geographic trend is evident in European grey partridges, with average clutch sizes increasing by about 20 per cent from England and Czechoslovakia north to northern Finland (Cramp and Simmons 1980). Similarly, there is a geographic trend in the clutch size of the European migratory quail, with six to seven eggs typical of southern and equatorial Africa, six to eight normal in India, and eight to twelve characteristic in Europe, and an additional tendency in Europe for clutch size to increase slightly from west to east. In this species there is no marked trend toward reduction in the sizes of later clutches (Wetherbee 1961), although very late (August) clutches from central Europe are significantly smaller than those laid earlier in the season (Glutz von Blotzheim 1973).

Roseberry and Klimstra (1983) noted a marked seasonal decline in clutch sizes of northern bobwhites that they largely attributed to reduced clutches of renesting efforts, and also observed significant year-to-year variations in average annual clutch sizes, which they believed may likewise have resulted from yearly variations in the incidence of renesting. No clear intraspecific geographic clines in clutch sizes are yet apparent in any New World quails, but should be looked for when good data on clutch sizes become available for the Central and South American populations of *Colinus*. A summary of some reported mean clutch sizes of wild populations of various quails and partridges for which fairly large samples are available is presented in Table 4.

Egg hatchability and hatching success under natural conditions

The incidence of infertile or inviable eggs in wild galliform populations is normally so small as to be relatively insignificant, judging from available data (Johnsgard 1973). In general, less than 10 per cent of the eggs laid in nests by wild females are incapable of hatching, with estimates of 90–95 per cent hatchability commonly reported for such species as the northern bobwhite (Roseberry and Klimstra 1983), grey partridge (Middleton 1936; Jenkins 1961b), and red-legged partridge (Jenkins 1957). However, the percentage of eggs (and clutches) that actually hatch is generally substantially less, owing to losses result-

Table 4. Reported clutch sizes of quails and partridges under natural conditions

Species	Normal range	Mean clutch size	References
Mountain quail	6–14	9.9 (29 nests)	*P.R. Quarterly* **10** (1950)
Scaled quail	5–22	12.7 (39 nests)	Schemnitz 1961
Gambel's quail	6–19	12.3 (40 nests)	Gorsuch 1934
California quail	9–17	13.7 (16 nests)	Lewin 1963
Northern bobwhite	7–28	14.4 (394 nests)	Stoddard 1931
		13.7 (347 nests)	Roseberry and Klimstra 1983
Montezuma quail	6–16	11.1 (24 nests)	Leopold and McCabe 1957
Chukar partridge	14–19	15.5 (4 US nests)	Mackie and Buechner 1963
Grey partridge	10–20	14.6 (4051 English nests)	Lack 1947
		15–18 (5 US studies)	Schulz 1977
		15.9 (104 Danish nests)	Paludan 1954
European migratory quail	8–13	10.2 (17 nests)	Westerskov 1947 (in Cramp and Simmons 1980)

Table 5. Egg hatchability and hatching success of quails and partridges under natural conditions

Species	Egg hatchability (per cent)	Percentage of nests hatching	References
Mountain quail	95.8 (82 eggs)	57 (14 nests)	*P.R. Quarterly* **8** (1948)
California quail	—	24.8 (83 nests)	Glading 1938
Gambel's quail	—	24 (44 nests)	Gorsuch 1934
Northern bobwhite	93.3 (00 eggs)	32.7 (793 nests)	Roseberry and Klimstra 1983
	86 (2874 eggs)	36 (602 nests)	Stoddard 1931
Chukar partridge	—	25 (16 nests)	Harper *et al.* 1958
Grey partridge	86 (average of four studies)	54 (average of eight studies)	Schulz 1977
	93 (59 203 eggs)	72 (7521 nests)	Middleton 1936

ing from predation, nest abandonment, trampling by large animals, and similar sources of nest and egg failure (Table 5).

Nesting losses are highly variable and often unpredictable as to their intensity and their effects on overall productivity. Roseberry and Klimstra (1983) reported annual nest success rates (percentage of initiated nests that hatch successfully) in northern bobwhites varying from 25 to 53 per cent during 12 years of study, and calculated an average overall nest loss rate of 2.75 per cent per day. The highest nest losses evidently resulted from predation, although variations in estimated annual predation rates did not correlate with overall nest success rates, apparently because of effective renesting efforts. Farming activities, nest abandonment, and unfavourable weather had progressively smaller effects on nesting success. In general, the numbers of nests found per hen (as an estimate of relative renesting efforts), and the proportion of nests judged to have hatched successfully correlated best with the estimated annual productivity index (ratio of observed chicks to the estimated abundance of hens the previous March), while annual clutch-size and egg-hatchability differences did not appear to contribute to observed yearly variations in female productivity. Similarly, Dimminck (1974) judged that annual differences in hen renesting efforts might have important influences in determining annual productivity of northern bobwhites in Tennessee. However, the environmental variables that might influence the bobwhite's optimum nesting season length, as well as potential factors that might regulate both renesting incidence and success, are still distinctly uncertain and may well vary geographically.

Interspecific clutch-size variations and their adaptive significance

The possible evolutionary significance of clutch-size variations in galliform birds was discussed earlier with reference to the New World quails (Johnsgard 1973), and it was then concluded that in temperate-breeding species of grouse and quails the upper limits of evolved clutch sizes might be adaptively related to the number of eggs an adult can effectively incubate and to the predation levels during the relatively long and vulnerable egg-laying period, at least in those species with large average clutch sizes. It was further concluded that energy drains on the females probably play only a subordinate role in influencing evolved clutch sizes, inasmuch as females can apparently largely compensate for these losses by increasing their food intake during the laying period (based on observations of captive birds). Although this conclusion is counter to arguments advanced by Lack (1968) supporting the idea that physiological factors associated with limited female energy reserves might have influenced the evolved clutch sizes of bird species having precocial young. Temperate-zone quails and partridges are remarkable both for their large average clutch sizes and their fairly high rates of pre-incubation nest losses. In northern bobwhites these pre-incubation nest losses may range from about 3 to 5.6 per cent per day (Roseberry and Klimstra 1983), or well above the levels theoretically required to place a limit on the most productive maximum clutch size (Johnsgard 1973).

The smaller average clutches characteristic of tropical species of quails and partridges evidently require another explanation, which may be related to such environmental factors as possibly reduced amounts of nutritious foods available around the time of nesting (Lack 1968), or perhaps to the shorter diurnal foraging periods but generally longer overall nesting seasons in tropical latitudes, making it more feasible to raise two or more relatively small broods during a single breeding season rather than necessarily try to raise a single brood of maximum size. Too few data on average clutch sizes in tropical species of the New World quails are available to estimate very accurately the actual degree of difference in clutch sizes of tropical and temperate species. However, limited data on *Odontophorus*, *Dactylortyx*, and *Dendrortyx* all suggest that the clutch sizes of these Central and South American forms average only about four to six eggs, or less than half those typical of North American quails.

The comparable smaller average size of clutches of the European migratory quail in equatorial Africa versus those of India and Europe has already been noted; likewise the other essentially tropical forms of *Coturnix* seem to have relatively small clutches of four to five eggs. Similarly, the eight- to twelve-egg clutch size of the most northerly ranging species of *Francolinus* (the black francolin) is larger than the reported clutch sizes of francolins from equatorial Africa, which usually average less than six eggs. The southernmost temperate-adapted francolin species (such as the red-billed and Cape) again have larger average clutches of six to eight eggs than such equatorial relatives as the Schlegel's, white-throated, and Latham's forest francolins, all of which have reportedly two- to four-egg clutches that are among the smallest of all the perdicines. Other perdicines reportedly having unusually small clutches of only two to four eggs include the long-billed partridge, the red, painted, and Ceylon spurfowl, the Snow Mountain quail, the stone partridge, several species of *Arborophila*, and a few additional francolins. Most of these are topical woodland or forest species, while the Snow Mountain quail is adapted to a tropical alpine habitat and the stone partridge to a tropical semi-desert.

I earlier (Johnsgard 1973) calculated collective average clutch weights as a percentage of adult female weights for six species of North American odontophorines plus the grey and chukar partridges. For the North American odontophorines these clutch weights average about 75 per cent of the female's weight. The highest averages, of above 90 per cent, were estimated for northern bobwhite and California quails, the two most temperate-adapted species of the entire odontophorine group. Among the perdicines for which reasonably good clutch-size and adult weight data are available, comparably high percentages (in excess of 80 per cent of the female's weight) occur only in four temperate-latitude forms; the European and Asian migratory quails and the chukar and rock partridges. Although the equally temperate-adapted grey partridge produces what is perhaps the largest average clutch of any perdicine, it lays a relatively small egg and the collective clutch weight is therefore not unusually high. Considering the opposite extreme, the available and still rather limited data suggest that at least nine species of perdicines produce clutches that collectively average less than 20 per cent of the female's estimated average weight (the latter sometimes necessarily indirectly determined from data on male weights and/or wing measurements). These species include six predominantly tropical African francolins (Schlegel's, Latham's forest, Shelley's, grey-breasted, yellow-necked, and red-necked), the swamp francolin of temperate India, the long-billed partridge of tropical Malaysian forests, and the alpine-adapted Snow Mountain quail of equatorial New Guinea. Of these, only the swamp francolin has a clutch seemingly smaller than might seem appropriate, based on their ecologies and distributions.

3 · Comparative ecology and population dynamics

Inasmuch as many of the species considered in this book are important game birds, a great deal of research on population densities, distributions, survivorship, and their environmental controls has been performed on at least some representatives of this group. For example, several books or symposia, for example, have been published on the biology and ecology of a single species, the northern bobwhite (e.g. Stoddard 1931; Rosene 1969; Morrison and Lewis 1972; Roseberry and Klimstra 1983; Lehmann 1984), and comparable volumes have dealt with the California quail (Leopold 1977) and the grey partridge (Kobriger 1977). Recently published bibliographies of the northern bobwhite (Scott 1985) and the grey partridge (in Peterson and Nelson 1980) comprise about 3000 and 1300 references, respectively, a large number of which are ecologically oriented. It would thus be impossible to summarize completely the ecological literature of even a single well-studied species such as either of these, and instead only a brief attempt will be made to update and expand the comparative information provided in my earlier (Johnsgard 1973) book on the North American quails.

Population densities and their effects on productivity

The observed population densities of quails and partridges are a dual reflection of each species' relative tendencies toward social gregariousness or spacing, and the capabilities of the environment to support a given biomass or population level of the species and its possible competitors for the same or similar resources. A large number of estimates have been published of such densities (which generally are based on counts during late winter, spring, or summer breeding populations, thus avoiding the abnormally high autumn densities brought about by juveniles), especially for the North American species (Table 6). In terms of biomass per acre, these densities often seem absurdly small, rarely exceeding a bird per acre of habitat, and sometimes as low as 20 acres per bird in marginal habitats, or generally far lower than mammalian herbivores of comparable size. This would suggest that indeed some inherent population restraints must be operating on these birds, and keeping them from reaching the levels at which resource exhaustion might occur.

Probably more work on environmental population controls has been done on northern bobwhites and grey partridges than on any other species of quails or partridges, and some results of these studies may be mentioned. Roseberry and Klimstra (1983) reviewed the data on population densities of the northern bobwhite throughout its range, finding that in 19 studies lasting at least six years the mean autumn densities per 100 ha. (247 acres) varied from 13.8 to 163.5 birds, with the highest levels generally associated with the southern states. These authors believed that on their Illinois study area the birds had shown a gradual decline in their average numbers or 'equilibrium density' over a 27-year period, as well as an apparent tendency for the density to oscillate over periods of 8–10 years, which they regarded as an underlying factor rather than a dominant force in population regulation. They attributed the gradual decline in average bobwhite numbers in their study area to increasing human populations and associated changes in bobwhite habitats during that period, and to long-term plant-succession trends. Annual population changes were evidently more closely related to variable summer recruitment than to variable winter survival, although Roseberry and Klimstra believed that the relative effects of these two phenomena might differ in other geographic regions, especially in those areas having severe winters. Lehmann (1984) concluded that adequate rainfall during spring and late summer is the most critical limiting factor controlling bobwhite populations in Texas, through its effect on influencing breeding effort and success. He apparently considered summer heat and lowered humidity to have more serious effects on juvenile and adult bobwhite survival than winter temperatures, at least when the cold temperatures are not accompanied by precipitation.

Leopold (1977) similarly concluded that the observed trends in California quail populations were largely the result of differential summer recruitment rather than being the result of compensatory adult mortality during any particular part of the year. Furthermore, in central and especially northern parts of the species' range its annual recruitment rates are

Table 6. Some reported population densities of quails and partridges in favourable habitats[1]

Species	Density/location	References
Mountain quail	2 acres per bird maximum spring density, California	Edminster 1954
Barred and elegant quails	Under 1 acre per bird locally, Mexico	Leopold 1959
Scaled quail	10.1 acres per bird in winter, Texas	Wallmo 1956
	0.84 acres per bird in winter, Oklahoma	Schemnitz 1961
California quail	1.7–3.9 acres per bird in late winter, California	Glading 1941
Gambel's quail	1.6 acres per bird in late winter, Nevada	Gullion 1962
Northern bobwhite	4–20 acres per bird in spring, good range (various states)	Edminster 1954
Singing quail	16.5 acres per breeding bird, Tamaulipas, Mexico	Warner and Harrell 1957
Montezuma quail	21–23 acres per bird in summer, Chihuahua, Mexico	Leopold and McCabe 1957
Rock partridge	0.75 acre per territorial pair, protected study area	Menzdorf 1976c
Grey partridge	3.5–5.3 acres per bird in winter, North Dakota	Hammond 1941
	7.47 acres per breeding bird in spring, Idaho	Mendel and Peterson 1980
	c. 1 breeding bird per acre, best European habitats	Glutz von Blotzheim 1973
European migratory quail	1.75 acres per breeding male, best European habitats	Glutz von Blotzheim 1973

[1] Expressed in acres per bird; to obtain hectares per bird multiply by 0.4.

apparently inversely related to variations in spring density. There no dominant relationship exists between annual weather parameters and recruitment rates, although in central California years of mild springs and sunny summers generally result in the highest quail productivity. However, in drier and more southerly areas adequate winter and spring rainfall is evidently of such critical importance in regulating recruitment that any influences of adult population density on recruitment rates are largely obscured. The role of precipitation in these southern areas may be fairly direct, by immediately providing increased insect food and water availability to newly hatched chicks. More probably it operates indirectly, through influences associated with differential stimulatory and nutritional values of important food plants.

In studies on the grey partridge, Potts (1973; and in Cramp and Simmons 1980) reported that winter losses exhibited only slight year-to-year variability, and these effects plus spring dispersion tendencies of pairs contributed little to annual variations in mean breeding densities. However, chick survival during the six post-hatching weeks was highly variable in different years, and was largely correlated with insect food availability. Weigand (1980) found in Montana that summer productivity estimates (mean August brood sizes) were inversely related to spring densities, but also that the percentage of females with young in August was positively correlated with spring densities. Weather conditions, especially timing and amounts of rainfall, furthermore, strongly influenced breeding success, with heavy rainfall during the period of nesting and hatching greatly reducing productivity. This result generally agrees with observations in England (Middleton 1950) suggesting that sunny weather during the weeks immediately following hatching results in maximum partridge productivity. Thus, as with the New World quails, population densities of breeding adults evidently do influence overall productivity rates, but these effects are often masked or completely overwhelmed by variable weather effects, particularly those occurring at the time of nesting and immediately thereafter. These weather effects greatly affect hatching success and chick survival rates.

Flocking and covey behaviour

Apparently virtually all quails, partridges, and francolins exhibit tendencies toward sociality (covey-formation or flocking) during the non-breeding period. Such sociality has several obvious advantages, such as facilitating pair-formation, increasing anti-predator vigilance levels, and sometimes providing thermal benefits through the birds' potential formation of circular 'roosting rings' during cold nights, with their consequent sharing of body heat. Flocking probably has other subsidiary social benefits as well, such as the sharing of information on roosting sites, escape, drinking or dusting locations, and prime foraging areas. Obvious disadvantages include the potential attraction of predators and increased probabilities of intraspecific competition and social strife. Not surprisingly, the intensity of coveying behaviour and the sizes of the associated groups vary considerably among and sometimes even within species (Table 7). In the northern bobwhite,

Table 7. Some reported group sizes of quails and partridges

Species	Group size	References
Mountain quail	Average of 21 coveys, 9.1 birds, range 3–30	Miller and Stebbins 1964
Barred quail	Average of 18 coveys, 12 birds, range 5–20 or 25	Leopold 1959
Scaled quail	Average of 325 coveys, 31.2 birds, range 4–150	Schemnitz 1964
Elegant quail	Coveys range from 6 to 20 birds	Leopold 1959
California quail	Average of 27 autumn coveys, 13.9 birds	Barclay and Bergerud 1975
	Coveys usually 30–70 birds, averaging about 50	Leopold 1977
Gambel's quail	Average of 40 coveys, 12.5 birds, range 3–40	Gullion 1962
Northern bobwhite	Average of 2815 winter coveys, 14.3 birds, range 6–25	Rosene 1969
	Average of 12 022 autumn-to-spring coveys, 11.4 birds	Lehmann 1984
Black-throated bobwhite	Usually 7–15 birds in covey	Leopold 1959
Spotted wood-quail	From 5 to 10 birds in covey	Leopold 1959
Montezuma quail	Average of 62 autumn and winter coveys, 7.6 birds, range 3–14	Leopold and McCabe 1957
Chukar partridge	From 10 to 40 or more birds per covey	Leopold 1959
Grey partridge	700 autumn and winter coveys, average 8.2 birds	Bishop et al. 1977
	520 winter coveys, average 5.9 birds	Weigand 1980

for example, the fairly consistent covey size is in part a reflection of the most efficient number of birds required for conserving body heat in a circular roosting group, while in the more arid-adapted New World quails large flocks often assemble in the general vicinity of reliable watering sites. In many New World as well as Old World species the covey's nucleus often comprises a single pair and its brood, plus varying numbers of unsuccessful breeders that attach themselves to such family units. Small covey sizes (averaging about three to seven birds) that evidently also reflect a familial organization have been observed in two African francolins (Mentis and Bigalke 1980). However, some coveys of North American quails may be made up of two or more different family groups or, less frequently, of entirely adult birds, such as in those years when breeding success has been unusually low. It may be semantically desirable to restrict the use of the term 'covey' to small, family-sized groups of quail- or partridge-like birds, and refer to much larger assemblages simply as flocks.

Home ranges and territories

Individual coveys and flocks of non-migratory quails and partridges typically exhibit definite home ranges, within which most or all of their annual activities occur (Table 8). Members of such coveys or flocks tend to remain within these home ranges, which often overlap with the home ranges of other coveys or flocks using the same general area. To a degree, each flock of grey partridges has a developed flock territorialism, so that little interchange occurs between members of different flocks occupying the same habitat (Blank and Ash 1956; Jenkins 1961a). Such flock-bonding is stronger in the grey partridge than in the *Alectoris* partridges, in which inter-flock hostility appears to be virtually nil. In the California quail there is also inter-covey hostility, but no indication of actual territorial defence of an area by one covey against another (Leopold 1977).

Studies of inter-covey shifts in the northern bobwhite by Lehmann (1984) indicated that most such shifts were made by single birds (usually males) or by two birds (often apparently pairs). 'Receiver' coveys seemed to assimilate one or two birds rather easily, but major changes resulting in distinctly oversized coveys either produced a splitting into two or more groups, or the covey 'expelling' individuals a few at a time. Similarly, 'donor' coveys were often not affected by the loss of a single member. However, losses of three or more birds from a covey tended to result in the disintegration of that covey within a few weeks, and no covey was known to survive the loss of six or more individuals. 'Stable' coveys tended to be on average slightly larger (12.9 vs 11 birds) than ones that were considered unstable, and also tended to be less mobile, but in both types of covey there was a trend from autumn to midwinter toward a common median covey size (12 birds) and an equal sex ratio.

Individual territorial behaviour in New World quails and some partridges is sometimes much harder to document than 'flock territoriality'. Although breeding pairs tend to disperse for breeding, actual territorial boundaries are often difficult to

Table 8. Some reported home ranges of quails and partridges

Species	Home range	References
Mountain quail	Nesting pairs occupied 5–50 acres, California	*P.R. Quarterly* 1951
Scaled quail	Winter covey home ranges average 52.3 acres, Oklahoma	Schemnitz 1961
	Winter covey home ranges average 360 acres, Texas	Wallmo 1956
Gambel's quail	Winter covey home ranges average 20 acres, Nevada	Gullion 1956
California quail	Winter covey home ranges average 26 acres, California	Emlen 1939
Northern bobwhite	Winter covey home ranges average 24 acres, Missouri	Murphy and Baskett 1952
	Winter covey home ranges (1154 coveys) average 13.2 acres and ranged from 4 to 77 acres	Rosene 1969
Grey partridge	Winter ranges (8 coveys) range from 12.1 to 97.3 acres	Schulz 1980
	Home ranges of three females range from 141 to 544 acres over 6 months	McCrow 1977
	Average winter home ranges of 22 coveys 3.4 acres, range 0.2–13.8 acres	Weigand 1980

establish, and instead the birds seem to exhibit varying degrees of proximity-tolerance toward other breeding pairs or unpaired males. In the grey partridge, as well as in the New World quails so far studied, pairing typically occurs prior to covey break-up, thus there is no correlation between the establishment and effective advertisement of a breeding territory and the probability of obtaining a mate, which is the usual case in territorial passerines. Menzdorf (1976c) observed marked year-to-year variability in the locations of rock partridge territories, and additionally noted that these territories typically were not contiguous and often were quite ill-defined, at least initially. Similarly, in the chukar there is no typical fixed territoriality, and the object of the male's defence is the female, rather than a geographically definable area (Mackie and Buechner 1963). Nevertheless, the rally call of the mated male chukar may tend to help disperse the breeding population.

Several francolins (black, grey-winged, and Levaillant's red-winged) have been reported as apparently territorial during the breeding season, and one might expect to find a positive correlation between such territorial tendencies and tarsal spur development (or other effective male dominance signals) among males of various species. The advertisement call of the black francolin may thus be a social dominance signal that helps assure effective spacing, since a common method of trapping this species is to place a tame decoy bird in the territory of a resident advertising male, who quickly rushes in when he hears the calls of the decoy, and is normally trapped by nooses. Occasionally such a decoy may be killed by a wild male who manages to avoid the nooses, as their sharp tarsal spurs can readily inflict fatal wounds (Baker 1930). Similar methods of trapping have long been used by natives in Africa for trapping various species of *Francolinus* and *Coturnix*, and elsewhere have been used for trapping *Arborophila*, *Perdicula*, and various other apparently typically territorial Perdicini.

In such New World quails as the northern bobwhite and the *Callipepla* forms, most advertisement calling is done by surplus unpaired males rather than by paired birds, which rarely if ever utter loud vocalizations other than separation calls following pairing. Such bachelor male 'crowing territories' are often as close to the locations of breeding pairs as the latter will permit, and often they are tolerated in fairly close proximity. These advertising calls by bachelors may be uttered persistently throughout the entire breeding season, and thus might have the disadvantageous effect of attracting the attention of predators to nesting pairs.

Sex ratios, age ratios, and annual mortality rates

The significance of sex ratios and age ratios in the understanding and management of upland game bird populations is very great, and the many potential applications of such data have been discussed earlier (Johnsgard 1973). In monogamous species such as quails and partridges it might be expected that adult (tertiary) sex ratios would ideally hover around unity, or have a slight excess of males for renesting species, so as to assure that sexually active males might be available to fertilize renesting females whose mates have died or otherwise become unavailable. Indeed, the available data (Table 9) suggest that sex ratios of immature quails and partridges are very close to 50:50, while in adults the proportion of males typically ranges from 55 to 60 per cent of the population. This slightly distorted adult sex ratio might result either from increased mortality of immature females during their first autumn and winter, or from differential mortality during the breeding season. At least for the northern bobwhite there is some evidence supporting both of these positions, as reported by Roseberry and Klimstra (1983). These authors summarized evidence indicating that the relatively small annual variations in sex ratios of this species are not easily correlated with variations in productivity, although years of declining populations tended to be those with the highest incidence of adult males in the autumn population.

These same authors also found no direct correlation between annual variations in sex ratios and age ratios, although as just noted a large excess of males in the autumn population was associated with a greater than expected reduction in the young to adult ratio (and reduced productivity) the following year. Autumn and winter age ratios in quails and partridges (Table 10) are fairly easily determined from

Table 9. Some reported sex ratios of quails and partridges (expressed as percentage of males in population)

Species	Age class	Males (%)	Sample	References
Scaled quail	Adults	56.5	1174	Campbell *et al.* 1973
California quail	Immatures	50.8	6335	Francis 1970[1]
	Adults	57.3	4347	Francis 1970[1]
Gambel's quail	Immatures	49.3	333	Raitt and Ohmart 1968
	Adults	57.8	154	Raitt and Ohmart 1968
Northern bobwhite	Immatures	50.3	104 896	Roseberry and Klimstra 1983
	Adults	59.8	17 727	Roseberry and Klimstra 1983
Montezuma quail	Mixed	63.0	502	Leopold and McCabe 1957[2]
Chukar	Mixed	48.8	176	Christensen 1954[1]
Grey partridge	Adults	58.0	115	McCabe and Hawkins 1946
	Mixed	51.0	30 208	Schulz 1977
	Mixed	51.0	14 167	Johnson 1964[1]

[1] Calculated from data presented by authors.
[2] From a museum sample.

Table 10. Some reported autumn and winter age ratios of quails and partridges (expressed as percentages of immatures in population)

Species	Immature (%)	Sample	References
Mountain quail	48	198	Leopold 1959[1]
Scaled quail	74	1219	Schemnitz 1961
	73.9	5624	Campbell *et al.* 1973[2]
California quail	63.3	5603	Emlen 1940[2]
	59.3	10 682	Francis 1970[2]
Gambel's quail	76	352	Raitt and Ohmart 1968[2]
Northern bobwhite	82.3	51 178	Bennitt 1951
	83	104 896	Roseberry and Klimstra 1983[2]
	80.8	65 026	Yoho and Dimmick 1972
	67	44 280	Lehmann 1984
Montezuma quail	61	57	Leopold and McCabe 1957[1]
Chukar	81.5	1716	Christensen 1970[2]
Grey-winged francolin	36	145	Mentis and Bigalke 1980[2]
Levaillant's red-winged francolin	35	168	Mentis and Bigalke 1980[2]
Grey partridge	79.5	14 167	Johnson 1964[2]
	71.9	1956	Weigand 1980[2]

[1] Based on museum specimens taken at various times of year.
[2] Calculated from author's data.

harvest surveys, and have proven to be one of the best and simplest methods of estimating the previous breeding season's success. Additionally, large samples of age ratios often prove to provide excellent estimates of annual mortality rates (the percentage of the immatures in the autumn population approximating to the annual adult mortality rates, and those of adults providing an estimate of annual survival rates), at least for those species that do not exhibit differential age vulnerability to the sampling technique, which is usually hunting or trapping. These data suggest annual mortality rates for quails and partridges that are generally well in excess of 50 per cent, with the exception of two fairly large African francolins, which approximated 35 per cent immatures in the populations sampled. It is probable that a loose positive correlation exists between a species' adult body size and its chances for survival, and therefore it might be expected that some of the highest percentages of immature birds (and mortality rates) will be typical of smaller and generally more vulnerable species, and vice versa.

Lehmann (1984) has presented age-ratio data for northern bobwhites extending over a 37-year period. His sample, totalling over 131 000 harvested birds, suggests an overall annual adult survival rate (percentage of adults in autumn population) of 30 per cent, and yearly variations ranging from 13 to 62 per cent. This average is fairly close to the 27 per cent adult component in age ratios based on a sample of over 104 000 birds taken in Illinois over a 30-year period. These and other similar age-ratio data strongly suggest that annual mortality rates of adult New World quails may typically range from 70 to 80 per cent, even in unhunted populations. Independent estimates of mortality rates, based on analysis of banded and subsequently recovered birds (Table 11), confirm these general results. Using complicated mathematical models, Roseberry and Klimstra (1983) determined for their northern bobwhite population that the effects of hunting-related deaths on overall mortality rates were somewhat intermediate between representing a purely additive mortality source (non-hunting mortality rates being independent of and supplemental to these hunting effects) and being completely compensatory (non-hunting mortality declining and productivity increasing as hunting mortality increases), but nearer the additive end of the spectrum.

Leopold (1977) summarized evidence for various western American quails such as the California quail and scaled quail to support the general position of wildlife biologists that moderate levels of hunting are not directly additive to non-hunting causes of death in determining overall mortality rates of game birds, but instead hunting losses tend to substitute for various other natural mortality factors. For example, the study of Campbell et al. (1973) indicated little difference in the local annual mortality rates of hunted and unhunted scaled quail populations, and in many other areas the densities of quails or partridges have been found to be apparently unaffected by differences in hunting or non-hunting practices within such areas.

Table 11. Some reported annual quail and partridge survival rates

Species	Survival rate(s) (%)	References
Scaled quail		
Adults (1–6 years)		
Hunted	29.7	Campbell et al. 1973[1]
Unhunted	32.7	Campbell et al. 1973
California quail		
Immatures	26.7	Raitt and Genelly 1964[1]
Adults	31.6	Raitt and Genelly 1964
Adults	23–41	Leopold 1977[2]
Gambel's quail		
Both sexes	28–40	Sowls 1960[1]
Northern bobwhite		
Both sexes	22	Marsden and Baskett 1958[1]
Both sexes	19	Roseberry and Klimstra 1983[2]
Both sexes	32.2	Lehmann 1984[2]
Grey partridge		
Both sexes	16–20	Paludin 1963[1]
Asian migratory quail		
Both sexes	10.6	Austin and Kuroda 1953[1]

[1] Calculated from banded birds.
[2] Calculated from age ratios.

Given these high adult mortality rates under natural, sometimes even fully protected, conditions, the average longevity of any quail and partridge tends to be very limited. The probable additional longevity ('mean afterlifetime') of a bird that has survived to its first autumn of life generally averages less than a year, and the maximum expected longevity or 'turnover rate' (the time required to remove by death at least 99 per cent of a single juvenile age-class from the population) averages only about 4–5 years in the species so far studied (Johnsgard 1973). In Table 12 some examples of a few individual birds reaching or surpassing this expected level of longevity are provided, one of the most remarkable of which is that of a banded European migratory quail (from a sample of 6175 ring recoveries) surviving for at least 8 years following ringing. However, data based on 1007 recoveries of Asian migratory quail ringed in Japan suggest that this species probably has an extremely high adult mortality rate of 89–90 per cent, and that only about 1.2 per cent of the birds survive into their third year of life (Austin and Kuroda 1953). Similarly, only 49 of the 6175 European migratory quail recoveries (or less than 1 per cent) were from birds surviving at least four years after ringing, suggesting that an approximate 3- to 4-year turnover rate is probably also characteristic of this species. A summary of 47 recoveries of Asian blue quails included 11 (23 per cent) that were recovered more than a year following ringing, and the oldest recovery was obtained 36 months after ringing (McClure 1974).

It would be of interest to learn whether the annual mortality rates of such wandering or nomadic species as the Asian and African blue quails and the more truly migratory black-breasted quail, harlequin quail, and European and Asian migratory quails average significantly higher than those of similar-sized but sedentary quails. At present far too little evidence is available on this point to make any conclusions possible, although the data mentioned above suggest very high mortality rates for the two most highly migratory quail species (*coturnix* and *japonica*).

Seasonal mortality rates

Given the high annual adult mortality rates of most quails and partridges, it is of interest to try to learn just when most of this mortality occurs, and how it might affect overall population biology. Several studies of North American quails have attempted to provide such data, as summarized in Table 13. There is no clear pattern that emerges from these studies—as might be expected, the seasonal distribution of mortality seems to vary both between and within species. Some of the best data derive from studies on the northern bobwhite. In Texas there seems to be a relatively low winter mortality rate (Lehmann 1984), whereas near the northern limit of the species' range in Illinois winter mortality comprises the majority of the annual mortality (Roseberry and Klimstra 1983). Contrariwise, spring and summer mortality of this species was apparently quite high in Texas, when during early spring there were substantial declines in population density during the time that coveys were breaking up and pairs were dispersing for nesting.

Losses attributable to hunting are similarly variable in different areas, and as noted above these losses seem to be only partially compensatory in terms of their relationship to overall annual mortality rates. It is of interest that hunting may remove as much as 40 (Leopold 1977) to 45 (Rosene 1969) per cent of the total autumn populations of various North American quails without reducing subsequent breeding potential, and hunting probably accounts for an average of about a quarter of the estimated total annual mortality in the introduced grey partridge (Schulz 1977). In this species the average annual adult mortality is about 70 per cent, with the most significant seasonal losses probably occurring during the period of late winter dispersal (Schulz 1980). Weigand (1980) judged that in Montana the annual kill by hunters was responsible for removing only about 3 per cent of the total grey partridge population.

Table 12. Some longevity estimates and maximum longevity records for quails and partridges

California quail	One male banded as an adult was recaptured when at least 80 months old. Mean life expectancy after September following hatching is 9.7 months (Raitt and Genelly 1964)
Gambel's quail	Four of 121 birds trapped as adults were alive 4 years later, and 10 of 321 birds trapped as juveniles were alive 4 years later (Sowls 1960)
Northern bobwhite	One of 1156 banded birds was recovered in its fifth year. Estimated life expectancy as of first fall after hatching 8.5 months (Marsden and Baskett 1958); 10.7 months (Lehmann 1984); 9.1–11.7 months (Rosene 1969)
Grey partridge	Average life expectancy of 1.8 years for adults (Weigand 1980). Estimated mean longevity for adults of 1.9 years and mean life expectancy for one-year-olds of 8.3 months. One banded bird reached age of five years and two months (Paludan 1963)
European migratory quail	One banded bird out of 6175 recoveries was at least eight years old (Toschi, cited in Cramp and Simmons 1980)

Table 13. Some estimated seasonal mortality rates in quails and partridges

California quail	Average (four-year) November–March mortality in a hunted area of 39 per cent, with hunting accounting for 82 per cent of latter; mortality in a non-hunted area averaged 30 per cent over same period (Glading and Saarni 1944)
	Average (two-year) winter mortality of 58.5 per cent (Barclay and Bergerud 1975)
Gambel's quail	Autumn (September–December) mortality averaged 25 per cent during four years, of which 80 per cent was caused by hunting (Gallizioli and Swank 1958)
Northern bobwhite	Autumn and winter (November–late March) mortality averaged 63 per cent over 27 years, of which 66 per cent was due to hunting (Roseberry and Klimstra 1983)
	Autumn and winter (October–late February) losses averaged 25 per cent, with 46 per cent of latter caused by hunting (Lehmann 1984)
	Autumn and winter (November–March) losses averaged 32.6 per cent in four study areas (Rosene 1969)
	Spring to autumn (late February–October) adult losses averaged 56 per cent (Lehmann 1984)
	Spring to autumn (March–November) adult losses averaged 57 per cent in four study areas (Rosene 1969)
	Spring to autumn (March–November) adult mortality averaged 39 per cent over 17 years (Roseberry and Klimstra 1983)
Grey partridge	Autumn hunting mortality averaged 26 per cent (range 6–55 per cent) in 11 different studies (Schulz 1977)
	Winter losses averaged 34 per cent (range 10–77 per cent) in 15 different studies (Schulz 1977)
	Pre-hunting (summer) adult losses ranged from 15 to 22 per cent in three different studies (Schulz 1977)

Mortality during the winter period averaged about 8 per cent per month, compared to an estimated overall monthly mortality rate of 6 per cent. The most significant spring mortality factor was believed by Weigand to involve predation, and the quality and quantity of protective cover during spring was thus postulated as the environmental factor most limiting the Montana partridge population.

Nest and brood mortality

Relatively high rates of losses of nests seem to be characteristic of quails and partridges, but balanced against this is the fact that persistent renesting behaviour is also a typical feature of these birds. Roseberry and Klimstra (1983) found no correlation between nesting density and the proportion of northern bobwhite nests lost to predation or abandonment, and also no strong relationship existed between annual breeding densities and rates of nest failures. They suspected but were unable to prove a positive relationship between the number of nests produced annually per hen (a reflection of relative renesting efforts) and the total percentage summer population increase; such a correlation was suggested in Tennessee data obtained by Dimminck (1974). Studies attempting to correlate yearly bobwhite productivity with annual variations in nesting chronology have not proven very convincing; in some cases early nesting initiation and success seemed to be correlated with high overall productivity, while in others a significant number of late-hatched birds in the autumn population apparently indicated an unusually long and successful nesting season, including possible double-brooding.

In any case, most field studies suggest that a relatively high rate of chick mortality normally occurs among newly hatched broods (Table 14), which frequently results in a 25–50 per cent reduction from the initial brood size during the first few months of life. Heavy chick losses are typically associated with cold, wet weather during the time immediately around hatching, but unusually hot and dry weather at that time can have similarly devastating effects on brood survival. Lehmann (1984) reported that summers of severe drought in Texas coincided with those of poorest northern bobwhite production, while studies in Illinois have not found that unusually dry years were clearly associated with reduced bobwhite production (Roseberry and Klimstra 1983). Other studies of bobwhites and grey partridges have correlated relative chick survival with the abundance and availability of live insect food during the first weeks of life. Not surprisingly, there is typically a strong positive correlation between the numbers of chicks successfully hatched per female and the number of juveniles in the subsequent autumn population (Roseberry and Klimstra 1983), although the probable role of variable post-hatching mortality factors in influencing autumn population densities should not be underestimated.

The presumed causal inverse relationship between adult breeding density and chick survival (Lack 1951), often called the 'inversity principle', also requires additional attention. Roseberry and

Table 14. Some estimates of early brood mortality in quails and partridges under natural conditions

Species	Mortality estimates	References
Mountain quail	Approximately 30 per cent (range 1–55 per cent) brood mortality	Edminster 1954
California quail	Approximately 45–60 per cent brood mortality by autumn	Edminster 1954
Gambel's quail	Average 48 per cent brood loss (range 42–51 per cent) during three years	Edminster 1954
Northern bobwhite	Approximately 28 per cent brood loss in first eight weeks	Lehmann 1984
	Approximately 40 per cent brood loss in first 16 weeks	Rosene 1969
	Brood mortality 28.6 per cent in first eight weeks	Klimstra 1950
Grey partridge	Average loss of 57 per cent of young first four weeks (average of three studies)	Schulz 1977
	Average chick mortality 38 per cent	Middleton 1936
	Average July–September brood reduction 26.6 per cent	Weigand 1980

Klimstra (1983) found some support for this view when they established a fairly strong negative correlation between rates of summer population gain and prior breeding northern bobwhite densities. They also determined a negative correlation between annual production and the relative number of days of snow cover the previous winter (presumably a reflection of the incidence of winter stress). Bobwhite productivity rates were less strongly positively correlated with variations in total spring precipitation, as well as with the numbers of crop fields present during the following autumn. Collectively these four variables seemed to account for over 70 per cent of the total variation in recorded rates of summer gain.

All told, it is obvious that year-to-year differences in breeding densities, nesting efforts, and brood-rearing success rates must play extremely important interacting roles in influencing autumn densities in quails and partridges, but it is unlikely that any one environmental factor can be positively identified as the usual operative determinant of annual productivity, even for a single species.

4 · Ontogenetic growth and development

The ontogenetic development of quails and partridges is externally marked by a series of rather predictable moults and plumages from hatching until the definitive adult plumage stage is attained, and corresponding changes in behavioural and anatomical development. The moulting sequence of New World quails was described in some detail in my earlier monograph (Johnsgard 1973), and only briefly needs to be summarized and amplified here.

Natal plumage

All galliform birds are hatched with a downy coat that is often distinctively patterned in such a way as to provide camouflage for the precocial chicks. During the first ten days of post-hatching life wing feather development proceeds about twice as fast as does body growth, with the result that initial flying often occurs when the birds are between one and two weeks of age. Fledging in 7–10 days has been reported for the rock partridge, at about 10 days for the red-legged partridge, at 10–15 days for the grey partridge, at 11–19 days for the European migratory quail, at 14 days for the northern bobwhite, and at 10–14 days for the California quail.

Post-natal moult and juvenal plumage

The first indication of the juvenal plumage is usually evident at the time of hatching, when the inner primaries, secondaries, and rectrices become evident. The first (innermost) juvenal primary typically is the first remix to emerge, with the others developing sequentially to the eighth. The two outermost juvenal primaries appear relatively late, and in apparently all New World quails and most Old World perdicines these feathers are retained throughout most of the first year of life. The upper coverts of these two outermost primaries actually begin to grow before their associated primaries, and possibly serve as functional substitutes for these feathers until they are well developed. The third secondary (from the outside) is typically the first of that group of remiges to appear, with the inner ones again appearing sequentially, but the two outermost secondaries not appearing until about the time the innermost secondaries erupt.

The juvenal remiges (flight feathers) and rectrices (tail feathers) are scarcely fully grown before they begin to be replaced by the remiges and rectrices of the next plumage, and during the short time they are present the rest of the body is also transformed from a down-covered condition to one covered with typical contour feathers. The feathers replacing the natal down are called 'juvenal' feathers, and the associated age-class is called the 'juvenile' stage. These feathers usually emerge first on the flanks, and moulting gradually proceeds forward toward the breast. Juvenal feathers also soon appear on the crown, base of the neck, scapular region, and upper legs, gradually spreading down the back and up the neck. As these feathers are developing, the associated preening behaviour also begins. At least in *Alectoris* this behaviour tends to occur in a somewhat predictable sequence, beginning with one side of the body, proceeding to the feather tracts of the other side of the body in reverse sequence, and terminating with the feathers of the breast and belly (Menzdorf 1976e).

Although in most species considered here the post-juvenal moult does not extend beyond the eighth primary, one important difference does exist between the New World quails and at least those Old World perdicines that have been studied. In the New World odontophorines both the ninth and tenth primaries as well as their associated greater coverts are apparently invariably retained, whereas in the Old World forms the outer greater coverts are often molted during the post-juvenal moult, regardless of whether or not their associated primaries are also replaced. Thus, since these covert feathers are typically marked with lighter or more buffy tips in odontophorines, they provide an alternative and often more obvious method of identifying first-year birds in this group. In the Old World perdicines it is usually necessary to examine the outermost one or two (rarely up to four) primaries for signs of unusual wear and fading as a basis for recognizing first-year birds (Fig. 5).

Use of the rates and timing of growth of the first eight juvenal primaries has additionally been found to be an excellent method of judging the age of young

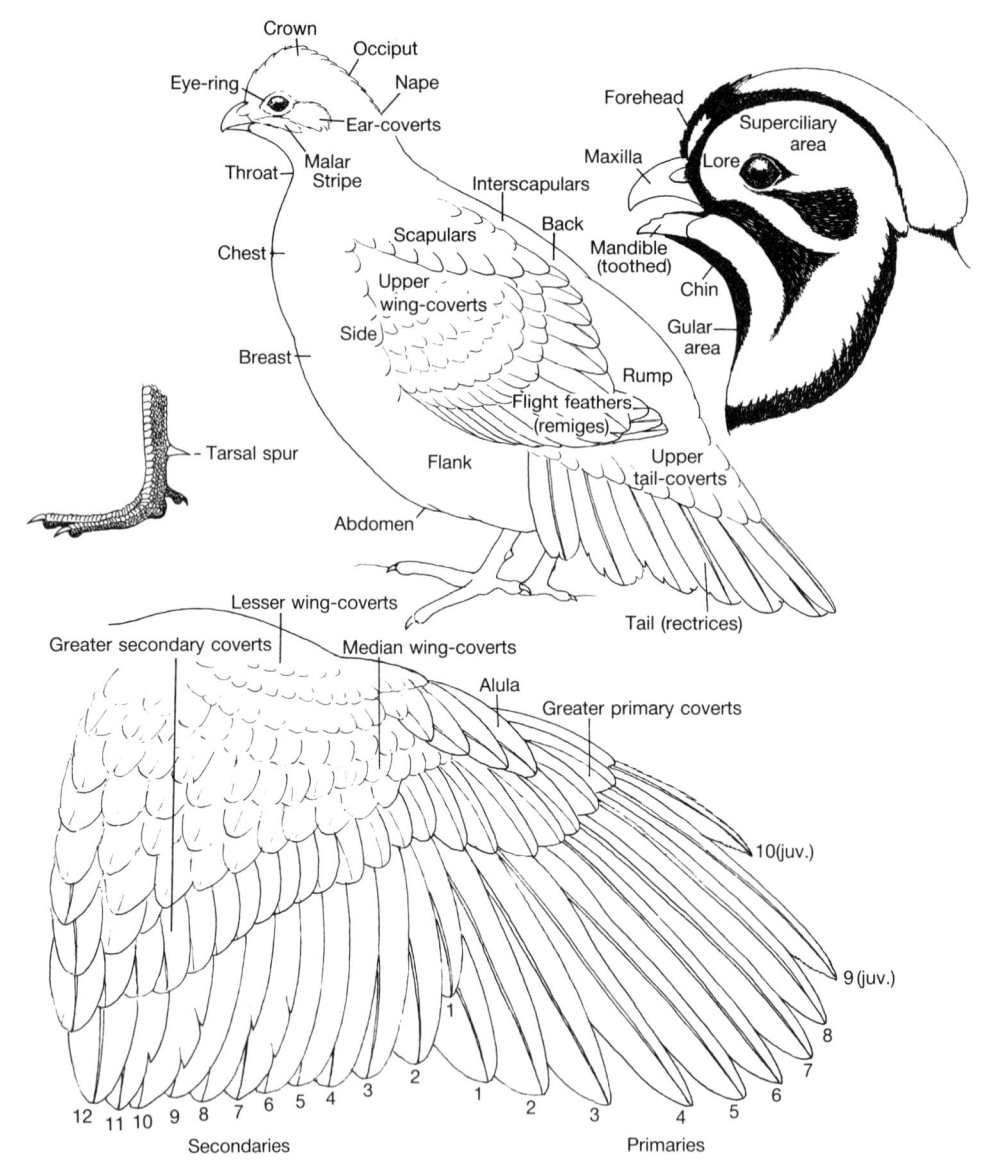

Fig. 5. Diagram identifying major external morphological features of galliforms, and method of numbering remiges (indicated condition of outer primaries is typical of immature birds).

phasianids for their first few months of life, which in turn provides a useful method for judging hatching dates in samples of first-year birds. Johnsgard (1973) summarized such primary-growth data for four species of New World quails plus the grey and red-legged partridges. A somewhat different and more comprehensive method for using this same kind of information is presented in Tables 15–18 as a means of estimating the ages of northern bobwhite, California quail, scaled quail, and grey partridge. Comparable information for the rock partridge has been provided by Menzdorf (1975e) and Disney (1978), and for the stubble quail by Crome et al. (1981). In this latter species only seven of the juvenal primaries are typically shed during the post-juvenal moult. Similarly, observations on both the European and Asian migratory quails indicate that the post-juvenal primary moult of different individuals may variously terminate at the sixth, seventh, or eighth primaries, and that the variable timing of the shedding of these outer primaries greatly restricts their value as an age-determination technique after about a month following hatching (Lyon 1962). A limited amount of variability on the degree of primary moult has also been reported in northern bobwhites, with some late-hatched birds retaining their seventh to tenth juvenal primaries, while others may have a precocious post-juvenal moult of the ninth, or rarely even the tenth primary (Johnsgard 1973). Lyon (1962) reported that although the Asian migratory quail's juvenal primary coverts are shed at the same time as their corresponding primaries, the replacement primary coverts of young birds are, nevertheless, different from those of older age-classes, with most young birds lacking the whitish rachis colour typical of the outer four or five primary coverts in adults. These feathers are also almost always more pointed and frayed in first-year birds than are the coverts of adults, and provide a useful device for estimating age exclusive of the condition of the outermost primaries.

Post-juvenal moult and first-winter or first-nuptial plumage

The post-juvenal (or 'prebasic') moult gradually replaces the juvenal body feathers during the first autumn of life, so that the bird gradually assumes its definitive first-winter (or 'first-basic') plumage. In the genus *Coturnix* an unusual second juvenal plumage is assumed with the post-juvenal moult, which is then followed by a second post-juvenal moult and assuming of the first-winter plumage (Wetherbee 1961). In many of the species considered here, little or

Table 15. Estimated age (in days) of northern bobwhite based on lengths of actively growing first-winter primaries[1]

Primary length (mm)	Primary no.							
	1	2	3	4	5	6	7	8
0–10	29–31	36–38	43–44	48–49	55–57	63–65	76–78	104–106
11–20	32–34	39–41	45–47	50–51	58–59	66–68	79–81	107–109
21–30	35–39	42–44	48–49	52–53	60–61	69–70	82–85	110–113
31–40	40–43	45–47	50–51	54–55	62–63	71–72	86–92	114–117
41–50	44–45	48–50	52–53	56–58	64–66	73–75	93–96	118–122
51–60	46–50	51–53	54–56	59–61	67–70	76–79	97–101	123–126
61–70	51–56	54–56	57–60	62–65	61–77	80–86	102–110	127–129
71–80		57–58	61–62	66–67	78–82	87–100	111–119	130–138
81–90					83–85	100–103	120–124	139–150+
Final average length	66	71	72	74	85	84	85	86
Age when fully grown (days)	56	58	62	67	85	103	124	150

[1] Adapted from Rosene (1969).

Table 16. Estimated age (in days) of California quail based on lengths of actively growing first-winter primaries[1]

Primary length (mm)	Primary no.							
	1	2	3	4	5	6	7	8
0–10	28–32	30–35	37–41	44–50	52–58	62–68	73–81	100–106
11–20	31–33	35–38	41–43	50–52	58–61	68–72	82–87	106–110
21–30	33–36	38–40	43–47	52–54	61–63	72–75	88–90	110–113
31–40		40–43	47–49	54–57	63–65	75–78	90–94	113–116
41–50			49–51	57–59	65–68	78–81	94–97	116–120
51–60				59–62	68–71	80–83	97–100	120–123
61–70					71–74	83–87	100–105	123–127
71–80						87–91	105–109	137–134
81–90							110–120	134+
Final average length	65	72	77	85	87	87	85	83
Age when fully grown (days)	55	62	70	80	90	108	121	141

[1] Estimated from graphic data presented by Raitt (1961).

Table 17. Estimated age (in weeks) of scaled quail, based on lengths of actively growing first-winter primaries[1]

Primary length (mm)	Primary no. 1	2	3	4	5	6	7	8
0–10	4	5	5	6	7	9	11	13
11–20	4	5	5	7	8	9	11	14
21–30	5	5	6	7	8	9	11	15
31–40	5	5	6	7	8	10	12	15
41–50	5	6	7	8	9	10	12	16
51–60	6	6	7	8	9	11	13	16
61–70	7	7	7	8	10	11	13	17
71–80	8+	8+	8	9	10	12	14	18
81–90	—	—	9+	10	11	13	15	19
91–100	—	—	—	12+	12+	14+	16+	20+
Final average length	73	78	86	94	94	95	93	88
Age when fully grown (days)	50	60	65	85	85	93	115	150

[1] Adapted from Ohmart (1967) and Smith and Cain (1984).

Table 18. Estimated age (in days) of grey partridge based on lengths of actively growing first-winter primaries[1]

Primary length (mm)	Primary no. 1	2	3	4	5	6	7	8
0–10	23–25	27–28	31–33	38–40	27–49	56–58	67–69	83–86
11–20	25–26	28–30	33–35	40–42	49–50	58–60	69–71	86–89
21–30		30–31	35–37	42–43	50–52	60–62	71–74	89–92
31–40			37–38	44–45	52–54	62–64	74–75	92–94
41–50			38–39	46–47	54	64–65	76–78	95–97
51–60				48	55	66–67	78–80	98–100
61–70							80–82	100–103
71–80							82–84	103–106
81–90							84–87	106–108
91–100								109–111
101–110								112–114
111–120								114–117
121–125								117+
Final average length	89	92	97	108	116	120	120	119
Age when fully grown (days)	49	56	63	70	77	91	105	126

[1] Adapted from Demers and Garton (1980).

no additional moulting occurs after the assumption of the first-winter plumage and the first breeding season, so that the first-winter plumage may in fact usually represent the bird's first breeding or 'nuptial' plumage. It appears that in some taxa of New World quails, such as the Cuban race of the northern bobwhite, a fairly extensive prenuptial (or 'pre-alternate') moult may occur prior to breeding (Watson 1962c), but this may well be an exception to the general rule. Sexual differences in plumage typically begin to appear following the post-juvenal moult, and are reflections of sex hormone effects on feather-follicle regulation of feather pigment deposition and, occasionally, feather length and shape. Similarly, the development of secondary sexual characteristics such as tarsal spurs, iris, and bill coloration, and the development of bare facial skin are also regulated by sex hormone activity.

Post-nuptial moult and succeeding plumages

Following the breeding season, apparently all species of the groups considered here normally undergo a complete moult, during which all of the body feathers, remiges, and rectrices are replaced. At this time any feathers still remaining from the juvenal plumage will be lost, and a plumage indistinguishable from older adult age-classes will be attained. The wing feathers are replaced in orderly sequence, in roughly the same order in which they had initially appeared. At this time the outermost primaries that typically have persisted from the juvenal plumage will normally be moulted. However, at least in the northern bobwhite a variable but often substantial percentage of adult females exhibit incomplete primary moults, especially during years of unusually wet or cold springs, and retain their outer two or three primaries at least through a second autumn (Stanford 1972b). The tail moult in all adult New World quails and Old World perdicines so far studied is centrifugal, proceeding from the middle pair outwardly.

At least in the California quail males begin their post-nuptial moult about a month in advance of females, but the rate of feather replacement is slower, so that both sexes complete their moult at about the same time. During those years when nesting may be terminated early because of unfavourable conditions the moult begins earlier in both sexes, but during years of protracted nesting the moult cycle may be delayed for as long as a month (Leopold 1977).

The second-year breeding plumage (termed the second annual or second nuptial, depending on whether one or two intervening moults are involved) is acquired in the same manner as the first, and later moults and plumages are likewise repetitions of earlier ones.

Ontogeny of behaviour and vocalizations

Although studies of the rock partridge by Menzdorf (1975c, 1982) indicate that the female lacks an innate recognition ability for her own eggs, the newly hatched chicks are eventually recognized individually by the female. They are guided from the nest shortly after hatching, and the physical abilities of the chicks thereafter tend to influence the daily activity pattern of the brooding female. Rock partridges are visually able to discriminate both shapes and colours; chicks show a preference for reds, whilst adults seemingly prefer greens. The chicks have an innate food-selection scheme (round objects preferred over oblong ones, and moving ones preferred to stationary ones), but must learn to distinguish edible from non-edible objects. Hand-reared chicks typically exhibit escape behaviour toward slow-moving and dark-coloured inanimate objects that resemble predators. This escape behaviour toward 'dummy predators' is evidently an innate response that is further modified during ontogeny. Experienced immature as well as adult birds are most prone to exhibit escape behaviour when they are exposed to models that move at a slow and constant speed. The size, shape, and colour of the models also influence the kind of escape behaviour that is evoked (Menzdorf 1975d).

Very young red-legged partridge chicks sometimes perform dustbathing behaviour; Goodwin (1953) observed a chick that began to dustbathe, using characteristic movements, when only about three hours old, after encountering some fine, dry earth. Similarly, downy chicks perform sunbathing behaviour as an apparent substitute for brooding when seeking a source of warmth. Chicks often cease to be brooded by adults when they are about 5 weeks old, and thereafter may huddle in groups when chilled. No doubt this kind of behaviour leads directly to the formation of circular roosting clusters among adults of various species.

Newly hatched rock partridge chicks can travel up to about 200 metres after 2–4 days, according to Menzdorf (1982). The female typically indicates the location of food items to chicks by pointing at such morsels and uttering a food call; tape-recorded food calls of a rock partridge were found to be more attractive when played to chicks of this species than were food calls uttered by a bantam hen foster mother. Chicks also respond to their mother's hawk-alarm call by looking for cover and crouching, and they maintain this posture until their mother calls them again. By their second week after hatching the chicks learn to distinguish flying objects visually (Menzdorf 1982).

The ontogeny of calling behaviour follows a pattern of gradual maturation of call number and function. Newly hatched rock partridge chicks utter four types of calls, including a lost call (which is similar to the sounds emitted by chicks just prior to hatching), a contact call, and a contentment call. Within three or four weeks after hatching a cackling call develops, which is the apparent precursor of the important adult rally call. The adult rally call, which initially occurs after the 'breaking' of the voice at about 50 days, also marks the beginning of the post-juvenal moult into the definitive adult plumage (Menzdorf 1976a). By the time the birds are adults they exhibit a total of 19 calls, of which eight are associated with sexual and/or aggressive behaviour, five are related to parental activities, and four are evoked by predators. The remaining two are the contact call, the most frequent of all vocalizations, and the louder rally call, which helps to reassemble scattered coveys in winter and also is used to help keep mates and families together during summer (Menzdorf 1977a). In this species as well as in *chukar* the rally call additionally serves as the primary male territorial advertisement call (Stokes 1961), while in other *Alectoris* species such as *rufa* and *barbara* the 'steam-engine' and advertising calls, respectively, evidently have this function (Cramp and Simmons 1980).

A study by Kochenderfer (1971) on the ontogeny of vocalizations in the northern bobwhite indicated a gradual development and diversification of adult vocalizations from three precursor calls in chicks, including a separation call (equivalent to the 'lost call' of rock partridges), a contact call, and a hand-held alarm or distress call. Juvenile birds gradually add two additional alarm notes while at the same time retaining their original three chick calls. As sexual maturity approaches the number of vocalizations increases, with the 'bobwhite' and copulation calls beginning at 15 weeks. The separation call takes on additional functions with maturity, such as serving to maintain pairs and attracting unmated birds as well as helping to reassemble groups, and various additional agonistic calls also appear with sexual maturity. Injection of testosterone in chicks resulted in an earlier-than-normal onset of reproductive and agonistic calls.

Observations by Stokes (1967) on the northern bobwhite as well as the California quail also indicated that the 'lost' call of chicks develops onto-genically into the adult separation call; Stokes thus regarded the latter as a 'primitive' call. Similarly, the 'lost' call of scaled quail chicks is analogous to the adult separation call (Anderson 1978). In both the California quail and the Gambel's quail the unmated male's single-syllable advertisement call is clearly acoustically derived from the multisyllable adult separation call, and thus seems to be functionally related to these more generalized distance-reducing calls of both chicks and adults. Similarly, the advertising 'shriek' of the unmated scaled quail is often interspersed with the 'chekar' (or 'pay-cos') separation call uttered by birds separated from their mates, and the former is acoustically very similar to the second syllable of the latter (Anderson 1978).

Annual cycles of seasonal reproductive activity

Perhaps because of the tropically oriented distribution of these birds, there is little evidence that annual photoperiodic changes are primary proximate factors in regulating gonadal cycles for most species. Kirkpatrick's (1955) work did indicate the presence of photoperiodic stimulation in the northern bobwhite, the most northerly of the New World quails, and there is similar evidence as to the effects of photoperiodism on gonadal stimulation and consequent possible regulation of breeding cycles in the equally northerly Asian migratory quail (Tanaka *et al.* 1965; Robinson 1980). However, unlike many northern hemisphere and migratory passerines, neither of these species appears to have an absolute photorefractory period during which reproductive activity cannot be induced by exposure to artificially increased photoperiods. Recent work by Sharp *et al.* (1986) on the grey partridge indicate that a typical long-day photorefractory period does occur in that species, which apparently is terminated by exposure to shorter daylight periods during autumn.

In Texas, rainfall rather than photoperiod appears to be a significant factor in seasonally stimulating male northern bobwhites to begin uttering advertising calls. Spring rainfall also seems to provide a general stimulus for females to begin nesting activity, perhaps through increasing the availability of vitamins or some other critical nutrient that is acquired from the consumption of actively growing green plants (Lehmann 1984). Roseberry and Klimstra (1983) judged that in southern Illinois the onset of nesting by northern bobwhites was not clearly influenced by annual variations in rainfall and their consequent effects on vegetative growth, although annual rainfall variations did influence overall productivity levels.

Leopold (1977) has reviewed the possible influences of rainfall both as a potential stimulus to breeding activity and as a determinant of reproductive success in the California quail. He judged that rainfall's stimulating effects on breeding, at least in arid or semi-arid areas, are most likely to be produced through correlated nutritional or other physiological mechanisms. Although recent evidence now has turned attention away from vitamins as a probable direct stimulus to breeding in the California quail, it is possible that some other annually variable inhibitory or stimulatory factor that is similarly associated with the consumption of green vegetation may be responsible. One of these might be the presence of phytoestrogenic compounds (isoflavones) that accumulate in leaves of some food plants during dry years and apparently tend to suppress quail reproduction through their negative feedback effects on the anterior pituitary's gonadotropic activity, but which are absent in rapidly growing leaves during moist years. Or, perhaps the large quantities of newly formed legume seeds that are typically consumed during years of favourable rainfall provide important nutritional benefits that could have stimulatory effects on the nesting physiology of females (Erwin 1975).

5 · Adult vocalizations and non-vocal behaviour

Adult northern bobwhites exhibit a minimum of 19 distinct calls (Johnsgard 1973), with five additional call variants being recognized as distinct by Stokes (1967). As with the similar 19-call repertoire of the rock partridge (Menzdorf 1977), the largest number of these (seven) are sexual and/or agonistic in function. Five others are associated with avoidance of enemies, five more with pair or group activities, and two are specifically associated with parental behaviour. Stokes (1961) described a very similar array of 14 adult calls in the chukar partridge, more than half of which are sexual or agonistic in function, and seven of which are apparently limited to males. A total of 14 adult calls and their functions have been reported for the California quail by Williams (1969), who classified six as reproductive in function, five as associated with alarm, two with social contact, and one with parental care. Only three of the calls were limited to males. Ellis and Stokes (1966) described 10 adult calls of the Gambel's quail, two of which are restricted to males and one to females.

Anderson (1978) has described 10 adult calls of the scaled quail, four of which are associated with aggregation and contact, three with alarm or distress, three with sexual attraction, and one with agonistic encounters. Only two of these (the advertisement shriek and the agonistic squeal) are uttered primarily or entirely by males; the majority of the vocal repertoire is shared by both sexes, a situation that appears to be fairly typical of New World quails but may be less typical of perdicines. Thus, in the European migratory quail the adult calls are mostly sexually dimorphic, with the male uttering five distinctive calls, the females two, and both sexes having three calls in common (Cramp and Simmons 1980). Observations summarized above on several *Alectoris* species indicate a considerable number of male-limited calls as well, suggesting that sexual dimorphism in vocalizations may be more important in proclaiming sexual identification and in asserting male dominance or facilitating spacing behaviour in the perdicines than in the odontophorines.

Several studies have been made on the behaviour of the Asian migratory quail, of which that by Eynon (1968) is perhaps the most complete. He determined that adults of both sexes have three calls in common (distress, a flushing call, and a 'flocking' call uttered by grouped birds), and that males have five additional calls while females have at least four. Three of the male calls (crowing, churring, and kekking) are concerned with aggressive or dominance assertion tendencies, a distress call is uttered by birds who are being pecked, and a weak call is uttered during heterosexual tidbitting behaviour. Females utter a 'long call' that sexually excites males, as well as an aggressive call, a greeting chatter (which was also heard in one hand-reared male), and an 'egg-laid' call associated with oviposition. Except perhaps for the flocking call, none of these is clearly associated with maintaining social aggregation and contact behaviour, and most are clearly agonistic in context, suggesting a low tendency toward social integration in this species. Similarly, nearly all of the postural displays of males are agonistic or dominance-related, including the most elaborate 'full strut' display, which not only is used between males during agonistic encounters but also sometimes precedes copulation and always follows it. Eynon did not observe the lateral 'waltzing' behaviour typical of dominance assertion in many male perdicines, but he did note that the male's full strut display may be either frontal or lateral in orientation.

Stokes (1967) provided a highly detailed inventory of the maintenance, agonistic, sexual, and parental behaviour and vocalizations of adult northern bobwhites, together with their situations of occurrence and the probable functions of many of these behaviours. He found that a confined group of seven males developed a straight-line social hierarchy that was established by display, calling, and actual attacks. Males who managed to maintain a position beside an introduced female rose in the social hierarchy, and the presence of such females in the group always increased the level of agonistic behaviour among the males. On the other hand, females confined as a group exhibited little agonistic behaviour among themselves except when a strange female was added.

Similarly, Ellis and Stokes (1966) found that much of the hostility they observed in Gambel's quail was associated with male sexuality, and that introducing a female into a pen of sexually deprived males would typically elicit fights. Such fighting was initially general and directed toward the female. However, one male soon would become the female's 'champion', thereafter defending her from attacks by the other males. Regardless of his social rank, no such male was found subsequently to lose his attachment to that female, even though more dominant males might later interrupt their courtship or copulation activities.

In a somewhat similar manner, Goodwin (1958) reported that pair formation in the red-legged partridge apparently requires some degree of mutual fear and hostility between the two birds; in his experience individual birds that have had prolonged social contacts with one another, such as those that have been together since they were chicks, do not form pair-bonds. Initial contacts between males and females are dominated by male displays that assert aggressive dominance, such as making the barred flank feathers more conspicuous by raising them and orienting them toward the bird evoking the displays, and simultaneously exposing the black-edged white throat by head-tilting and feather-ruffling. Goodwin judged (1953) that the typical lateral 'waltzing' display of this (and many other galliform) species, during which the near-side flank feathers are maximally exposed and the opposite wing lowered (perhaps for balance) as the displaying bird walks past or circles another, was generally or always aggressive in character, with overt attack being inhibited by fear or sexual tendencies. However, he also recognized the importance of food-calling by the male to the female, and associated symbolic courtship-feeding (or 'tidbitting') as an important part of sexual behaviour. This behavioural pattern seems to be widespread in the Phasianidae and probably serves a similar sexual or pair-bonding function in all the groups where it occurs (Johnsgard 1986). Illustrations of several of the major postural components of calls and displays important in the social behaviour of *Alectoris* are provided in Fig. 6, based primarily on sketches of *rufa* by Goodwin.

Menzdorf's (1976b, 1982) extensive studies of the rock partridge included descriptions of four forms of stationary postures, four postural forms of aggressive behaviour, three submissive postures, overt fleeing by walking, running, or flight, and two defensive responses to terrestrial and aerial enemies. He stated that courtship consists of specific sequences of behaviour patterns that typically begin with aggressive chasing of the hen by the male and terminate with copulation. A nearly identical sequence of courtship behaviour was described for the chukar partridge by Stokes (1961), who believed that 'tidbitting' by the male has an important female-appeasement function, and observed that copulation was most likely to occur after a series of tidbitting displays, but rarely occurred during waltzing sequences. Stokes also observed a 'nest ceremony' display, during which the male enters a clump of vegetation and performs nest-scraping movements. This ceremony apparently serves a valuable function in keeping the male closely oriented to his mate during the egg-laying and incubation phases of reproduction, and may be even more important in orienting the male to the future nest location in the event that he takes over incubation of the first clutch. Stokes found that pair formation was a subtle process and early in the pair-forming season was apparently quite gradual, but late in the season a male might pair with a hen within 24 h of being placed in the same pen. Although Stokes' penned birds were not obviously territorial, his observations of birds in the wild suggested that under such conditions they hold well-defined territories that are defended vigorously.

Stokes later (1963) listed 8 calls and 14 postural signals or behavioural sequences that occur during agonistic and sexual interactions of the chukar partridge. He found that penned males have a straight-line dominance hierarchy that is rapidly established after initial contacts by aggressive displays and fighting, whereas females are highly tolerant of one another. Initial male aggressive tendencies toward both sexes serve to disperse rival males and allow displaying males to recognize females by the latter's failure to respond with similar aggressive displays. Specifically, the male's waltzing display thus identifies his sex and establishes his dominance over females, while tidbitting serves to appease the female and induces her to crouch for copulation. The later nest ceremony probably helps to strengthen the social bond between the pair.

The best descriptions of social behaviour in *Perdix* are those of Jenkins (1961a), who studied almost 400 individually marked wild birds. He observed that coveys usually remained distinct, but with some interchange of individuals occurring in autumn, especially among unsuccessful breeders, and again in January, if yearlings were left alone as a result of early pairing by the adults in their coveys. Fighting occurred most often in late winter, when as many as 50 birds would engage in fighting, both between and within coveys. Then males would chase and fight with hens, brothers would chase brothers, and sisters chase sisters. Males would sometimes peck potential mates, and parents would even attack their own offspring. Much of this threat and fighting activity eventually actually led to pair formation, suggesting a strong agonistic element in sexual bonding. Thus, several 'courtship' displays by males toward females in this species are little if at all different from purely agonistic displays performed toward other males

Fig. 6. Postures associated with social behaviour in *Alectoris*, including (A) flight-intention call; (B) male advertising ('steam-engine') call; (C) separation ('rally') call; (D) cautious approach; (E) fleeing; (F) submissive crouching; (G) aggressive head-tilting; (H) head-tilting with wing-lowering (from far side); (I) waltzing (from far side); (J) tidbitting; (K) copulation; and (L) brooding. Adapted from sketches of *rufa* by Goodwin (1953), excepting I and K, which are after photos of *chukar* by Stokes (1961).

(Fig. 7.) Male threat displays are marked by an erect posture, a ruffled neck, and exhibition of the chestnut abdominal pattern. This posture is often accompanied by repeated harsh *tit* or more querulous *kee-erick* notes. The latter 'rusty gate' call is typical of unmated males, and is often uttered from an elevated site, thus serving both as a threat and a self-advertisement display (Cramp 1980). This 'upright-threat' posture was not observed by Jenkins in females, although alert adults of both sexes assumed a similar but silent upright and watchful posture, with the breast not thrust forward nor with the neck

34 *The Quails, Partridges, and Francolins of the World*

feathers so ruffled. According to Jenkins, 'courtship' postures performed by males toward females closely resembled those associated with male threat, but were usually seen only during and after the pair-forming period, and additionally the male's behaviour was generally reciprocated by the female. The barring on the male's flanks appeared to Jenkins to be of special visual significance during courtship. The female would direct her display toward this area or toward the breast patch by extending her neck, waggling her head, and making sinuous movements of the neck. Eventually the two birds might stand breast-to-breast, the female pointing her bill skyward, rubbing her neck along his, and the two birds sometimes rubbing their beaks together. Copulation was not preceded by specific displays, but instead the female would squat and the male, if responsive, would approach and mount her. Copulation usually occurred when the pairs were alone, but sometimes was performed while still in coveys.

A descriptive and pictorial inventory ('ethogram') of the range of adult egocentric and social behaviour

Fig. 7. Postures associated with social behaviour in the grey partridge, including: (A) male threat-upright; (B) sinuous-neck display of female; (C) male courtship-upright display; (D) intense display by male; (E) upright-alert posture; and (F) pre-copulatory behaviour. After Jenkins (1961a).

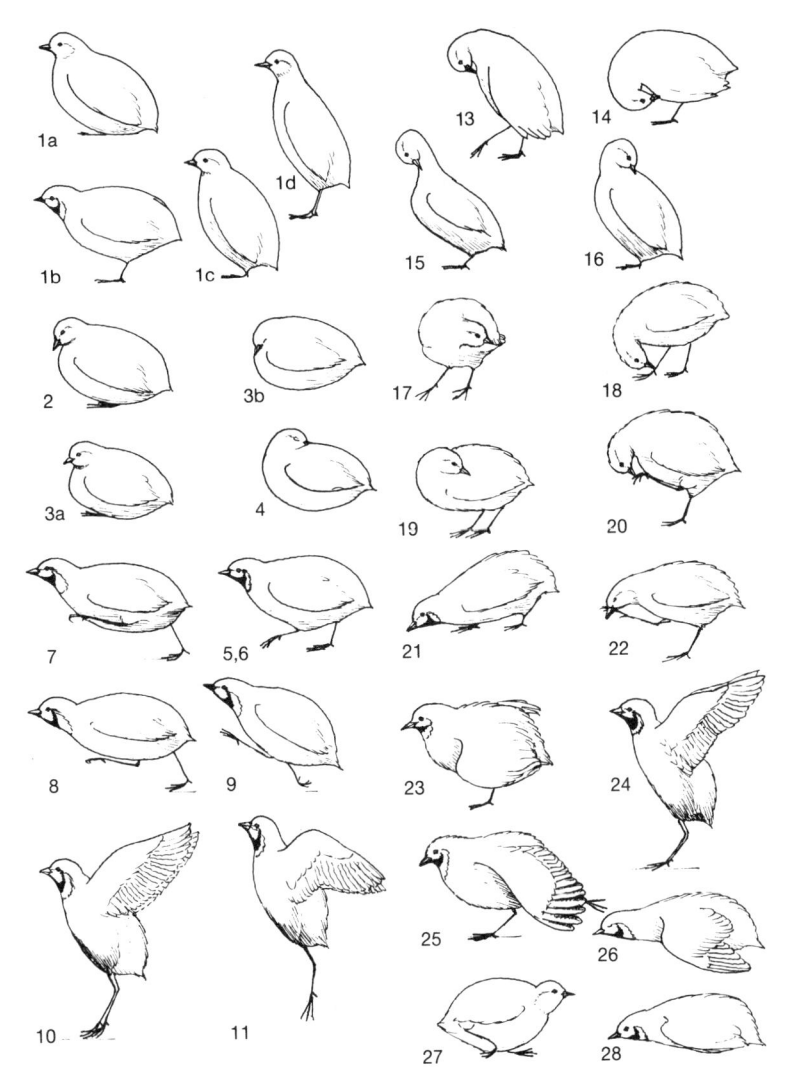

Fig. 8. Postures associated with egocentric behaviour in the Asian blue quail. Numbers refer to postures as listed in associated text. After sketches by Schleit *et al.* (1984).

patterns of the Asian blue quail has been presented by Schleidt *et al.* (1984), who recognized the following categories of posturing (here slightly condensed and reorganized):

Egocentric ('individual') behaviour
    Stationary postures (both sexes)
        1a–d. Standing (four postural subtypes)
        2. Sleep-like resting while standing
        3a, b. Sitting (resting or sleeping)
        4. Back-sleeping while sitting (bill in scapulars)

    Locomotory movements (both sexes)
        5. Walking
        6. Pacing
        7. Running
        8. Dashing
        9. Leaping
        10. Hovering
        11. Take-off
        12. Airborne (sustained flight)

    Comfort activities (both sexes)
        13–18. Preening (neck-, breast-, belly-, back-, tail, and wing-preening)
        19. Leg-picking
        20. Toe-picking
        21. Beak-wiping
        22. Head beak-scratching
        23. Body wing-shaking
        24. Wing back-stretching
        25. Wing- and leg-stretching
        26. Sunning
        27. Dustbathing with scratching
        28. Dustbathing with substrate-rubbing

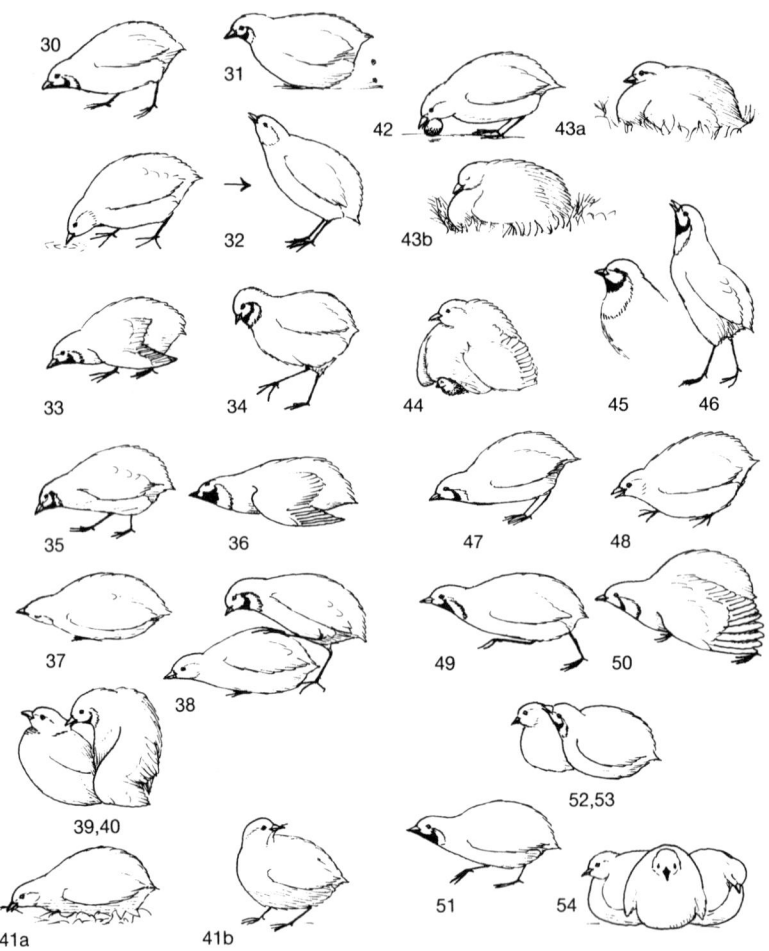

Fig. 9. Postures associated with egocentric and social behaviour in the Asian blue quail. Numbers refer to postures as listed in associated text. After sketches by Schleit *et al.* (1984).

    29. Dustbathing with wing-beating

Ingestion–egestion activities (both sexes)
    30. Foraging (food ingestion)
    31. Defecation
    32. Drinking

Social ('interindividual') behaviour
  Mating and courtship
    33. Invitation to preen (with call) (males)
    34. Strutting (males to females)
    35. Tidbitting (with call) (males to females)
    36. Body-down, wings-out posture (males to females)
    37. Invitation to mate (with call) (females to male)
    38. Mounting (males)
    39. Copulation, active partner (males)
    40. Copulation, passive partner (females)
  Nesting behaviour
    41a, b. Nest-building: standing or sitting (females)
    42. Egg-rolling (both sexes)
    43a, b. Incubation posture, alert or sleep-like (females)
  Parental behaviour
    44. Brooding posture (females)
  Agonistic behaviour
    45. Hissing posture (and call) (males)
    46. Crowing posture (and call) (males)
    47. Head-down, tail-up posture (both sexes)
    48. Crouching posture (females to males)
    49. Wings-normal attack (males)
    50. Wings-down attack (males)
    51. Defensive walk (males to males)
  Miscellaneous social behaviour (both sexes)
    52. Mutual preening, active bird
    53. Mutual preening, passive bird
    54. Huddling (contact resting)

Outline drawings of nearly all of the behavioural categories listed here are presented in Figs 8 and 9. The original listing recognized a total of 60 distinct and commonly observed adult postures and activities exclusive of additional non-posturally defined vocalizations. Each of these was morphologically described by 17 postural variables, providing an almost astronomic number of descriptive permutations. Consideration of associated vocalizations, when present, would, of course, increase the number of potentially measurable variables even more. Even the somewhat condensed listing shown here provides an indication of the difficulties of documenting all the behavioural variations and possible social communication signals ('displays') of a single species, not to mention the attendant problems of describing and differentiating them as well as trying to determine their possible social functions.

In spite of the very large number of species included in this book, there is seemingly a fairly limited degree of variation in adult social behaviour patterns discernible within both the odontophorines and the perdicines. This situation seems to be a reflection of the general similarities of sociality and mating systems occurring within these groups, at

Fig. 10. Posture associated with high-intensity frontal display in the northern bobwhite, compared with lateral wing display (waltzing) in the Asian blue quail. Above: original based on author's photo. Below: after Kruijt (1962).

least as compared with the pheasants, in which diverse spacing mechanisms and mating systems have led to a myriad of conspicuously variable heterosexual social signals, sometimes even within fairly closely related groups. Among the odontophorines and perdicines species-specific visual signalling devices important in sexual and agonistic interactions appear to be concentrated on the head, throat, breast, and flanks, but few if any such signal elements are situated in the wing and tail (or tail-covert) areas, which in pheasants are often modified for display function. (The chestnut or otherwise distinctively coloured outer tail feathers of some perdicines are probably associated with signals concerned with take-off and flight rather than being used primarily during ground displays, when they are often mostly or entirely hidden.) It is also interesting that in the perdicines lateral wing display typically involves the lowering of the more distant wing and associated maximum exposure of the near-side flank feathers [Fig. 10(b)], whereas in most but not all pheasants the near-side wing is lowered and the tail is correspondingly tilted, exposing both of these frequently colourful feather surfaces to the visual field of the 'target' bird but often hiding the near-side flanks. Among the odontophorines, which seemingly lack lateral wing displays but often perform frontal displays during agonistic interactions (including initial sexual encounters), there is a comparable concentration of frontally visible male signals in the form of elaborate crests and contrasting throat patterns [Fig. 10(a)]. Similarly, conspicuous and functionally analogous male throat patterns exist in various perdicines such as *Alectoris* and *Coturnix*, but a definite crest is present in only one perdicine genus (*Rollulus*). Finally, the taxonomically widespread tidbitting behaviour of phasianids (Stokes and Williams 1972) is seemingly not associated with the exhibition of any specific plumage specializations (although the bright bill and/or eye coloration of some phasianid taxa may represent significantly correlated visual signals) and instead may have been ritualized primarily through the evolution of associated, sometimes apparently species-specific, vocalizations.

II · SPECIES ACCOUNTS

1 · Subfamily Odontophorinae

The New World quails are a group of small[1] to medium-sized gallinaceous birds comprising nine genera and 31 species that range from southern Canada south to southern Brazil and northern Uruguay, in temperate to tropical habitats. Introductions have also been made well beyond the group's native range. They differ from the Perdicini in that the lower mandible edge is always serrated or slightly toothed, the tarsus is never spurred in adults of either sex, and the juvenal primary-coverts are not moulted and replaced during the post-juvenal moult. In all species so far studied the post-juvenal moult of the remiges is incomplete, with two juvenal primaries typically retained in the first-winter (first-basic) plumage. Plumage sexual dimorphism ranges from none to moderate, and size dimorphism of the sexes is slight. Irisdescent plumages and tarsal spurs are lacking in all species, although a few have eye-rings or larger areas of colourful bare skin around the eyes. Vocalizations are apparently important social signals, especially in forest-dwelling species, which in some cases are known to utilize duetting for pair interactions. As with the Perdicini, ritualized courtship-feeding or 'tidbitting' behaviour is apparently an important pair-bonding mechanism, but the 'waltzing' display typical of the Phasianinae is either absent or very poorly developed. Monogamous pair-bonding is typical of all the species so far studied. No species is known to undertake long migrations, although considerable seasonal altitudinal movements do occur in at least one montane form (*Oreortyx*). Most species are gregarious out of the breeding season, forming assemblages ranging from small coveys to large flocks. Most are primarily seed-eaters, but the more forest-adapted forms may be largely insectivorous. Nests are usually open scrapes, but in some tropical forest forms (*Odontophorus*) may be well-hidden, roofed-over structures. The average clutch-size ranges from 5 to 15 eggs, and the incubation period is from 22 to 30 days. The young are precocial, often fledging in about two weeks, and breeding during the first year of life is regular.

KEY TO THE GENERA OF ODONTOPHORINAE

A. Tail long (over 105 mm), considerable bare red skin present behind eye: *Dendrortyx* (3 spp.)
AA. Tail less than 95 mm; little or no bare skin evident behind eye
  B. Tail less than half the length of the folded wing; a bushy crest of soft, broad feathers usually present; tips of extended feet reach beyond tail
    C. Claws elongated with tips of lateral claws extending beyond base of middle claw
      D. Tail feathers soft, narrowing toward tips; flattened crest at nape: *Cyrtonyx* (2 spp.)
      DD. Tail feathers firm, broad, and with rounded tips; no definite crest: *Dactylortyx* (1 sp.)
    CC. Claws not elongated; tips of lateral claws not reaching base of middle claw
      D. Tail over 60 mm long, of 12 rectrices: *Odontophorus* (14 spp.)
      DD. Tail under 50 mm long, of 10 rectrices: *Rhynchortyx* (1 sp.)
  BB. Tail longer than half the length of the folded wing; virtually crestless or bearing a distinct crest near the front of head; tips of extended feet not reaching end of tail
    C. Scapulars and inner secondaries spotted; tail under 70 mm long
      D. With an erect, narrow crest; sides and flanks vertically barred: *Philortyx* (1 sp.)
      DD. Either crestless, or with a short, upturned, and unbarred crest; not vertically barred on flanks: *Colinus* (4 spp.)
    CC. Scapulars and inner secondaries unspotted; tail over 70 mm long
      D. Crest long (over 60 mm), folded wing over 120 mm: *Oreortyx* (1 sp.)
      DD. Crest shorter, wing under 120 mm: *Callipepla* (4 spp.)

---

[1] In this and the following descriptive taxonomic diagnoses, 'small' indicates those species with average wing-lengths of less than 125 mm, medium those with wing-lengths from 125 to 175 mm, 'large', those with wing-lengths from 175 to 225 mm, and 'very large' those over 225 mm.

2 · Genus *Dendrortyx* Gould 1844

The tree-quails are medium-sized montane and forest-adapted species of Mexico and Central America with long tails of 12 rectrices that are sometimes as long as the wing. The wing is rounded, with the fifth and sixth primaries the longest. Some juvenal greater primary-coverts and the two outer primaries are retained in the post-juvenal moult. The sexes are similar as adults, with relatively stout bills, bushy crests, and areas of bare skin around the eyes. Three largely allopatric species are recognized:

D. macroura (Jardine and Selby): long-tailed tree-quail
    D. m. griseipectus Nelson 1897
    D. m. macroura (Jardine and Selby) 1828
    D. m. diversus Friedmann 1943
    D. m. striatus Nelson 1897
    D. m. inesperatus Phillips 1966
    D. m. oaxacae Nelson 1897
D. barbatus Gould 1846: bearded tree-quail
D. leucophrys Gould: buffy-crowned tree-quail
    D. l. leucophrys Gould 1844
    D. l. hypospodius Salvin 1896

KEY TO THE SPECIES OF *DENDRORTYX*

A. Chin and throat grey; tail under 120 mm: bearded tree-quail (*barbatus*)
AA. Chin and throat not grey; tail over 120 mm
  B. Chin and throat white: buffy-crowned tree-quail (*leucophrys*)
  BB. Chin and throat black: long-tailed tree-quail (*macroura*)

LONG-TAILED TREE-QUAIL (Plate 1)

D. macroura (Jardine and Selby) 1828
Other vernacular names: long-tailed partridge; colin à longue queue (French); Langschwanzwachtel (German); codorniz coluda (Spanish).

Distribution of species (see Map 1)

Resident of highland forests of central and southern Mexico from Jalisco south-east to Oaxaca and Veracruz.

Distribution of subspecies

D. m. diversus: north-western Jalisco.
  D. m. griseipectus: Pacific slope mountains, in Distrito Federal, Mexico, and Morelos.
  D. m. macroura: valley of Mexico and Veracruz.
  D. m. striatus: southern Jalisco to Michoacan and Guerrero.
  D. m. inesperatus: vicinity of Chilpancingo, Guerrero.
  D. m. oaxacae: eastern Oaxaca.

Measurements

Wing, males 151–166 mm, females 141–158 mm; tail, males 138–169 mm, females 119–151 mm (Ridgway and Friedmann 1946). Weight, both sexes 350–465 g (Leopold 1959); males average 433 g, females 390 g (Johnsgard 1973). Egg, average 49.2 × 33.5 m (Warner 1959), estimated weight 30.5 g.

Identification

In the field (12–15 in.)

Associated with dense undergrowth of mountain forests of Mexico, and more often heard than seen. Its song, heard mostly at dawn and dusk, consists of a series of about five grunting, hooting notes that rise in volume and are followed by a loud, ringing *kor-eee'-oh* call that is repeated several times, and a similar series of phrases, spaced about a second apart, often uttered by a chorus. If seen, the long tail and the black throat markings set it apart from all other New World quails.

Map 1. Distribution of the bearded (Be), buffy-crowned (Bu), and long-tailed (L) tree-quails.

In the hand

The combinations of a long tail (almost half the length of the bird), a black throat, ear-coverts, and forehead, and a bare area of red skin around the eye, allow easy in-hand recognition.

Adult female. Like the male, but averaging smaller, and with a shorter tail.

Immature. Apparently with the two outer primaries frayed, and with buffy tips on the greater primary-coverts (Petrides 1942).

General biology and ecology

Like the other tree-quails, this species is associated with dense montane forests, where it is highly elusive and rarely seen. In Colima it normally occurs in the higher cloud forests but has been found in pine–oak forests of the Volcanes de Colima, and occasionally extends to the lower edge of the arid pine–oak zone. Similarly, in Oaxaca it has been found in the humid pine–oak and cloud forest zones, between 5800 and 9000 ft (1500–2700 m), and in Michoacan has been reported from fir forests and less commonly pine–oak forests at an elevation of nearly 9000 ft (2700 m) on Cerro Moluca. South of Mexico City the birds have been found in the least disturbed humid fir–pine–oak forests at elevations of 2800–3300 m. Even in favoured cloud forest habitats the density is probably quite low, in the vicinity of only about two pairs per hundred acres (Johnsgard 1973).

The only specific information on the foods of this species comes from Leopold (1959), who observed that a specimen he studied had a crop full of legume seeds, mostly tick trefoils (*Desmodium*). Warner (1959) stated that it consumes flowers, flower buds, and small fruits from arboreal perches, but also scratches for food in the leafy forest litter.

Local overlap with the bearded tree-quail occurs on Pico de Orizaba and Cofre de Perote in eastern Puebla and Veracruz (Leopold 1959), suggesting that possible competition may occur there, but otherwise little if any competition from other New World quails is likely.

Social behaviour

Nothing is known of possible territoriality in this species, but like other tree-quails it engages in daily bouts of 'singing' at dawn and dusk, apparently from pairs or coveys roosting in the same tree. Warner (1959) described these calls (see Field identification section above) as being uttered from February until July, which corresponds to the probable major breeding season.

Reproductive biology

Three nests of long-tailed tree-quails have been described to date. Warner (1959) found a nest during July in a semi-open pine–fir forest at 2900 m elevation, on a very steep mountain slope and in a tangle of brush. Dead branches jutted out over a 2-ft high rock face, forming a roof for a cavity about 2 ft wide and 3 or 4 ft long. A mat of leaves and woody materials formed a light-impervious roof, and a single opening about 6 in. wide led from the nest into the forest. A total of six eggs were found, the last two of which were laid within an approximate 25-h period. One of the nests, found by Rowley (1966), was also associated with a rock outcrop, while the other was in a slight depression at the base of a small shrub, but neither nest was well concealed. The first nest was found in mid-April (the time of locating the second was not reported) and each of the two nests contained four incubated eggs, suggesting to Rowley that the six-egg clutch found by Warner might have been the work of two females.

There is no information on the incubation period or on the growth of the young. Young birds have been collected as early as late May, and immatures (three-quarters grown) have been obtained as late as 1 December, suggesting a relatively prolonged nesting period and possibly the occurrence of double-brooding.

Evolutionary relationships

The view of Holman (1961) that *Dendrortyx* is the most anatomically generalized of all the New World quails is an attractive one from an ecological and zoogeographical standpoint. The nearest relatives of this genus are probably *Philortyx*, as suggested by Holman, or *Odontophorus*, as is implied by ecological and natal plumage similarities in these two genera (Johnsgard 1973). Within the genus *Dendrortyx* there is no clear evidence of relationships; all three species seem to be quite distinct from one another and the available evidence provides no overt clues as to their phyletic history.

Status and conservation outlook

Leopold (1959) stated, 'As long as the high mountain forests remain intact, tree-quails will be found in the thickets.' This assessment holds equally true today, although the rate of forest destruction in Mexico does not make this a particularly optimistic statement.

BEARDED TREE-QUAIL (Plate 2)

D. barbatus Gould 1846

Other vernacular names: bearded wood-partridge;

colin barbu (French); Bartwachtel (German); chiviscoyo (Spanish).

Distribution of species (see Map 1)
Resident of highland forests of eastern Mexico from Veracruz to eastern San Luis Potosi and eastern Hidalgo. No subspecies recognized.

Measurements

Wing, males 147–166 mm, females 148–152 mm; tail, males 117–121 mm, females 110–119 mm (Ridgway and Friedmann 1946). Estimated weight, males 459 g, females 405 g. Egg, average 46.6 × 31 mm (Johnsgard 1973), estimated weight 24.7 g.

Identification

In the field (9–13 in.)
Associated with the montane cloud forests of eastern Mexico, where no other similar species is present except in a few areas of Puebla and Veracruz, where the long-tailed tree-quail also occurs. The calls of these two species are very similar, and visual distinction may be necessary. The bearded tree-quail lacks black on the throat and head, and is smaller, with a proportionately shorter tail (about a third the total length of the bird).

In the hand
The grey throat and anterior portions of the head separate this species from the other two forms of *Dendrortyx*, and the long (over 105 mm) tail easily distinguishes *Dendrortyx* from all other Odontophorini genera.

Adult female. Like the male, but averaging smaller, and usually with a shorter tail.

Immature. Older juveniles are barred and spotted with dark brown on the flanks and undersides, and are less richly coloured with chestnut on the breast. The two outer primaries may be slightly more pointed in the first-basic plumage than in adults.

General biology and ecology

Very little has been written on the habitats of this species, but it has been reported from the cloud forests of Veracruz and Puebla, and also reportedly occurs at elevations of about 5000 ft (1500 m) in the latter state. In Hidalgo I obtained live specimens that had been captured in cloud forests at an elevation of about 5500 ft (1650 m), and was told by local residents that the species occurred in primary forest remnants at elevations of less than 3000 ft (900 m).

There may be limited contacts between this species and one or more other forest-adapted quail species, particularly the singing quail. Local contact also occurs with the long-tailed tree-quail. Almost nothing is known of its foods and foraging behaviour, but reportedly bearded tree-quails visit planted fields in forest openings when the black beans are ripening. I have observed (Johnsgard 1973) that captive birds seem particularly to relish soft fruits, such as grapes and bananas, and they also readily eat fairly large seeds such as soaked black beans and whole kernels of corn.

Social behaviour

Nothing has been written on the social behaviour of this species under natural conditions. I kept five birds (all of which had been caught as chicks from a single brood) in a single large cage after purchasing them from a peasant in Mexico, and had them with me in my truck for several weeks during field work in Mexico. At that time the birds were highly social, remaining close to one another, and when two females were removed from the group and placed out of sight from me and from the others. The separated birds soon began to try to regain contact with the others by uttering progressively louder calls. These calls were immediately answered by two others that I believed to be males, based on their somewhat larger size and slightly louder and lower-pitched voices, and the four birds soon produced a tremendous clamour that did not stop until I reappeared and returned the separated individuals. Most or all of the group would occasionally produce dawn and dusk chorus when they were kept in quiet and undisturbed surroundings. These choruses would last about 15–20 minutes and sounded very much like the calls generated by separation. Since there were no other tree-quails available to initiate or respond to these calls, nor were the calls the result of separation, I concluded that such dawn and dusk calling is essentially a group location-announcement signal, and inasmuch as two apparent pairs were present in the same cage without fighting I judged that territorial spacing is probably not the major function of the call. One of these pairs later laid eggs and produced chicks after being brought to the US and placed in a large breeding cage.

Reproductive biology

No nests of this species have yet been found under natural conditions and described. One of the females that I obtained laid eggs in a fairly simple nest that was constructed of palm leaves placed around a depression that the birds had dug in the corner of their cage. The female laid a total of 16 eggs (which

Plate 1. Long-tailed tree-quail, pair. Painting by H. Jones.

Plate 2. Bearded tree-quail, pair. Painting by H. Jones.

Plate 3. Buffy-crowned tree-quail, pair. Painting by H. Jones.

Plate 4. Barred quail, pair. Painting by H. Jones.

Plate 5. Mountain quail, pair. Painting by H. Jones.

Plate 6. Scaled quail, pair. Painting by H. Jones.

Plate 7. Elegant quail, pair. Painting by H. Jones.

Plate 8. Gambel's quail, pair. Painting by H. Jones.

Plate 9. California quail, pair. Painting by H. Jones.

Plate 10. Northern bobwhite (*virginianus*), pair. Painting by H. Jones.

Plate 11. Black-throated bobwhite, pair. Painting by H. Jones.

Plate 12. Spot-bellied bobwhite (*hypoleucus*), pair. Painting by H. Jones.

Plate 13. Crested bobwhite (*sonnini*), pair. Painting by H. Jones.

Plate 14. Marbled wood-quail (*gujanensis*), pair. Painting by H. Jones.

Plate 15. Spot-winged wood-quail, pair. Painting by H. Jones.

Plate 16. Rufous-fronted wood-quail (*erythrops*), pair. Painting by H. Jones.

Plate 17. Chestnut wood-quail, pair. Painting by H. Jones.

Plate 18. Dark-backed wood-quail, pair. Painting by H. Jones.

Plate 19. Rufous-breasted wood-quail, pair. Painting by H. Jones.

Plate 20. Black-breasted wood-quail, pair. Painting by H. Jones.

Plate 21. Gorgeted wood-quail, pair. Painting by H. Jones.

Plate 22. Black-fronted wood-quail (right) and Tacarcuna wood-quail (left), adults. Painting by Mark Marcuson.

Plate 23. Venezuelan wood-quail, adult and immature. Painting by H. Jones.

Plate 24. Stripe-faced wood-quail, pair. Painting by H. Jones.

Plate 25. Starred wood-quail, pair. Painting by H. Jones.

Plate 26. Spotted wood-quail, pair. Painting by H. Jones.

Plate 27. Singing quail, pair. Painting by H. Jones.

Plate 28. Montezuma quail, pair. Painting by H. Jones.

Plate 29. Ocellated quail, pair. Painting by H. Jones.

Plate 30. Tawny-faced quail, pair. Painting by H. Jones.

Plate 31. Verreaux's monal-partridge, adult. Painting by H. Jones.

Plate 32. Szechenyi's monal-partridge, adult. Painting by H. Jones.

were removed as they were laid) at intervals from 1 to 18 days. Five of these hatched after incubation periods of between 28 and 30 days. None of the chicks survived to fledging.

Evolutionary relationships

See the long-tailed tree-quail account for comments on the relationships of this species.

Status and conservation outlook

This species has a very limited range in an area of Mexico where the primary forests are rapidly being destroyed, and as a result little hope can be held out for any long-term survival.

BUFFY-CROWNED TREE-QUAIL (Plate 3)

D. leucophrys Gould 1844
Other vernacular names: buffy-fronted wood-partridge, Guatemala partridge; highland partridge; colin à sourcils blancs (French); Guatemalawachtel (German); gallina del monte; perdix de los altos (Spanish).

Distribution of species (see Map 1)
Resident of highland forests of Central America from Chiapas, Mexico, south to Costa Rica.

Distribution of subspecies
D. l. leucophrys: mountains of Chiapas south to Nicaragua. Includes *nicaraguae*.
 D. l. hypospodius: mountains of Costa Rica.

Measurements

Wing, males 138–160 mm, females 130–153 mm; tail, males 113–149 mm, females 108–157 mm (Ridgway and Friedmann 1946). Estimated weight, males 397 g, females 340 g. Egg, average 44 × 33 mm, estimated weight 26.4 g.

Identification

In the field (12–14 in.)
This is the only tree-quail found from Chiapas southward, and like the others it is largely associated with montane cloud forests. Its calls are loud and rollicking, and consist of four rather than three syllables. Its long tail, which is usually held slightly cocked and somewhat vaulted, will visually confirm it as a tree-quail.

In the hand
The white to buffy forehead, throat, and superciliary stripe of this species is unique, and together with a black bill should easily serve to provide identification. As in the other *Dendrortyx*, there is a fairly large area of red skin around each eye.

Adult female. Like the male, but averaging smaller, and with a shorter tail.

Immature. Similar to adults, but with the outer two primaries more frayed and pointed, and with buffy tips on the greater primary-coverts (Johnsgard 1973).

General biology and ecology

Over the relatively broad range of this species it has been reported to occur from as low as 1000 ft (300 m) to as high as about 9500 ft. It has been reported both from cloud forests at these higher elevations and from drier forest habitats such as oak forests of the arid lower tropical zones. Most observers suggest that cloud forests are the primary habitat, but in both El Salvador and Honduras the species has been reported as preferring drier forests to the higher and moister montane forests. The elusive nature of the birds makes it difficult to be certain of their most characteristic habitats, but heavy primary forests or the thick undergrowth associated with secondary forests would seem to be preferred. The only available estimate of population density is of a single pair in a 15-acre study area of mature pine–oak forest at an elevation of 7700 ft (2300 m) in Chiapas, and it seems likely that relatively low densities are probably typical of this forest-adapted species (Johnsgard 1973).

Social behaviour

Like the other tree-quails this species is usually to be found in small groups of four to six birds that probably represent family units, but occasionally as many as about a dozen have been seen together. These coveys or small flocks break up as the breeding season approaches. Like the other *Dendrortyx* species the birds resort to tall roosting trees for the night-time hours. It is quite possible that all the birds from the surrounding area converge on such trees, where regular dawn and dusk singing occurs. It seems possible that this behaviour may facilitate the meeting of birds from different family units and thus help bring about pairing by relatively unrelated individuals. The calls are evidently quite similar to those of the bearded and long-tailed tree-quails, and their names – *guachoque, guachoco* and *chirascua'* – all provide various interpretations of such vocalizations.

Reproductive biology

No detailed descriptions of nests are available for this species, but it reputedly nests on the ground and lays four or five eggs (Dickey and van Rossem 1938). Evidence of the breeding season comes from a female with a brood patch that was collected during May in Guatemala, and well-grown chicks collected in Honduras between April and July. These limited data suggest a spring and summer breeding season similar to that of the other tree-quails.

Evolutionary relationships

See the long-tailed tree-quail account for comments on this.

Status and conservation outlook

This species has the broadest range of any of the tree-quails and the only one extending beyond Mexico into parts of five other countries. As such it has the greatest promise of survival in remnant protected primary montane forest areas such as parks and sanctuaries, and it also apparently does well in the relatively drier habitats of secondary forest and arid upper tropical forest types that are more widespread than true cloud forests.

3 · Genus *Philortyx* Gould 1846

The barred quail is a small, arid-adapted Mexican species with a moderately long tail of 12 rectrices, about 75 per cent as long as the wing. The wing is rounded, and the seventh and eighth primaries are the longest. Some juvenal greater primary-coverts and the two outer primaries are retained in the post-juvenal moult. A distinctive black-throated juvenal plumage stage is present. The sexes are alike as adults, with strong vertical flank-barring and straight, narrow crests. One monotypic species is recognized: *P. fasciatus* (Gould) 1844, barred quail.

BARRED QUAIL (Plate 4)

Other vernacular names: banded quail; Mexican barred partridge; colin barre (French); Bandwachtel; Bindenwachtel (German); codorniz listada (Spanish).

Distribution of species (see Map 2)

Resident of arid scrub and woodlands in western Mexico from Jalisco south to Guerrero. No subspecies recognized.

Measurements

Wing, males 95–102.5 mm, females 94–104 mm; tail, males 58–66 mm, females 59–68 mm (Ridgway and Friedmann 1946). Weight, both sexes 115–160 g (Leopold 1959); males average 130 g, females 126 g (Johnsgard 1973). Egg, average 30 mm × 23.7 mm (Johnsgard 1973), estimated weight 9.3 g.

Identification

In the field (7.5–8.5 in.)

Associated with arid scrub and weedy overgrown pasturelands of western Mexico, and usually not seen until flushed from heavy cover. The calls are distinctive, and consist of a series of loud *ca* or *cow* notes, as well as a more prolonged series of *ca-UT-la* calls that sometimes are terminated by a descending whistled note. This same squealing whistle is also typically uttered every time a covey is flushed, and seems to then serve as an alarm call.

In the hand

Rather easily identified in the hand by the combination of strongly barred black and white flanks and a

Map 2. Distribution of the barred (B) and mountain (M) quails.

short (under 30 mm), straight crest that is blackish brown and tipped with rufous.

Adult female. Like the male, but with a shorter (less than 25 mm) crest (Johnsgard 1973).

Immature. Up to almost five months of age, varying amounts of black facial feathers are present (at maximum covering the entire lores, superciliary area, and most of the chin and throat), the upper wing-coverts have buffy tips and buffy shaft-streaks, and the outer two primaries are buffy-tipped and relatively frayed and pointed (Johnsgard 1973).

General biology and ecology

The preferred habitat of this species consists of tropical scrub and thorn forest areas of western Mexico lying between sea-level and about 1600 m, with almost no records of birds occurring at higher elevations, although the Mexican 'highlands' have usually been part of the general descriptions of its range. Open lowland thorn forest close to weedy, overgrown pastures or cultivated fields seem to represent optimum habitat, and these conditions are widespread in the vicinity of Acapulco. There the birds found shade and protection in brushy clumps of mimosa and staghorn acacia, and weedy herbs offered an abundance of seeds and green leafy materials for food. In such locations coveys were often flushed within 100 yards of one another, and the flushed birds would typically land in brushy thickets after quite short escape flights.

Observations by Leopold (1959) indicated that such legumes as *Desmodium* and *Crotalaria*, plus seeds from such weeds herbs as sunflowers, thistles, corn cockle, and doveweeds were all consumed, and crop plants including beans and sesame seeds were also eaten. Small seeds, such as sesame seeds, were preferred over larger ones by the captive birds that I had. Although the birds often, perhaps regularly, roost in trees, they probably do not forage in them to any significant degree. They sometimes also fly into trees upon being flushed. In heavy cover very little running seems to be done, but in more open habitats they may scurry into thicker cover after landing. In connection with this the birds appear to be highly sedentary, and what seemed to be the same covey could be repeatedly found within a few dozen yards of where it had been located the day before.

Social behaviour

The coveying behaviour of these birds is much like that of bobwhites, with the usual covey numbering under 20 birds and averaging 12 (Leopold 1959) or as low as 5.8 (Johnsgard 1973). This latter estimate of covey size was obtained in June, just prior to adult breeding dispersal, and probably reflects a substantial reduction in covey size from those seen by Leopold, which were made up of approximately 50 per cent immature birds. The breeding season in the Acapulco area is evidently later than that of northern bobwhites breeding in the same region, and the available information suggests that barred quail nesting probably does not begin there until August.

Very little is known of the vocalizations and displays of this species; I was able to capture only a single bird (probably a male), and its most frequently uttered vocalization was a long series of somewhat nasal *ca* or *cow* notes that are similar to those of a *Callipepla* quail; these notes seemed to serve as a location-announcement signal. This call was most often produced in late afternoon or early morning hours, but also could be induced at other times by playback of a tape recording of the call. It closely resembled the first syllable of a three-noted call, sounding to me like *ca-ut'-la*, that was uttered by birds that had been scattered following flushing and that apparently served as a separation call.

Reproductive biology

Thus far no nests from the wild have been described, but captive birds in California have bred, and these laid between late July and September. The eggs were deposited in simple grass nests, which were slightly roofed-over. Although natives in Mexico had told me that the clutch of wild birds is usually 14–16 eggs, four clutches produced by two pairs ranged from three to seven eggs, averaging 5.5 eggs each. The eggs were laid at approximate two-day intervals, and the incubation period averaged 22.6 days, ranging from 21 to 23 days. The downy young are rather distinctive in appearance, with some similarities to those of *Dendrortyx* and others to those of *Colinus*, *Callipepla*, and *Oreortyx*. There is an interesting juvenal plumage phase during which the cheeks and throat are black. This condition is lacking in adults, and presumably reflects an ancestral trait that has been selected out of the adult plumage (Johnsgard 1973).

Evolutionary relationships

Although this species bears a strong outward similarity to *Callipepla*, its skeletal anatomy suggests that it is quite generalized and that its nearest living relatives are the tree-quails. In its plumage it has some strong similarities to the scaled quail. These two species have closely allopatric distributions in somewhat similar arid habitats, although the scaled quail is more grassland-adapted and less shrub-dependent. I believe that *Philortyx* and *Callipepla* had independent evolutionary origins from a *Dendrortyx*-like ancestor, probably during late Tertiary times as arid lands were expanding in western coastal areas and the interior highlands of Mexico, respectively.

Status and conservation outlook

This species thrives on marginal agriculture or ranching, and in various natural habitats too dry to support many human activities. Such habitats are widespread in western Mexico, and thus the species should pose no conservation problems in the foreseeable future.

4 · Genus *Oreortyx* Baird 1858

The mountain quail is a medium-sized montane North American species, with a rounded tail of 12 rectrices and about 60 per cent as long as the wing. The wing is rounded, and the sixth and seventh primaries are the longest. Some juvenal greater primary-coverts and the two outer primaries are retained in the post-juvenal moult. The sexes are alike as adults, with long, straight, and narrow crest plumes, and strong vertical flank-barring. One polytypic species is recognized:

O. *pictus* (Douglas): mountain quail
  O. p. *palmeri* Oberholser 1923
  O. p. *pictus* (Douglas) 1829
  O. p. *confinis* Anthony 1889
  O. p. *eremophila* van Rossem 1937
  O. p. *russelli* Miller 1946

MOUNTAIN QUAIL (Plate 5)

Oreortyx pictus (Douglas) 1829
Other vernacular names: mountain partridge; painted quail; plumed quail; colin des montagnes (French); Bergwachtel; Berghaubenwachtel (German); cordorniz de montana (Spanish).

Distribution of species (see Map 2)

Resident of brushy North American woodlands from western Washington (introduced north to British Columbia) south through California to northern Baja California. Also locally introduced in western Idaho.

Distribution of subspecies

O. p. *palmeri*: south-western Washington south to north-western San Luis Obispo County, California. Introduced locally on Vancouver Island, British Columbia, and in western Washington.

O. p. *pictus*: extreme western Nevada west to the west slope of the Cascade Range in southern Washington and south to the Sierra Nevada and inner Coast ranges of California.

O. p. *confinis*: Baja California in the Sierra Juarez and Sierra San Pedro Matir ranges.

O. p. *eremophila*: southern and west-central California in the Sierra Nevada south to extreme northern Baja California and extreme south-western Nevada.

O. p. *russelli*: Little San Bernadino Mountains in Riverside and San Bernadino counties, California.

Measurements

Wing 125–140 mm, females 126–138 mm; tail, males 73–92 mm, females 71–86 mm (Ridgway and Friedmann 1946). Weight, both sexes 200–290 g (Leopold 1959); males average 235 g, females 230 g (Johnsgard 1973). Egg, average 34.5 × 26.5 mm, estimated weight 13.5 g.

Identification

In the field (9–11.5 in.)

This fairly large and strongly crested quail is easily recognized by its slender head-plumes and strongly barred flank pattern. During spring, a loud, whistled *guee-ark'* or *plu-ark'* call is produced by advertising males. Generally associated with forest edges, dense brush, and disturbed forest habitats.

In the hand

This is the largest of the quails found north of Mexico, with a wing of more than 120 mm. It has a distinctive crest of two narrow plumes that are usually over 60 mm in length.

Adult female. Like the male, but usually with shorter and more brownish plumes, and with a more brownish rather than bluish grey hind-neck. The plume length is not a reliable means of sexing birds in most areas (Brennan and Block 1985).

Immature. With buff-tipped greater primary-coverts, and with the two outermost primaries more pointed and frayed than the inner ones (Johnsgard 1973).

General biology and ecology

Gutierrez (1980) reported that of nearly 600 observations of individuals or groups of mountain quail in California, 85 per cent were associated with mixed evergreen forest, about 12 per cent were in chaparral habitat, and a few were in foothill or riparian woodlands or ecotone situations. This habitat association persisted throughout the entire year with but minor variations. Compared with data on the California quail in the same area, the mountain quail was found on average to be associated with greater slope, more ground, shrub and crown cover, less herbaceous cover, more and larger trees, a shorter distance to cover, and a lesser degree of habitat 'edge' present. In general the birds show a strong aversion to open ground and grasslands by comparison with the California quail.

Gutierrez also discussed the foods of the mountain quail, based on examination of 96 specimens. He found that seeds, followed progressively by bulbs and acorns, were the most important categories of diet, with flowers, fruits, greens, animal materials, and fungi of increasingly subsidiary importance. Six genera of predominantly perennial plants (*Lithophragma*, *Quercus*, *Stellaria*, *Erodium*, *Trifolium*, and *Rhus*) comprised over 60 per cent of the total volume of foods found in the sample. In association with their bulb-eating diet, a fair amount of digging is done by the birds, as is some plucking of fruits or seeds from shrubs and also some acorn-hulling. Like the other North American quails these birds are diurnal foragers, with most foraging done in early morning and late afternoon. Fruits were also found to be important food sources in Washington and Idaho, as were various seeds, acorns, tubers, and roots, with animal materials making up a very small proportion of the diet even in young birds (Yocom and Harris 1952; Ormiston 1966).

Social behaviour

Social groups in mountain quails tend to be small; average numbers observed in various studies have ranged from five to nine birds, with maxima of about 20 individuals. A covey made up of a pair and its well-grown young might be expected to average about seven birds, assuming a 50 per cent brooding loss. Although they avoid crossing open grasslands the birds are actually quite mobile, tending to escape by running uphill or flying if necessary. In some areas rather substantial seasonal migrations to lower altitudes during cold months are undertaken, producing annual movements that may be in excess of 20 miles on western slopes of the Sierra Nevadas. However, daily movements rarely exceed a few hundred metres, although flights of about half a mile have been observed in disturbed birds. Evidently the birds must obtain water at least once a day, and typically they visit such watering areas in late morning or late afternoon. They usually arrive on foot in single file, and where such watering sites are extremely limited groups of hundreds or rarely thousands of individuals may concentrate (Johnsgard 1973).

The vocalizations of the mountain quail are still incompletely known, but the most familiar call is the separation or rally call, which is an extended series of loud *kow* or *cle* notes. The male's advertising *plu-ark* call is a loud, clear whistle that is mainly uttered during spring but can be heard occasionally in early autumn. In California male crowing usually begins in February, as the coveys are starting to dissolve.

Reproductive biology

During February and March the coveys of mountain quails start to break up, as males begin to show antagonism toward one another and unpaired birds begin uttering advertising calls. Individual breeding pairs spread out to occupy areas of between 5 and 50 acres, and most nesting probably occurs in April. Although clutch-size data are still not very satisfactory, it appears that 9 or 10 eggs represent the normal clutch (the average of 29 clutches reported by Grinnell *et al.*, 1918, was 8.7 eggs, exclusive of two probably artificially large clutches). The nests are usually well concealed in thick vegetation, and typically are within a few hundred yards of a source of water. Males assist in nest defence, and also help to defend the brood. They may occasionally even assist with incubation, judging from the frequent presence of male brood patches. The incubation period has been variously reported as from 21 to 25 days, with the longer periods more likely than the shorter extreme, and no information is available on growth rates or moulting sequences. There is no indication that multiple brooding occurs in this species, but it is common for unsuccessful females to make a second or even third nesting effort (Leopold 1959).

Evolutionary relationships

Holman (1961) considered the scaled quail to be the nearest relative of the mountain quail, and the other crested quails ('*Lophortyx*') to be somewhat more distant, as are the *Colinus* forms. As Gutierrez *et al.* (1983) have suggested, it seems likely that *Oreortyx* diverged perhaps as long as 12 million years ago from an ancestor that later separated to form *Callipepla* and *Colinus*. It is not difficult to imagine such a forest-adapted ancestral type, not markedly different from the modern tree-quails, with *Oreortyx* retaining a larger number of forest-related adaptations and *Callipepla* becoming progressively more adapted for open-country existence.

5 · Genus *Callipepla* Wagler 1832 (including *Lophortyx* Bonaparte 1838)

The crested quails are small grassland to arid-adapted species of North America with moderately long tails of 12–14 rectrices that are about 75 per cent as long as the wing. The wing is rounded, with the seventh primary usually the longest. Some juvenal greater primary-coverts and the two outer primaries are retained in the post-juvenal moult. The sexes are very similar or dimorphic as adults; both have erect bushy or recurved (comma-shaped) crests that are usually larger or more conspicuous in males. Four mostly allopatric species are recognized:

C. squamata (Vigors): scaled quail
    C. s. castanogastris Brewster 1881
    C. s. pallida Brewster 1881
    C. s. hargravei Rea 1973
    C. s. squamata (Vigors) 1830
C. douglasii (Vigors): elegant quail
    C. d. douglasii (Vigors) 1829
    C. d. languens (Friedmann) 1943
    C. d. bensoni (Ridgway) 1887
    C. d. teres (Friedmann) 1943
    C. d. impedita (Freidmann) 1943
C. gambelii (Gambel): Gambel's quail
    C. g. sana (Mearns) 1914
    C. g. gambelii (Gambel) 1843
    C. g. ignoscens (Friedmann) 1943
    C. g. pembertoni (van Rossem) 1932
    C. g. fulvipectus (Nelson) 1899
    C. g. stephensi (Phillips) 1959
    C. g. friedmanni (Moore) 1947
C. californica (Shaw): California quail
    C. c. brunnescens (Ridgway) 1884
    C. c. californica (Shaw) 1798
    C. c. catalinensis (Grinnell) 1906
    C. c. canfieldae (van Rossem) 1939
    C. c. orecta (Oberholser) 1932
    C. c. plumbea (Grinnell) 1926
    C. c. decoloratus (van Rossem) 1946
    C. c. achrustera (Peters) 1923

KEY TO THE SPECIES OF *CALLIPEPLA*

A. Crest of brown or black feathers that curve forward and are enlarged toward tips
  B. Abdomen feathes edged with darker colour in a scalloped pattern; flanks marked with olive-brown: California quail (*californica*)
  BB. Abdomen feathers extensively blackish, or buffy with mottling and streaking; flanks marked with chestnut: Gambel's quail (*gambelii*)
AA. Crest feathers neither recurved nor enlarged toward tip
  B. Crest bushy and buff-coloured; body feathers marked with dark scallops: scaled quail (*squamata*)
  BB. Crest pointed and brownish or cinnamon-coloured; pale, rounded spots on sides and abdomen: elegant quail (*douglasii*)

SCALED QUAIL (Plate 6)

Callipepla squamata (Vigors) 1830
Other vernacular names: blue quail; chestnut-bellied quail (*castanogastris*); cottontop quail; Mexican blue quail; scaled partridge; colin écaille (French); Schüppenwachtel (German); codorniz escamosa (Spanish).

Distribution of species (see Map 3)

Resident of arid North American grasslands from eastern Colorado and western Kansas south through Arizona, New Mexico, and western Texas to central Mexico. Introduced in central Washington and Nevada.

Distribution of subspecies

C. s. castanogastris: southern Texas south through Tamaulipas, Nuevo Leon, and eastern Coahuila, Mexico.
    C. s. pallida: southern Arizona, southern New Mexico, and western Texas south to northern Sonora and Chihuahua.
    C. s. hargravei: western Oklahoma, south-western Kansas, south-eastern Colorado, northern New Mexico, and north-western Texas.
    C. s. squamata: northern Sonora and Tamaulipas south to the valley of Mexico.

Measurements

Wing, males 109–121 mm, females 109.5–120 mm; tail, males 75–90 mm, females 75–88 mm (Ridgway and Friedmann 1946). Weight, males 151–202 g, average of 52, 179.4; females 130–274 g, average of 58, 173.4 (Lehmann 1984). Egg, average 32.5 × 25 mm, estimated weight 11.2 g.

Map 3. Distribution of the elegant (E) and scaled (S) quails.

Identification

In the field (10–12 in.)
Found only in arid grasslands and semi-deserts, these birds are a distinctive bluish grey throughout, except for a whitish crest that is variably exposed but always partially visible. The birds are more prone to run than to fly, and utter distinctive *pey-cos* separation or rally calls, especially when a covey has been dispersed. In spring a loud *whock!* note is produced by advertising males.

In the hand
The greyish plumage, with black edging on the feathers that produces a unique 'scaly' patterning, is distinctive, as is the short, bushy white crest.

Adult female. Similar to the male, but with dusky shaft-streaks on the throat, a more buffy crest, a black marginal line inside the white margins to the inner webs of the secondaries, and the outer margins of the secondaries with indistinct buff mottling.

Immature. With buff-tipped greater primary-coverts, and probably slightly more pointed and frayed outer primaries.

General biology and ecology

Studies by Schemnitz (1961, 1964) indicate that areas having opportunities for ground-level movement, but which also provide some overhead protection, are most often used by this species. Of more than 2000 observations, habitats dominated by shrubs 3–20 ft high provided the majority of sightings, while those having man-made 'cover' such as structures present contributed 30 per cent, and the remainder were more or less equally divided among habitats comprised of forb clumps, cropland, or open grassland. These last two habitat types were used much less frequently than their availability might have suggested, whereas areas having shrubs such as skunk-brush (*Rhus*) or man-made structures received higher usage than expected.

During the spring and summer the birds move to less heavy cover, but shrubs and forbs nevertheless provided the primary nesting cover, with very few nests found in purely grassy situations. A source of midday shade and loafing cover, together with dusting sites, are probably important during summer, and it is probable that the birds prefer areas near a surface source of water, although this may not be required. Foods in the form of weed or grass seeds are primarily consumed, while tree seeds or fruits such as those of mesquite (*Acacia*) and hackberry (*Celtis*) are less often eaten, and animal materials are taken in small quantities, at least by adults. Compared with the Gambel's quail, a smaller reliance on herbaceous legumes appears to be typical. A sample of autumn and winter crops from Texas suggested that grass seeds were eaten less by scaled quails than by northern bobwhites, and green materials eaten substantially more (Lehmann 1983). Evidently the birds are quite opportunistic and catholic in their food choice, which may be a desirable adaptation for survival in a semi-arid environment.

The scaled quail is fairly sedentary, with estimated winter home ranges of 10 birds varying from 24 to 84 acres, and home ranges averaging about 52 acres in one study. Winter home ranges of separate coveys generally overlap only slightly or not at all, and there is a very limited degree of covey-shifting by covey members (Schemnitz 1961).

Social behaviour

The fairly large winter coveys remain intact until the males begin to exhibit increasing hostility toward one another and begin pairing with females. At about the same time unmated males begin uttering their shriek-like 'whock' advertisement calls, which are probably produced through the entire period that such males remain in reproductive condition. This call is probably limited to unmated birds (Wallmo 1956; Anderson 1978). The equally familiar 'pey-cos' or 'chekar' call is a separation call, and is uttered both by birds separated from their covey as well as by either sex when visually separated from its mate (Johnsgard 1973). It is probable that paired males can individually recognize the calls of their mates (Anderson 1978). Both sexes share nearly all calls in this species, the shriek of unmated males being the major exception that is completely limited to males, while an agonistic squeal is primarily uttered by females. Anderson (1978) did not consider any of the calls to be specifically territorial in function, and it is doubtful that true territories exist in this species.

Reproductive biology

Nesting records in both the United States and Mexico suggest that breeding is fairly late, extending from June through to September or even October, and is apparently timed to coincide with the usual period of summer and autumn rainfall. The long breeding season may also facilitate renesting and possibly even double-brooding. The best evidence on this comes from Wallmo (1956), who noted one case of a male raising the pair's first brood while the female began a second clutch. Average clutch estimates range from 13 to 14 eggs, and males apparently assist in incubation only rarely. As with the other *Callipepla* species, the incubation period averages about 23 days. The young fledge fairly rapidly, and by the time the young are 11–15 weeks old they weigh nearly as much as adult birds (Lehmann 1984).

The high percentage of young in hunter kills during autumn (Schemnitz 1961 reported ratios indicating 74.1 per cent immatures) suggests that there is a high annual mortality rate in adults, and there is probably an even higher rate of mortality in young birds.

Evolutionary history

I have suggested earlier (Johnsgard (1973) that the current distribution of the scaled and elegant quails suggest that they might have speciated following isolation from a common ancestral population by the Sierra Madre Occidental mountains, and both species are now desert-adapted and, depending upon the presence of shrubby or brushy vegetation, occur in relatively scattered (for the scaled quail) or more continuous (for the elegant quail) groupings. Curiously, the only known hybrids between these two species have apparently been sterile (Banks and Walker 1964), although varying degrees of hybrid sterility are not uncommon among congeneric hybrids of the New World quails (Johnsgard 1973).

Status and conservation outlook

Although the range of the scaled quail has retracted in historical times as desert grasslands have become degraded (Rea 1973), the species is able to exist fairly well in overgrazed, weedy habitats that are too dry for the northern bobwhite to survive in.

ELEGANT QUAIL (Plate 7)

Callipepla douglasii (Vigors) 1829
Other vernacular names: Benson's quail (*bensoni*); crested quail; Douglas' quail (*douglasii*); Yaqui quail; colin élégant (French); Douglaswachtel (German); codorniz gris (Spanish).

Distribution of species (see Map 3)
Resident of arid scrub habitats from western Chihuahua south to Jalisco.

Distribution of subspecies
C. d. douglasii: extreme southern Sonora south through Sinaloa and north-western Durango.
   C. d. languens: endemic in western Chihuahua.
   C. d. bensoni: endemic in Sonora.
   C. d. impedita: endemic in Nayarit.
   C. d. teres: endemic in north-western Jalisco.

Measurements

Wing, males 101–115 mm, females 98–115 mm; tail, males 66–94 mm, females 65–87 mm (Ridgway and Friedmann 1946). Weight, both sexes 160–196 g (Leopold 1959); males average 175 g, females 169 g (Johnsgard 1973). Egg, average 34 × 24 mm (Johnsgard 1973), estimated weight 10.8 g.

Identification

In the field (9–10 in.)
The pale, rounded flank and underpart markings are distinctive, as is the fairly long crest of orange-buff (males) or brown (females) that is not recurved or enlarged toward the tip. The birds are found in arid coastal and upland habitats of the Sonoran desert, and their separation call is a rasping *ca-cow!* call much like the *pey-cos* note of the similar but allopatric scaled quail. Males also utter a sharp, nasal whistle, likewise sounding much like the *wock!* note of advertising male scaled quails.

In the hand
This is the only New World quail that has the flanks, breast, and abdomen marked with whitish round

spots, and with a long, straight crest of graduated feathers that are uniformly brown or orange-buff.

Adult female. Differs from the male in having the top and sides of the head and nape olive-brown, the crest usually dark brown, the mantle washed with greyish and mottled with whitish buff; the rest of the upperparts similarly marked; the inner wing-coverts, scapulars, and outer secondaries deep blackish brown, more or less mottled with rufous and margined with buff; the chin and throat white with dark shafts; and the rest of the underparts greyish brown, covered with rounded white spots.

Immature. Birds in first-winter plumage have frayed and pointed outer primaries, and at least some of their greater primary-coverts are tipped or edged with buffy.

General biology and ecology

This species is still virtually unstudied, but appears to be associated with deciduous forests, thorn forests, and scrub thickets in tropical areas. Leopold (1959) reported that dense second growths of tropical forests are favoured, but that open fields and pastures are avoided. In areas of such favoured habitats the population density may locally exceed a bird per acre, although limited census data suggest that usual breeding densities are far lower than this (Johnsgard 1975). Evidently a wide assortment of weed seeds, fruits, and insects are eaten, as might be expected in a desert-adapted form, with the seeds of legumes possibly being a preferred food (Leopold 1959).

Social behaviour

Leopold (1959) observed that coveys of this species ranged from 6 to 20 birds, and that they began to break up in mid-April, when males were then crowing from low perches. The male's advertising call is a sharp, nasal whistle that acoustically closely resembles the 'shriek' of the scaled quail. Although not yet studied in the wild, the apparent separation call as uttered by birds in my collection consisted of a somewhat nasal *ca-cow'* that usually occurred in groups of two to five notes, averaging about three, separated by intervals of about one-quarter second, and with the midpoints of the notes almost exactly a half-second apart. Like the *chi-ca-go* call of the California and Gambel's quails the frequency of the call is variable, rising and falling with each note. When disturbed, both sexes would utter typical *Callipepla* chipping notes, and the male would utter sharp whistled *wheet* notes when alarmed. Both sexes produced sharply down-slurred distress whistles when held in the hand, also like those of other *Callipepla* species.

Reproductive biology

Leopold (1959) judged that nesting in Nayarit probably begins in early May, and in southern Sinaloa nesting also probably occurs in April and May, or substantially earlier than the barred and scaled quails of the same general region of western Mexico. Although no good descriptions are available, the nests are reportedly placed on the ground. They usually have 8–12 eggs, but sometimes as many as 20 have been found, perhaps reflecting the efforts of two females. At least in captivity the maximum rate of egg-laying is one per day; one captive female laid seven eggs in 8 days. The incubation period is 22–23 days (Johnsgard 1973). I was unable to raise any of the chicks successfully, and no information is otherwise available on growth or development of the young, either under captive or natural conditions.

Evolutionary history

As noted in the preceding account, I believe that the scaled quail represents the nearest living relative of the elegant quail. Hubbard (1974) hypothesized that a widespread ancestral *Callipepla* type was initially disjoined into three refugial isolates, perhaps by the Illinoian glaciation, which included an ancestral *douglasii* type in western Mexico, separated from ancestral *squamata* in central upland Mexico, and ancestral *californica–gambelii* in the Baja area. The subsequent Wisconsin glaciation may have brought about separation of these last two forms. Holman's (1961) osteological data indicated that the elegant, Gambel's, and California quails show greater interspecific differences from one another than do the Recent species of *Colinus*.

Status and conservation outlook

No specific or even general information exists on the status of this species, but its adaptation to a variety of thorn forest and shrubby tropical woodland habitats would seem to give it reasonable protection from exploitation or habitat loss.

GAMBEL'S QUAIL (Plate 8)

Callipepla gambelii (Gambel) 1843
Other vernacular names: Arizona quail; desert quail; fulvous-breasted quail (*fulvipectus*); Tiburon Island quail (*pembertoni*); colin de Gambel (French); Gambels Wachtel; Helmwachtel (German); codorniz de Gambel (Spanish).

Map 4. Distribution of the Gambel's quail.

Distribution of species (see Map 4)

Resident of western North American deserts from Nevada, Utah, and Colorado south to Baja California, Sonora, and Chihuahua, Mexico. Locally introduced in Idaho.

Distribution of subspecies

C. g. sana: endemic in western Colorado.

C. g. gambelii: southern Utah and southern Nevada south to the Colorado and Mojave deserts and north-western Baja California.

C. g. ignoscens: southern New Mexico and extreme western Texas.

C. g. pembertoni: endemic on Tiburon Island, Gulf of California.

C. g. stephensi: southern Sonora, near the Sinaloa border.

C. g. friedmanni: coastal Sonora, from Rio Fuerte to Rio Culican.

Measurements

Wing, males 106–122 mm, females 105–118 mm; tail, males 81–107 mm, females 83–102 mm (Ridgway and Friedmann 1946). Weight, both sexes 160–176 g (Leopold 1959); males average 161 g, females 156 g (Johnsgard 1973). Egg, average 31.5 × 24 mm, estimated weight 10.0 g (Johnsgard 1973).

Identification

In the field (9.5–11 in.)

Associated with desert habitats, the Gambel's quail is easily identified by its forward-tilting teardrop- or comma-like crest, its underparts, which are streaked or blotched with black but are devoid of scaly black markings, and its fairly bright chestnut flank streaking. A loud *chi-ca'-go-go* call is uttered by birds separated from their mates or coveys, and a loud *kaa!* or *cow!* note is uttered by advertising males.

In the hand

Only this species and the California quail have forward-tilting, comma-shaped crests, and distinction between these two species may be achieved by the comments provided above in the description section.

Adult female. Similar to *californica*, but with the crest slightly better developed, the back of the neck without white spots, the chest with dark shafts, the feathers of the breast and abdomen not margined with black, and the flanks margined with chestnut. Differs from the male in its shorter and more brownish crest, lack of a black and white facial and throat pattern, less rufous crown, and lack of a black abdomen.

Immature. Similar to adults, but with buff-tipped greater primary-coverts, and with the outer two primaries usually more pointed and frayed than the inner ones.

General biology and ecology

Over its range the Gambel's quail occurs in three major climatic and habitat types. The first two of these are warm desert habitats, including rather low desert valleys dominated by mesquite (*Prosopsis*) and various warm-desert shrubs, where annual precipitation is extremely low. The second consists of upland desert habitats where the desert vegetation is mostly of cat's-claw (*Acacia*), yucca (*Yucca*), prickly pear (*Opuntia*), and various desert shrubs adapted to fairly mild winters and annual precipitation up to about 10 in. The third habitat type is in the cool-desert habitats of the Colorado River basin, where sagebrush (*Artemisia*) and other alkaline-adapted shrubs dominate and winter temperatures are sub-freezing, with snow cover often persisting for considerable periods, producing marginal survival capabilities for the northernmost populations (Gullion 1960).

At least in some areas fairly high densities of birds of about a bird per two to five acres have been reported. The birds apparently do best in areas where annual precipitation is more than 5 in. or where subsurface moisture allows for adequate vegetational growth to provide nutritional needs for breeding. Where moist and succulent plants are available, drinking water is evidently not needed, although springs or other sources of water are highly attractive to the birds in extreme desert areas.

Several studies indicate that the birds consume a wide variety of plants, with leafy materials, especially from legumes, and the seeds of both legumes and non-leguminous plants being of primary importance. In general, forbs are the most important plant life form, and seeds comprise the single most important food item category (Judd 1905; Hungerford 1962). Observations by Campbell (1957) led him to believe that the species' capacity to show foraging flexibility by utilizing a wide variety of food sources helped to explain its ability to survive in agricultural as well as non-developed desert areas, where the vegetational complexes are so different.

These birds occur in fairly small coveys that in one study averaged 12.5 birds, 10 of which had home ranges that varied from 19 to 95 acres, averaging 35.7 acres. These coveys moved about quite variably, with most movement in coveys of adult birds, and especially during March and early April during prenesting activities. During spring there was a good deal of covey-shifting, especially by young males presumably searching for mates, and also by a few young females (Gullion 1962).

Social behaviour

Winter coveys are basically comprised of family units or their multiples, and have relatively little overlapping of home ranges. During late winter the coveys begin to dissolve because of increased male hostility toward other males as well as toward females. Raitt and Ohmart (1966) believed that chases of females by males might be associated with pair formation under natural conditions. Fighting among males during this time is associated with social dominance rather than territorial activity, although *cow* calling by unmated males begins about the time that coveys are breaking up and the pairs dispersing. Ellis and Stokes (1966) confirmed that only unmated males utter *cow* calls, advertising their sexual availability. The birds are strongly monogamous, although the mortality rate is so high that probably few pairs remain intact from one year to the next.

Beside the *cow* call, the separation call is the most familiar call of the species. It is sufficiently complex acoustically to allow for individual recognition. Both sexes produce this call, which sounds like *cow-cow!-cow-cow* or *chi-ca'-go-go*, and it can be easily induced among captive paired birds by visually separating them.

Reproductive biology

In New Mexico egg-laying begins in late April, with nest sites being scratched out under vegetational cover. The eggs are deposited on a nearly daily basis, with a clutch typically consisting of about three cycles of laying four to six eggs daily, followed by a skipped day. Thus, a total of 12–14 eggs usually comprise a complete clutch. Although as many as 19 eggs were found in a sample of 44 nests, most of these nests had 10–16 eggs present (Gorsuch 1934). Incubation is by the female alone, and usually requires 23 days. The male typically sits on a perch some distance away and serves as a 'look-out'. Upon disturbance by a dangerous mammal the male will perform a nest-distraction display, and on some occasions the male may even take over complete care of the first brood, leaving the female free to begin a second clutch. Or, more probably, the chicks may be 'weaned' when only about a month old and left in the care of other birds in the area, thus allowing the pair to start a second clutch (Gullion 1956).

Evolutionary history

I have suggested (Johnsgard 1973) that isolation between the ancestral populations of the Gambel's and California quails may have been achieved by the Sierra Nevada range, which still effectively serves to prevent extensive contact between these two species. Where limited contact does occur (as in Joshua Tree National Monument) hybridization has been reported. This suggests that separation of these gene pools has been fairly recent; Hubbard (1974) hypothesized that it might have occurred as recently as during the Wisconsin glaciation.

Status and conservation outlook

The importance of this species as a sporting bird in the arid south-west ensures that it will be managed carefully in that region, and additionally it survives well where natural desert habitats are converted to agricultural lands through irrigation.

CALIFORNIA QUAIL (Plate 9)

Callipepla californica (Shaw) 1798
Other vernacular names: San Lucas quail

(*achrustera*); Santa Catalina quail (*catalinensis*); valley quail; Warner Valley quail (*orecta*); colin de Californie (French); Kalifornische Schopfwachtel; Schopfwachtel (German); codorniz Californiana (Spanish).

Distribution of species (see Map 5)

Resident of grasslands and shrublands of western North America from Oregon to Baja California. Locally introduced elsewhere in North America (British Columbia, Washington, Idaho, Utah, and Colorado), and in New Zealand, Norfolk Island, Hawaii (Kauai and Hawaii), central Chile, and on King Island (Bass Strait).

Distribution of subspecies (excluding most introductions)

C. c. orecta: south-eastern Oregon (Warner Valley).

C. c. brunnescens: coastal areas from extreme northern California to southern Santa Cruz Country.

C. c. californica: interior northern Oregon and western Nevada south to southern California and Los Coronados Islands, Baja California.

C. c. catalinensis: endemic on Santa Catalina Island; introduced on Santa Rosa and Santa Cruz islands, southern California.

C. c. canfieldae: east-central California (Owens Valley).

C. c. plumbea: San Diego County, California, south through north-western Baja California.

C. c. decoloratus: Baja California between about 25 and 30°N latitude

C. c. achrustera: southern Baja California.

Measurements

Wing, males 102–119 mm, females 101–117 mm; tail, males 77–99.5 mm, females 78–92 mm (Ridgway and Friedmann 1946). Weight, males average 176 g, females 162 g (Johnsgard 1973). Egg, average 32 × 25 mm, estimated weight 11.0 g (Johnsgard 1973).

Identification

In the field (9.5–11 in.)

Except in the very limited area (in southern California) of contact between this species and the Gambel's quail, identification is simple, and requires only seeing the comma-shaped or teardrop-like crest. Where there is danger of confusion with the Gambel's quail, the rather 'scaly' underpart pattern and the more olive-brown flank streaking serve to separate these otherwise similar species. Their calls are also similar, but the California quail produces a three-noted (occasionally four-noted) *cu-ca-cow!* separation call, with the last rather than second note the loudest, and a single-noted *cow!* is used as a male advertisement signal.

In the hand

See the Gambel's quail account for in-hand distinction from that species.

Adult female. Differs from the male in having the crest shorter and browner, no black and white pattern on the head and throat, most of the feathers being dirty white with dark centres, the general colour of the neck, mantle, and chest brownish grey, and of the underparts white, tinged with buff on the abdomen. Differs from the female of *gambelii* as noted under that species.

Immature. Similar to adults, but with the outer two primaries relatively pointed and frayed, and with buff-tipped greater primary-coverts.

General biology and ecology

The rather broad distribution pattern of this species encompasses several climatic and vegetational

Map 5. Distribution of the California quail.

zones. Leopold (1977) reported that four major ecological zones are encompassed, including arid ranges (mostly in southern California and Baja California), transitional ranges in the Sacramento Valley, humid forest ranges associated with the Coast and Cascade ranges, and interior Great Basin and Columbia Basin ranges (mostly recently occupied). Of these the transitional ranges in the Sacramento Valley foothills provide the most stable quail habitats, and are characterized by mild winters, moderately dense ground vegetation, and generally adequate nesting and escape cover. The species' habitat needs have been analysed by Gutierrez (1980), who found that among nearly 500 sightings the most common habitat association was grassland (39 per cent), followed progressively by mixed evergreen forest and ecotone or transitional habitats, and with small usage of foothill woodlands, riparian woodlands, and chaparral vegetation. Compared with mountain quail of the same area, the birds used less steep slopes, areas with less crown, shrub, and ground cover, but those with more herbaceous cover and 'edge' present, and were observed at a somewhat greater average distance to dense, protective cover. These findings may not apply equally to all subspecies, but do suggest the importance of transitional habitats having an interspersion of woody and more open and herbaceous vegetational types.

Foods of the California quail include the usual array of forbs, with legumes and their seeds probably being the single most important food source. Green materials are consumed in largest quantities during the winter and spring months, with legumes such as bur clover (*Medicago*), clover (*Trifolium*), deervetch (*Lotus*), and lupines (*Lupinus*) all of importance in various areas for their leaves or seeds (Leopold 1977). Various weedy species of filaree (*Erodium*) are also sources of seeds in the California foothills area. In some areas acorns are consumed in quantity, and the fruits of a few other shrubs or trees are also consumed locally, especially in winter. In general, food studies suggest that the need for brushy habitat by the California quail is primarily a reflection of its requirements for protective cover, while most of its foods comes from herbaceous forbs, especially legumes (Johnsgard 1973).

Coveys of California quail are often fairly large and tend to be fairly stable, with little interchange by covey members, at least during autumn and winter. Four coveys studied by Emlen (1939) had home ranges of 17–45 acres, which were related to the distribution of brushy cover. By late February these coveys began to break up as pairs formed and unmated males began to leave the covey, probably in search of females. Unmated males began to *cow* call in late April, and often associated with mated pairs, but sometimes assumed a fairly nomadic existence. Mated birds had rather small home ranges of only 12–25 acres prior to the start of nesting, and even smaller ranges of about 3–10 acres thereafter.

Social behaviour

Evidently most pairing by California quail occurs before the testes of males are much enlarged, and pairs can be formed among late-winter coveys in the absence of conspicuous displays or copulatory behaviour. However, when sexually active males are initially exposed (in captivity) to females their immediate response is a seemingly highly aggressive or dominance-assertion frontal display, and probably only the female's submissive behaviour and her lack of male plumage features inhibit overt attack (Johnsgard 1973). As male-to-male aggression in the covey increases during spring, unmated males are probably forcibly excluded from the group. Evidently the only actual territorial defence is that of unmated males, who self-advertise with *cow* calls and may fight with other unmated males that attempt to enter their crowing territories (Genelly 1955). Such advertising behaviour probably increases the chances of these males by allowing unmated or recently widowed females to find mates rapidly, but the localization of crowing areas in the vicinity of nesting females may tend to increase the predation rate on such nesting birds.

Reproductive biology

Nest sites in this species are in various sheltered locations such as among dry grass or weeds, at the base of shrubs, in rock piles, and the like. Eggs are deposited at the rate of about five per week. In various studies the average clutch-size has been estimated at from 10.9 to 14.2 eggs, with the higher figure probably closer to the typical situation. Incubation periods are 22 or, more commonly, 23 days, based on incubator data (Johnsgard 1973). Although incubation is normally done only by the female, the male will take over if his mate is killed, and in any case is attentive during the chick-raising phase. Only a few established cases of double-brooding have been verified, but the substantial number of young broods that are tended only by males in some years suggest that in favourable years double-brooding may not be uncommon (Anthony 1970). Erwin (in Leopold 1977) also believed that by turning over the chicks to the male after they are about two weeks old and starting a second clutch females may attain a significant augmentation of productivity during favourable breeding years, at least in the drier portions of this species' range.

The chicks grow rapidly, initially fledging at about two weeks of age, and completing their juvenal plumage by about 10 weeks. By 21 weeks all of the juvenal flight feathers except for the outer two will have been replaced and an adult-like plumage attained (Raitt 1961).

Evolutionary relationships

A discussion of this can be found in the Gambel's quail and mountain quail accounts.

Status and conservation outlook

With over two million California quail harvested each year by hunters in the United States, the population status of this species is clearly excellent, and does not warrant special attention.

6 · Genus *Colinus* Gouldfuss 1820

The colins or bobwhites are small grassland to savannah or woodland edge-adapted species of North, Central, and South America, with rounded tails of 12 rectrices that are more than half the length of the wing. The wing is rounded, with the seventh primary the longest. Some juvenal greater primary-coverts and the two outer primaries are retained in the postjuvenal moult. The sexes are dimorphic as adults, and males are slightly to distinctly crested, with highly variable black and white facial and throat patterning. Both sexes have blackish blotches or spots on the rump, scapulars, and inner secondaries. Four highly variable allopatric species are recognized here, but some authorities have recognized as few as two (by merging *leucopogon* with *cristatus* and *nigrogularis* with *virginianus*), and larger numbers have often been recognized at various times:

C. virginianus (L.): northern bobwhite
  C. v. marilandicus (L.) 1758
  C. v. virginianus (L.) 1758
  C. v. floridanus (Coues) 1872
  C. v. cubanensis (G. R. Gray) 1846
  C. v. mexicanus (L.) 1766
  C. v. texanus (Lawrence) 1853
  C. v. taylori Lincoln 1915
  C. v. ridgwayi Brewster 1885
  C. v. maculatus Nelson 1899
  C. v. aridus Aldrich 1942
  C. v. graysoni (Lawrence) 1867
  C. v. nigripectus Nelson 1897
  C. v. pectoralis (Gould) 1843
  C. v. godmani Nelson 1897
  C. v. minor Nelson 1901
  C. v. atriceps (Ogilvie-Grant) 1893
  C. v. thayeri Bangs and Peters 1928
  C. v. harrisoni Orr and Webster 1968
  C. v. coyolcos (P. L. S. Müller) 1776
  C. v. salvini Nelson 1987
  C. v. insignis Nelson 1897
  C. v. nelsoni Brodkorb 1942
C. nigrogularis (Gould): black-throated bobwhite
  C. n. caboti Van Tyne and Trautman 1941
  C. n. persiccus Van Tyne and Trautman 1941
  C. n. nigrogularis (Gould) 1843
  C. n. segoviensis Ridgway 1888
C. [*cristatus*] *leucopogon* (Lesson): spot-bellied bobwhite
  C. l. incanus Friedmann 1944
  C. l. hypoleucus Gould 1860
  C. l. leucopogon (Lesson) 1842
  C. l. leylandi Moore 1859
  C. l. sclateri Bonaparte 1856
  C. l. dickeyi Conover 1932
C. cristatus (L.): crested bobwhite
  C. c. mariae Wetmore 1962
  C. c. panamensis Dickey and van Rossem 1930
  C. c. decoratus Todd 1917
  C. c. littoralis Todd 1917
  C. c. cristatus (L.) 1766
  C. c. badius Conover 1938
  C. c. leucotis Gould 1844
  C. c. bogotensis Dugand 1943
  C. c. parvicristatus Gould 1843
  C. c. continentis Cory 1913
  C. c. horvathi Madarasz 1904
  C. c. barnesi Gilliard 1940
  C. c. mocquerysi Hartert 1894
  C. c. sonnini Temminck 1815

KEY TO THE SPECIES OF *COLINUS*

A. A short upturned crest present; throat white to (usually) cinnamon-tinted, sometimes spotted with black; the underparts spotted or barred: crested bobwhite (*cristatus*)
AA. No distinct crest; the throat rarely tinted with cinnamon; underparts highly variable
  B. Chin and throat black or mostly black (males)
    C. Underpart feathers with black edges, producing a scalloped effect: black-throated bobwhite (*nigrogularis*)
    CC. Underpart feathers mostly cinnamon to chestnut and unspotted northern bobwhite (*virginianus*, primarily Mexican races)
  BB. Chin and throat not uniformly or mostly blackish (both sexes)
    C. Throat entirely white (males)
      D. Underparts brown spotted with white, or entirely white: spot-bellied bobwhite (*leucopogon*)
      DD. Underparts mainly white, with variable black barring or spotting: northern bobwhite (*virginianus*, primarily non-Mexican races)
    CC. Throat not entirely white (both sexes)
      D. Throat with some black present (males): spot-bellied bobwhite (*leucopogon*)
      DD. Throat mostly or entirely buffy, with little or no black on head and neck (females)
        E. The breast, sides, and flanks conspicuously spotted: spot-bellied bobwhite (*leucopogon*)
        EE. The breast, sides, and flanks not conspicuously spotted

F. Breast feathers with two terminal spots: black-throated bobwhite (*nigrogularis*)
FF. Breast feathers lacking terminal white spots: northern bobwhite (*virginianus*)

NORTHERN BOBWHITE (Plate 10)

Colinus virginianus (L.) 1758

Other vernacular names: black-breasted bobwhite (*pectoralis*); black-headed bobwhite (*atriceps*); common bobwhite; Coyolcos bobwhite (*coyolcos*); Cuban bobwhite (*cubanensis*); eastern bobwhite (*virginianus*); Florida bobwhite (*floridanus*); Godman's bobwhite (*godmani*); Grayson's bobwhite (*graysoni*); Guatemalan bobwhite (*insignis*); Jaumave bobwhite (*aridus*); least bobwhite (*minor*); masked bobwhite (*ridgwayi*); Nelson's bobwhite (*nelsoni*); plains bobwhite (*taylori*); Puebla bobwhite (*nigripectus*); Salvin's bobwhite (*salvini*); Thayer's bobwhite (*thayeri*); Virginian colin; colin de Virginie (French); Virginiawachtel; Virginische Baumwachtel (German); codorniz comun (Spanish).

Distribution of species (see Map 6)

Resident of brushy and woodland edge habitats of North America from Maine to Florida, west to South Dakota, eastern Wyoming, Colorado, and New Mexico, and south through Mexico to extreme northern Guatemala. Also present on Cuba and Isle of Pines (possibly introduced). Introduced elsewhere in northwestern North America (British Columbia, locally to Oregon, Idaho, and central Wyoming), the West Indies (New Providence, Andros, Hispaniola, Puerto Rico, and St Croix), and New Zealand. Other introductions (Hawaii, Europe, etc.) have been attempted.

Distribution of subspecies (excluding most introductions)

C. v. marilandicus: south-western Maine south to Delaware, Maryland, and central Virginia, and west to Pennsylvania. Often included in *virginianus*.

C. v. mexicanus: eastern United States west of the Atlantic seaboard, west to the range of *taylori*. Often included in *virginianus*.

C. v. virginianus: southern Atlantic seaboard from Virginia south to northern Florida and south-eastern Alabama.

C. v. floridanus: endemic to peninsular Florida.

C. v. cubanensis: Cuba and the Isle of Pines (where possibly introduced).

C. v. taylori: South Dakota south to northern Texas and east to western Missouri and north-western Arkansas.

C. v. texanus: south-western Texas south to northern Mexico including parts of Coahuila, Nuevo Leon, and Tamaulipas.

C. v. ridgwayi: previous resident of Arizona (where extirpated); currently limited to a small relict population in north-central Sonora. Listed as 'endangered' in the *ICBP bird red data book* (King 1981).

C. v. maculatus: central Tamaulipas south to northern Veracruz and west to south-eastern San Luis Potosi.

C. v. aridus: central and west-central Tamaulipas south to south-eastern San Luis Potosi.

C. v. graysoni: south-eastern Nayarit and southern Jalisco south to the valley of Mexico, Morelos, southern Hidalgo, and south-central San Luis Potosi.

C. v. nigripectus: Puebla, Morelos, and Mexico.

C. v. pectoralis: eastern slopes of mountains in central Veracruz.

C. v. godmani: lowlands of Veracruz.

C. v. minor: north-eastern Chiapas and adjacent Tabasco.

C. v. atriceps: interior of western Oaxaca.

C. v. thayeri: north-eastern Oaxaca.

C. v. harrisoni: south-western Oaxaca.

C. v. coyolcos: Pacific coast of Oaxaca and Chiapas.

C. v. salvini: southern coastal Chiapas.

Map 6. Distribution of the northern bobwhite.

C. v. insignis: southern Chiapas and adjacent Guatemala (Rio Chiapas valley).

C. v. nelsoni: extreme southern Chiapas. Questionably distinct from *insignis*.

Measurements

Wing (US forms), both sexes 98–119 mm; tail, both sexes 49–70 mm. Wing (Mexican forms), both sexes 90.5–114 mm; tail, both sexes 45–67 mm (Ridgway and Friedmann 1946). Weight (eastern US), males average 173 g, females 170 g (Johnsgard 1973); (Mexico) both sexes of various races average 129–159 g (Leopold 1959). Weights tend to increase from south to north and from east to west in the US (Rosene 1969). Egg (US forms), average 31 × 25 mm, estimated weight 10.7 g.

Identification (see Fig. 11)

In the field (7.5–9.5 in.)

Although northern bobwhites vary greatly in appearance throughout their range, they have little or no contact with quails of comparable size, except for limited contact with the dissimilar scaled quail. Throughout their broad range the birds are associated with fairly heavy cover, and unlike *Callipepla* species are more prone to hide and flush upon disturbance than to run from danger. When separated, birds utter a series of *hoy*, *hoy-poo*, and *hoyee* notes of increasing loudness. Males of probably all races of northern bobwhites utter the distinctive *bob-white!* or *ah-bob-white!* note that is heard most commonly in unpaired males.

In the hand

Male northern bobwhites are easily distinguished from the more southern species of *Colinus* by their lack of a crest, absence of scalloped or spotted underparts, and their white or black throat markings, but different races vary greatly in the amount of white vs black on the throat, and the underpart coloration, which may be mostly white, barred and marked with black, or extensively cinnamon to chestnut, usually liberally marked with black. The buffy-throated females are less easily distinguished. However, their breast feathers lack definite terminal white spots, and their sides and flanks are also more distinctly streaked or barred than spotted.

Adult female. Differs mainly from the male in having a buff throat; the black areas of the face, throat, and foreneck are represented only by black tips to the feathers; the black bars on the undersides are also much less distinct and usually absent on the middle of the breast and abdomen.

Immature. Similar to adults, but with the other two primaries relatively pointed and frayed, and with buffy edging on most of the greater primary-coverts.

General biology and ecology

The total distributional range of this species is remarkably broad, and encompasses many climatic and habitat types, as summarized earlier (Johnsgard 1973). However, in most or all areas certain habitat features are constant, including the presence of grassy or herbaceous nesting cover, cultivated crops or a similar source of natural vegetative foods, and brushy or woody cover. Grasslands provide nesting cover, some foraging cover, and limited roosting cover. Croplands offer feeding, loafing, dusting, and limited roosting cover. Brushy areas or woodlands are used throughout the year for escape and roosting, and in some areas provide vital autumn and winter feeding habitats. Areas that have all three habitat types in close proximity, namely, those that provide a good deal of 'edge', are probably ideal. Nesting cover is typically open herbaceous vegetation under 20 in. high, having some nearly bare ground below, and with enough space between the plant stems to allow easy walking. Available surface water for drinking is needed, as the birds apparently cannot get enough moisture from plant materials to fill their needs, and dusting sites having dry and rather powdery soils are also important habitat components (Johnsgard 1973).

The foods of northern bobwhites also vary greatly in different regions (Rosene 1969) but seeds, especially legume seeds, are universally eaten, with cultivated grains tending to replace seeds in some areas, and particularly in winter such crops as maize, wheat, and soybeans are especially important. Likewise in some wooded areas acorns are important autumn foods. The use of insects varies regionally and seasonally, but is always highest during the period of reproduction, and likewise is an important source of proteins for young chicks. Bobwhites typically forage twice a day, in early morning and again in late afternoon. Birds of a covey forage together without aggression, and grit may be picked up at the same time or obtained separately along roadways or cuts.

Social behaviour

The coveys of bobwhites tend to be fairly small, usually of about 10–15 birds, and averaging about 12. This consistency in covey size is at least partly a reflection of the optimum number of birds needed for forming efficient roosting circles. The coveys tend to be distinctly sedentary in most areas, often having home ranges of less than 20 acres, and rarely these are

Fig. 11. Adult male plumage variation in the northern bobwhite, subspecies (A) *floridanus*, (B) *virginianus*, (C) *coyolcos*, (D) *ridgwayi*, (E) *godmani*, (F) *graysoni*, (G) *insignis*, and (H) *atriceps*.

more than 50 acres (Rosene 1969). The greatest instability of coveys probably occurs in autumn, as unsuccessful pairs attach themselves to pairs with well-grown young, members of broods may break up and become attached to new coveys, and the like. During the autumn period individual movements are sometimes fairly considerable, sometimes of several miles, and in very rare instances may cover more than 20 miles. However, after this 'autumn shuffle', most coveys become fairly uniform in size and composition, and remain so until spring (Lehmann 1984). As spring approaches, members of individual coveys become more aggressive toward one another, and fighting may break out periodically, especially among males. Males perform a strong frontal display (see Fig. 10) toward other males and initially also toward females as a social dominance signal. They only later begin to respond selectively toward females by performing tidbitting behaviour, bowing, and assuming a lateral display posture. During lateral display the male walks slowly about the female with his flank feathers ruffled, his tail fanned, and his upper body surface tilted toward her, but with the wings not distinctly drooped (Stokes 1967). This display is silent, is usually brief, and perhaps corresponds to the circling or 'waltzing' display of many perdicines. Probably tidbitting represents the major method of pair-bonding in bobwhites, as it continues well beyond the pair-forming period (Johnsgard 1973).

Vocalizations in the bobwhite are numerous and complex (Stokes 1967), but as in the other New World quails the two most familiar vocalizations are the separation call and the unmated male advertisement call. The separation call actually consists of a series of progressively louder *hoy*, *hoy-poo*, and *hoyee* notes that not only serve to reunite separated pairs and reassemble scattered coveys but may also help to space coveys, attract unmated males to females, and repel intruding birds. During fierce male-to-male aggressive encounters a loud and raucous 'caterwaul' or a series of whining *squee* notes usually replace these less strictly agonistic calls. The familiar 'bobwhite' call is limited almost exclusively to males during the breeding season, particularly unmated ones.

Reproductive biology

Nests of bobwhites are typically built in dead grass or herbaceous vegetation from the prior breeding season, usually in fairly open surroundings. The site is probably chosen by the pair, and both typically help build the nest. The first egg is usually laid a day or two after the nest is completed, and thereafter an egg is laid on a nearly daily basis until the clutch is complete. Clutch-sizes average from about 12 to 16 eggs in various parts of the US range, with a slight tendency for larger clutches in more northerly areas (Rosene 1969). The incubation period averages 23 days. Nest losses tend to be quite high, often up to 60 or 70 per cent, and thus some females must begin up to as many as four nests before successfully hatching a brood. However, such persistent renesting behaviour is typical, and probably most females do eventually succeed in hatching a brood. There is only very limited evidence for double-brooding in wild US populations (Stanford 1972a). The hatching curves from birds in southern Texas as shown by Lehmann (1984) do not exhibit an extended or secondary late peak as might be expected if regular double-brooding occurred there, but the possibility of double-brooding in more tropical parts of the range remains.

The young chicks fledge in about two weeks, and by 150 days of age the outermost first-winter primaries (the eighth) have attained their maximum length (Rosene 1969). By this age, too, captive hand-raised birds can be stimulated into reproductive activity by increased photoperiods, but under wild conditions sexual activity would not begin for several additional months.

Evolutionary relationships

I have suggested (Johnsgard 1973) that *Colinus* may have originated in the highlands of southern Mexico, moving gradually northward to form the *virginianus* group, eastward to the Caribbean lowlands to form the *nigrogularis* group, and progressively southward to form a series of populations now represented by the *leucopogon–cristatus* superspecies. Great regional variations in bodily feather pigmentation gradually evolved, which are generally but not always related to local variations in humidity and vegetational cover. Some of the most southerly forms also acquired fairly conspicuous crests. The timing of this period of speciation must remain speculative, but Holman (1961) commented that the skeletal differences among the *Colinus* types are fairly small, suggesting a fairly recent evolutionary divergence. Hubbard (1974) suggested that a recent glaciation may have fragmented the ancestral ranges of at least the US and Mexican forms of *Colinus* and accounted for the resulting speciation process and their present-day distributions.

Status and conservation outlook

The general status of the northern bobwhite is certainly favourable over much of its US range, although the 'masked' race of northern bobwhite has been extirpated from Arizona and has markedly declined since the initiation of population surveys during the late 1960s within its very limited remaining Sonoran range (Brown and Ellis 1977). Attempts at reintroduction of this race into Arizona have not proven to be successful, and its fate in Sonora remains questionable.

BLACK-THROATED BOBWHITE (Plate 11)

Colinus nigrogularis (Gould) 1843
Other vernacular names: black-throated quail; Honduras bobwhite (*nigrogularis*); Progreso bobwhite (*persiccus*); Yucatan bobwhite (*caboti*); colin à gorge noire (French); Schwarzkehlwachtel (German); codorniz garganta negra (Spanish).

Distribution of species (see Map 7)
Resident of brushy, savannah, and woodland edge habitats of Yucatan, the Lake Petan area of Guatemala, and the savannahs of Belize, eastern

Map 7. Distribution of the black-throated (B), crested (C), and spot-bellied (S) bobwhites.

Honduras, and adjacent Nicaragua. Allopatric with both *leucopogon* and *virginianus*, and occasionally considered conspecific with the latter.

Distribution of subspecies

C. n. caboti: eastern Campeche and Yucatan Peninsula except for the Progreso area.

C. n. persiccus: endemic to the Progreso area of the Yucatan Peninsula.

C. n. nigrogularis: Belize and adjacent northern Guatemala.

C. n. segoviensis: north-eastern Honduras and northern Nicaragua.

Measurements

Wing, males 95–103.5 mm, females 95–103.5 mm; tail, males 50–58.5 mm, females 50–59 mm (Ridgway and Friedmann 1946). Weight, both sexes 126–144 g (Leopold 1959); males average 137 g, females 139 g (Johnsgard 1973). Egg, average 30.5 × 23 mm (Johnsgard 1973), estimated weight 8.9 g.

Identification

In the field (7.5–8.5 in.)

The very limited range of this species is isolated from that of all other bobwhites, and additionally the males have distinctive black throats and eye-stripes, both of which are conspicuously bounded by white. Advertising males produce *bob-white!* calls, but this call is uttered much more rapidly (lasting about a second) than in the northern bobwhite, and often lacks the preliminary syllable.

In the hand

The black blotches on the scapulars help to identify this species as a bobwhite, and the lack of a crest separates males from crested bobwhites. Females are distinguished from the more southern forms of northern bobwhites by their relatively tawny underparts, and from the Central and South American bobwhites by their bright buff superciliary stripes and throats, which lack any black markings, and the absence of any crest.

Adult female. Differs from the male mainly in having a buffy chin and upper throat, and in having less markedly scalloped underpart patterning. Very similar to females of some races of *virginianus*, but the breast is generally less tawny-coloured or cinnamoneous.

Immature. Similar to adults, but with buffy-tipped greater primary-coverts. The outer two primaries are also usually more pointed and frayed than the others.

General biology and ecology

Like the northern bobwhite, this species seems to need a weedy seed supply, brushy or woody escape cover, and fairly open grassy or herbaceous nesting and foraging cover. In the Yucatan Peninsula these needs are well met by the weedy fields of *henequin* (*Agave*) that are widespread in areas cleared of natural vegetation. In this region there is virtually no surface water, even during the wet season, and thus moisture must be obtained from herbaceous leafage. Not many specimens have been examined, but it is known that such weedy species as tick trefoil (*Desmodium*) are consumed by the birds (Leopold 1959). Much feeding is done by the birds in the *henequin* fields, where the bayonet-like leaves of this plant doubtless provide a good deal of overhead protection from large predators but still allow for unimpeded ground-level movement. During the dry season the birds are strongly attracted to sources of water, suggesting that in some areas it may be limiting. There is no detailed information on movements or on population density, but at least in some areas of Yucatan it must approach or exceed a bird per acre (Johnsgard 1973).

Social behaviour

The coveys of this species tend to be of about the same size as the northern bobwhite, namely of about a dozen birds. Probably these coveys are maintained until the start of the rainy season, which usually begins between April and June. Observations on captive birds suggest that females may also be stimulated by increased photoperiods. At about the same times males begin uttering their *bob-white!* calls, which are given from elevated sites and repeated about every 15–25 s. These calls are fairly brief, each lasting about a second. Probably pair-forming behaviour is much like that of the northern bobwhite; observations of captive birds (Cink 1971) indicated that many of the vocalizations of these two species are comparable, although they tend to be higher in pitch and uttered more rapidly in *nigrogularis*. However, the multi-syllable caterwaul call of the northern bobwhite is replaced in this species by a brief and essentially single-element *churr* note similar to that found in *cristatus*. The most commonly observed agonistic behaviour observed in captive birds was overt attacking and pecking; frontal display was not seen. Lateral display was also brief by comparison with the northern bobwhite.

Reproductive biology

Indirect evidence from Mexico suggests that nesting occurs over a fairly long period, probably extending from April to August or even later. The nests in the Yucatan area are reportedly often placed under *henequin* plants, but no information on clutch-sizes is available from wild birds. Eggs have been laid by birds I have maintained in captivity at irregular rates, but the incubation period has proven to be the same as that of the northern bobwhite, i.e. 23–24 days (Johnsgard 1973). The young have also grown at essentially comparable rates, as have hybrids between these species, although specific weight or mensural measurements have not been made.

Evolutionary relationships

Inasmuch as Cink (1971) concluded that the vocalizations of *nigrogularis* have more in common with *virginianus* than with *cristatus* I am inclined to consider these two as allospecies, but not as subspecies as has sometimes been suggested. There is no natural contact between these forms at present, and captive unmated males of each species showed virtually no response when they were exposed to female separation calls of the other, suggesting that at least a degree of intrinsic isolation has occurred since the two gene pools separated. Comments on the possible timing of this separation have been made in the account of the northern bobwhite.

SPOT-BELLIED BOBWHITE (Plate 12)

C. leucopogon (Lesson) 1842
Other vernacular names: Dickey's quail (*dickeyi*); Leland's quail (*leylandi*); Sclater's quail (*sclateri*); white-breasted bobwhite (*incanus* and *hypoleucus*); white-faced bobwhite (*leucopogon*); white-throated quail (*leucopogon*); colin à huppe blanche (French); Salvadorhaubenwachtel; Tupfenwachtel (German); codorniz vientrimanchada (Spanish).

Distribution of species (see Map 7)
Resident of brushy fields and woodland edges of Central America from southern Guatemala to Costa Rica. Recognized as specifically distinct by Blake (1977) but allopatric with and often included in *C. cristatus*, which it replaces geographically north of Panama.

Distribution of subspecies

C. l. incanus: southern Guatemala.
C. l. hypoleucus: western El Salvador.
C. l. leucopogon: eastern El Salvador.
C. l. leylandi: Honduras.
C. l. sclateri: Nicaragua.
C. l. dickeyi: Costa Rica.

Measurements

Wing, males 95–107 mm, fameles 95–102 mm; tail, males 50–64 mm, females 47–63 mm (Ridgway and Friedmann 1946). Estimated weight, males 144 g, females 115 g. Egg, average 31.1 × 25 mm (Schönwetter 1967), estimated weight 10.7 g.

Identification (see Fig. 12)

In the field (8–8.5 in.)

In its limited Central American range this is the only bobwhite species, and there it occupies similar kinds of habitats to those of the northern bobwhite. Males also utter a *bob-white!* call, but it and the species' other calls have not yet been compared sonagraphically with those of the other bobwhites. The birds are quite variable in appearance throughout their range, but the more northerly races are strongly white on

Fig. 12. Adult male plumage variation in the spot-bellied bobwhite, subspecies (A) *leucopogon* and (B) *leylandi*, and the crested bobwhite, subspecies (C) *cristatus*, (D) *leucotis*, (E) *parvicristatus*, and (F) *sonnini*.

the face, and the underparts are also white or heavily spotted with white.

In the hand

Separation from the crested bobwhite is not always easy, especially among females, but adults of this species lack a definite crest, and in males the throat varies from white to brown or sometimes mostly black, while in females it is always brownish, with a black and white border and with dusky speckles or streaks. Northern forms (*leucopogon* northward) have the forehead, superciliaries, and throat white, with the underparts entirely white or spotted with white. Southern races (*leylandi* southward) have the throat brownish with a black border, or blackish with a white to buffy border; a white or buffy post-ocular stripe; hindneck and sides of neck spotted black and white; the underparts deep reddish brown to pale greyish brown, spotted with white or buff. Females are closely similar to females of *cristatus*, but generally more greyish rather than brownish on the upperparts, and with black vermiculations rather than black blotches on the upper back.

Adult female. Similar to the male, but with buffy superciliaries and throat, the latter speckled with dusky, the underparts pale buff, variably streaked and barred; the breast, sides, and flanks heavily spotted with white.

Immature. Juvenile males have the feathers of the breast, abdomen, sides, and flanks spotted with buff and brown much like those of the females, but these areas soon become intermixed with adult-like first-winter feathers. Older birds are closely similar to adults, but the outer two primaries more pointed and frayed than are the others.

General biology and ecology

This species is generally described as occurring in open grassy and scrubby areas, especially shrubby borders and fencerows, and the thickety stands and overgrown ravines of scrubby ranchlands. There seems little doubt that it has the same general ecological needs as the better-known northern and black-throated bobwhites.

Social behaviour

The covey sizes of this species have not been described, but Slud (1964) noted that the coveys he encountered tended to escape on foot rather than fly to safety. Coveys make twittering sounds that change to a *prrrit* when they are disturbed, and after flushing the birds utter soft repeated *chit* noises resembling the whistling of wings. Flushed birds sometimes alight on slender branches in thickets or on shrubs, but this is probably exceptional. The male's advertising call may have two or three notes.

Reproductive biology

In El Salvador male 'bob-white' calls have been heard as early as mid-March, and egg-laying probably commences by about the end of May. Although most young are well feathered by late September, some breeding occurs quite late in the year, since half-grown juveniles have been collected as late as the latter part of January (Dickey and van Rossem 1938). There is no information on egg-laying rates, clutch-sizes, or incubation periods, but these are likely to be similar to those of the northern bobwhite, as are growth and maturation rates.

Evolutionary relationships

General comments on the evolution of *Colinus* were made in the northern bobwhite account. I am inclined to think that speciation and range expansion in the *leucopogon–cristatus* group proceeded from Central to South America rather than in the reverse direction, and that the genus *Colinus* is subtropical to warm-temperate rather than tropical in origin, which might help account for the apparent sensitivity of *nigrogularis* females to increased photoperiods in stimulating reproduction. In that sense the South American forms of crested bobwhites might be considered the most derived types, whereas most other South American odontophorine quails fall in the more generalized and seemingly more primitive group.

CRESTED BOBWHITE (Plate 13)

C. cristatus (L.) 1766
Other vernacular names: crested quail; crested colin; Horvath's quail (*horvathi*); Mocquery's quail (*mocquerysi*); short-crested quail (*parvicristatus*); Sonnini's bobwhite (*sonnini*); white-eared quail (*leucotis*); colin zonecolin (French); Haubenwachtel (German); codorniz crestuda; codorniz chico (Spanish).

Distribution of species (see Map 7)

Resident of brushy Neotropical fields, savannahs, and woodland edge habitats from Panama to western Colombia and east to the Guianas; also on Aruba, Curacao, and Margarita islands, and introduced on Mustique (Grenadines) and St Thomas (Virgin Islands).

Distribution of subspecies

C. c. mariae: Panama (Chiriqui).
   C. c. panamensis: Panama (western lowlands).
   C. c. decoratus: Caribbean coast of Colombia.
   C. c. littoralis: northern Colombia (northern base of Santa Marta Mountains).
   C. c. cristatus: extreme northern Colombia (Guajira Peninsula and eastern base of Santa Marta Mountains).
   C. c. badius: Cauca Valley to Pacific slope of western Andes, Colombia.
   C. c. leucotis: northern Colombia (Magdalena and Sinu valleys).
   C. c. bogotensis: east-central Colombia (Boyaca and Cundinamarca).

C. c. parvicristatus: north-eastern Colombia (east slope of eastern Andes) and Venezuela (northern Amazonas, north-western Bolivar, and upper Apure Valley).

C. c. horvathi: north-western Venezuela (Merida Andes).

C. c. continentis: north-western coastal Venezuela; Aruba, Curacao.

C. c. barnesi: north-central Venezuela.

C. c. mocquerysi: north-eastern Venezuela.

C. c. sonnini: coastal northern Venezuela, the Guianas, and extreme northern Brazil (upper Rio Branco).

Measurements

Wing, males 93–120 mm, females 92–119 mm; tail, males 46–74 mm, females 46–72 mm (Blake 1977). Weight, males 132–153 g, females 131–141 g (Haverschmidt 1968). Egg, average 30 × 40 mm, estimated weight 9.5 g.

Identification (see Fig. 12)

In the field (8–8.5 in.)

This is the only species of bobwhite found from Panama southward, which simplifies identification. Both sexes exhibit short crests, and are usually found near fairly heavy brushy cover, into which they run or fly when disturbed. The male's *bob-white!* call is relatively high-pitched and rapidly uttered, and the species' other calls are correspondingly higher and faster than those of the northern bobwhite.

In the hand

The presence of a short crest, which is slightly curved upward, is quite distinctive, but its length varies considerably in different races. In females the crest is shorter and is a darker brown, and in both sexes the underparts tend to be rather heavily spotted and barred with buff, cinnamon, and black, at least as compared with *leucopogon*.

Adult female. Differs from the male mainly in having the crown brown or buff, speckled or finely streaked with black.

Immature. jeveniles are spotted with brown and white on the underparts and have pale yellow (in skins) bills. Similar to the adults after moulting the juvenal plumage, but retaining the outer two juvenal primaries, and also retaining some greater primary-coverts that are edged and tipped with buff.

General biology and ecology

In Panama this species is mainly found in thickets and along the edges of woodlands bordering fields and savannahs. The birds occur in small coveys, and when startled they rarely fly more than about 10 m, dropping out of sight in the nearest cover, or simply disappearing on foot into low shrubbery. They feed on the ground (Wetmore 1965), probably consuming the usual array of weedy seeds that the other bobwhites are known to prefer.

Social behaviour

Judging from observations of Wetmore (1965), coveys begin to break up and nesting begins at the start of the rainy season. He heard males in Panama starting to call as early as late January, and as late as the latter part of June. Nearly grown young were observed by the end of May. The advertising whistle of the male is a *bob-white!* of two or three notes, with isolated birds prone to add a preliminary third note, at least in captivity (Cink 1971). This species appears to have only one type of separation call, the *hoy-poo*, with the notes much more closely spaced and the intervals shorter than in the corresponding call of the northern bobwhite. The caterwaul call is a staccato *whirrr-churr* that sounds like a single phrase. It is much more abbreviated than in the northern bobwhite but is more like that of the black-throated bobwhite. The most commonly observed agonistic behaviour that Cink observed was bill-fighting, which was invariably accompanied by caterwauling. Lateral display was shorter than in the northern bobwhite, averaging 3 s rather than 8–20 s, and frontal display was not observed at all (Cink 1971).

Reproductive biology

To my knowledge no nests have been described from wild birds, but captive females in my laboratory appeared to be much like the other *Colinus* forms as to their reproductive biology. The eggs had the same incubation period (22–23 days) under artificial incubation conditions, and the chicks seemed to exhibit roughly the same rates of growth and plumage development, although specific measurements were not made. The low hatchability that occurred in crosses between the northern bobwhite and crested bobwhite, together with a high rate of embryonic deaths and deformities, suggest that some genetic barriers to hybridization may exist between them. However, some first generation hybrids that did survive proved to be fertile and produced vigorous offspring when back-crossed to the northern bobwhite (Cink 1971; Johnsgard 1973).

Evolutionary relationships

See the previous accounts of the white-bellied and northern bobwhites for comments on this.

Status and conservation outlook

The many widely scattered but little-studied populations of this species are probably fairly secure, as the birds thrive in second-growth and marginal agricultural situations, and generally benefit from forest clearing activities.

7 · Genus *Odontophorus* Vieillot 1816

The wood-quails are medium-sized forest-adapted species of Central and South America, with short tails of 12 rectrices that are up to half the length of the wing. The wing is rounded, with the sixth and seventh primaries the longest. Some juvenal greater primary-coverts and the two outer primaries are retained in the post-juvenal moult. The sexes are alike or very similar as adults, with relatively heavy bills, bare eye-rings or larger areas of exposed facial skin, generally inconspicuous and often sexually dimorphic bushy crests, and mostly rufous to olive-toned or dark brown plumages. Fourteen mostly allopatric species are recognized here; several of these—the forms here indicated (with brackets) as allospecies—perhaps might be better classified as subspecies:

O. gujanensis (Gmelin): marbled wood-quail
    O. g. castigatus Bangs 1901
    O. g. marmoratus Gould 1844
    O. g. medius Chapman 1929
    O. g. gujanensis Gmelin 1789
    O. g. buckleyi Chubb 1919
    O. g. pachyrhynchus Tschudi 1844
    O. g. rufogularis Blake 1959
    O. g. simonsi Chubb 1919
O. capuiera Spix: spot-winged wood-quail
    O. c. plumbeicollis Cory 1915
    O. c. capuiera Spix 1825
O. erythrops Gould: rufous-fronted wood-quail
    O. e. verecundus Peters 1929
    O. e. melanotis Salvin 1865
    O. e. parambae Rothschild 1897
    O. e. erythrops Gould 1859
O. [*speciosus*] *melanonotus* Gould: dark-backed wood-quail
O. [*speciosus*] *hyperythrus* Gould 1858: chestnut wood-quail
O. speciosus Tschudi: rufous-breasted wood-quail
    O. s. soederstroemii Lönnberg and Rendahl 1922
    O. s. speciosus Tschudi 1843
    O. s. loricatus Todd 1932
O. leucolaemus Salvin 1867: black-breasted wood-quail
O. strophium Gould 1844: gorgeted wood-quail
O. [*strophium*] *dialeucos* Wetmore 1963: Tacarcuna wood-quail
O. [*strophium*] *columbianus* Gould 1850: Venezuelan wood-quail
O. atrifrons Allen: black-fronted wood-quail
    O. a. atrifrons Allen 1900
    O. a. variegatus Todd 1919
    O. a. navai Aveledo and Pons 1952
O. balliviani Gould 1846: stripe-faced wood-quail
O. stellatus Gould 1843: starred wood-quail
O. guttatus Gould 1838: spotted wood-quail

KEY TO THE SPECIES OF WOOD-QUAILS (*ODONTOPHORUS*)

A. Breast not spotted with white
  B. Breast uniformly rusty red
    C. Superciliary area rust to rufous chestnut
      D. Chin and throat black: rufous-fronted wood-quail (*erythrops*)
      DD. Chin and throat rusty red: chestnut wood-quail (*hyperythrus*)
    CC. Superciliary area not rusty red
      D. Throat black; a distinct superciliary stripe present: rufous-breasted wood-quail (*speciosus*)
      DD. Throat white; no distinct superciliary stripe: dark-backed wood-quail (*melanonotus*)
  BB. Breast grey, black, or barred with brown and buff
    C. Breast black; throat white: black-breasted wood-quail (*leucolaemus*)
    CC. Breast grey to buffy; throat not white
      D. Breast uniformly grey: spot-winged wood-quail (*capueira*)
      DD. Breast barred with brown and buff: marbled wood-quail (*gujanensis*)
AA. Breast spotted with white, at least laterally
  B. Chin and throat brown or grey to buffy
    C. Underparts uniformly reddish; chin and throat grey: starred wood-quail (*stellatus*)
    CC. Underparts spotted with white; chin and throat brown to buffy: stripe-faced wood-quail (*balliviani*)
  BB. Chin and throat mostly or entirely white, or black and white
    C. Chin and throat mostly or entirely black
      D. White shaft-streaks present on throat: spotted wood-quail (*guttatus*)
      DD. No white shaft-streaks on throat
        E. Chin and throat entirely black: black-fronted wood-quail (*atrifrons*)
        EE. Throat-band black, dividing white chin and foreneck: Tacarcuna wood-quail (*dialeucus*)
    CC. Chin and throat white, sometimes with black markings
      D. Throat variably barred or spotted with black: Venezuelan wood-quail (*columbianus*)
      DD. Throat white, banded below with black: gorgeted wood-quail (*strophium*)

MARBLED WOOD-QUAIL (Plate 14)

Odontophorus gujanensis Gmelin 1789

Other vernacular names: Buckley's partridge (*buckleyi*); Chiriqui wood-quail (*castigatus*); Duida partridge (*medius*); Guianan partridge (*gujanensis*); Simon's partridge (*simonsi*); thick-billed partridge (*pachyrhynchus*); tocro de Guyane (French); Guayana-Zahnhuhn; Guayanawachtel (German); gallito del monte jaspeado; perdiz común (Spanish).

Distribution of species (see Map 8)

Resident of tropical forests of the Neotropics from southern Costa Rica to Colombia, eastward to Venezuela and the Guianas; thence southward, mostly east of the Andes, to eastern Bolivia and central and north-eastern Brazil.

Distribution of subspecies

O. g. *castigatus*: Costa Rica and western Panama.

O. g. *marmoratus*: eastern Panama, northern Colombia, and north-western Venezuela.

O. g. *medius*: southern Venezuela to north-western Brazil.

O. g. *gujanensis*: south-eastern Venezuela east to the Guianas and south through Brazil to the northern Mato Grosso and east to north-eastern Paraguay.

O. g. *buckleyi*: eastern Colombia south through eastern Ecuador to northern Peru.

O. g. *pachyrhynchus*: east-central Peru (Junin and Ayacucho).

O. g. *rufogularis*: north-eastern Peru (upper Rio Javary).

O. g. *simonsi*: eastern Bolivia.

Measurements

Wing (chord), males 130–154.5 mm, females 130–149 mm; tail, males 51–77.5 mm, females 51–75 mm (Blake 1977). Weight, males 313–349 g, females 298 g (Haverschmidt 1968). Egg, average 36.8 × 27.5 mm, estimated weight 15.3 g.

Identification

In the field (11–12 in.)

This is the most widespread of the *Odontophorus* species, and over most of the Amazonian drainage is the only one likely to be encountered. In the upper Amazon it is replaced by *stellatus*, which has distinctive white breast-spotting. Farther north it is in possible contact with several other congeners, where its marbled greyish to tawny underparts, relatively unmarked upperparts, and similarly unpatterned greyish to rufous head colour provide for identification. Its usual call is a repeated, liquid multi-noted *k-wuck-oo* or *corcoro-vado* sequence that is actually a duet.

In the hand

The lack of black or white on the throat, or of white spotting on the underparts, and instead a rather uniformly tawny to dusky 'marbled' coloration, with a grey mantle, provide for species identification.

Adult female. Like the male but slightly smaller.

Immature. Crest and top of head dark reddish brown, very faintly vermiculated with dusky, the mantle is faintly or almost imperceptibly vermiculated, the outer secondaries are almost devoid of vermiculations, the inner webs of the remiges are mottled with buff, and the bill is orange-red. The two outermost juvenal primaries and the greater primary-coverts are retained in the post-juvenal moult (Petrides 1945).

General biology and ecology

This forest-adapted species is usually found in regions of irregular terrain, particularly in hilly

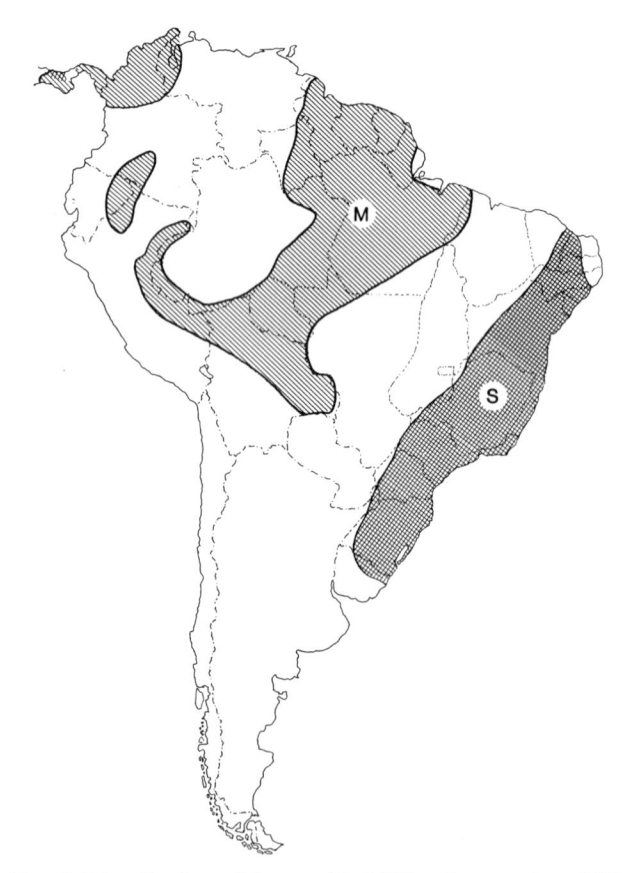

Map 8. Distribution of the marbled (M) and spot-winged (S) wood-quails.

country, where they inhabit undergrowth vegetation. In Costa Rica they are most abundant in the rainforests at altitudes of 2000–3000 ft (600–900 m), but sometimes occur in tall second-growth vegetation (Skutch 1947). They take shelter around fallen trees, or on steep and broken slopes, where they are usually overlooked, as they typically crouch rather than flush when disturbed. When frightened they may take-off with a flurry of wings and fly some 60–80 m, after which they land and begin running. They have been observed feeding under berry-producing trees on the fruits that have fallen to the ground, and sometimes go to the edges of rivers to drink. Most of the crop and stomach contents of one specimen consisted of starchy seeds, but there were also the remnants of such invertebrates as millipedes, ants, cockroaches, spiders, and beetles (Wetmore 1965). Skutch (1947) observed the birds feeding on small, dried bananas and possibly also on associated insects.

Social behaviour

The usual covey size is of six to eight individuals, which almost certainly represent family groupings. These birds move about through the forest in single file, and members of the covey hunt for food in close cooperation and preen one another's plumage. Birds in these coveys communicate almost constantly by nearly inaudible calls, but at some times such as dawn and dusk, and occasionally during moonlit nights or early forenoon hours, they sing loudly enough to be heard for half a kilometre or more. Wetmore (1965) described the call as sounding like a *perro-mulato* sequence that might be uttered several hundred times over a period of several minutes. Skutch (1947) stated that it sounded like *burst the bubble*, and said that the evening chorus usually occurs in the twilight hours when most other avian sounds have hushed. Along the Caribbean slope the call has been described as sounding like *corcorovado*, and Chapman (1929) determined from a captive pair of birds that this is actually a duet, with the two birds singing antiphonally the two portions of the call. Wetmore (1965) observed that when he collected the female of a pair, the remaining male called with a *corcoro* call for several evenings, apparently until a new mate was obtained about a week later and the full call was heard once again. This would suggest that the call at least in part may serve as a male advertising call, as in the North American odontophorines, but has been supplemented by an integrated female response, perhaps as possible mating-bonding or pair-location announcement mechanism.

Reproductive biology

One nest was found in Colombia in early March. It was at the foot of a large forest tree, and consisted of only a single egg. There is also a record of an oviducal egg obtained from a female in the latter part of March. In Panama chicks or juveniles have been collected between late February and early April (Wetmore 1965). Skutch (1947) found three nests in Costa Rica, all of which were well-enclosed chambers roofed over with dead vegetation, and with lateral round entrances. They were situated on sloping ground in second-growth woodland, at the base of a mound produced by the roofs of a fallen tree, and among the protruding roots of a standing tree. Eggs in these nests were laid in January, April, and probably May. Two completed clutches had four eggs, but the other clutch was lost before being completed. Only one bird, apparently the female, was observed incubating, and she incubated almost continuously every day, being on the nest about 75–86 per cent of the daylight hours. Each morning the putative male would call his mate from the nest for a short time, and later escort her back to its vicinity. Toward the end of incubation a third grown bird of unknown relationship to the others was also present. The incubation period of one nest was between 24 and 28 days, and the chicks were led from the nest by the female when they were less than 22 h old. At that time the male joined the group. Downy chicks were also observed travelling with coveys of five or six grown birds.

Evolutionary relationships

This is the most widely distributed of all the *Odontophorus* species, and one of the most plainly coloured. Its plumages do not offer any clear evidence of its relationships, although I have suggested (Johnsgard 1973) that it may represent a superspecies with the spot-winged wood-quail on the basis of the two species' allopatric distributions in similar tropical lowland habitats and their overall general plumage similarities.

Status and conservation outlook

Inasmuch as this species has a very broad distribution that is centred in some of the most remote Amazonian forests, it is likely to survive longer than some of the other *Odontophorus* types with much smaller distributions in areas that are rapidly being affected by human activities.

SPOT-WINGED WOOD-QUAIL (Plate 15)

Odontophorus capueira Spix 1825
Other vernacular names: capueira partridge; capueira

wood-quail; Ceara partridge (*plumbeicollis*); tocro uru (French); Brazilien-Zahnhuhn; Capueirawachtel (German); perdix capueira (Spanish).

Distribution of species (see Map 8)
Resident of tropical forests of South America from north-eastern Brazil south to Paraguay and extreme north-eastern Argentina.

Distribution of subspecies
O. c. plumbeicollis: north-eastern Brazil (Ceara and Alagoas).
O. c. capueira: eastern Brazil, eastern Paraguay, and north-eastern Argentina.

Measurements

Wing, males 151–172 mm, females 144–153 mm; tail, males 75–82 mm, females 70–77 mm (Blake 1977). Estimated weight, males 457 g, females 396 g. Egg, average 40.1 × 29.1 mm (Schönwetter 1967), estimated weight 18.7 g.

Identification

In the field (11 in.)
This is the only wood-quail found in eastern South America south of the Rio Parnaiba, which simplifies field identification. It calls at dawn and again at twilight, although the calls are still only poorly described. By one early account the call is of only two notes, sounding like *ura*, but another early observer states that it is comprised of three or four notes, rapidly repeated (Gould 1850), suggesting that perhaps the longer vocalizations are duets.

In the hand
This species can be separated from other *Odontophorus* by the combination of dark greyish feathers on the throat and underparts, and having the outer webs of the primaries barred with white.

Adult female. Very similar to the male, but slightly smaller.

Immature. Pale shaft-streaks on the mantle wider and more conspicuous than in adults, the black spots on the lower back and rump much coarser, and the grey underparts washed with reddish brown.

General biology and ecology

Almost nothing is known of the biology of this species under natural conditions, but it is associated with dense tropical forests and probably is much like the marbled wood-quail in its ecology. Its foods in the wild are unstudied, but captive virds were found to thrive on game bird 'crumbles' and high protein 'trout chow' (Flieg 1970).

Social behaviour

No detailed information is available on this. See In the field section above for some comments on vocalizations.

Reproductive biology

Flieg (1970) reported that three adults (two males, one female) of this species were bred during 1965 at the St Louis Zoo, which is the first and only known case of any wood-quail nesting in captivity. Nesting activity began by the three birds individually picking up materials and tossing them backward toward the next individual, thus gradually transporting the materials to the nest site. The nest they constructed was domed, and measured about 40 × 50 cm. It took about three days to complete, and five eggs were laid at daily intervals. These eggs were removed, and a second clutch of three eggs was produced about two weeks later. Three eggs hatched after incubation periods of 26–27 days, although in one case pipping began four days earlier. The young birds reportedly grew very rapidly at first, but after reaching about half adult size they reportedly slowed down, and full adult size was reached at about two months of age. This, none the less, represents a surprisingly rapid growth rate for a tropical species of quail.

Evolutionary relationships

This is the southernmost species of *Odontophorus*, and its range is nearly in contact with that of the marbled wood-quail, its most likely nearest living relative.

Status and conservation outlook

This species occurs in three countries (Brazil, Paraguay, and Argentina), and its status is essentially unknown in all of them.

RUFOUS-FRONTED WOOD-QUAIL
(Plate 16)

Odontophorus erythrops Gould 1859
Other vernacular names: black-eared wood-quail (*melanotis*); chestnut-eared partridge (*erythrops*); Honduran partridge (*verecundus*); Paramba partridge (*parambae*); rufous-breasted wood-quail; Veraguan partridge (*coloratus*); tocro à front roux (French); Rotstirnwachtel (German); gallito del monte pechicastaño; perdiz rojiza (Spanish).

Genus *Odontophorus* Vieillot 1816　75

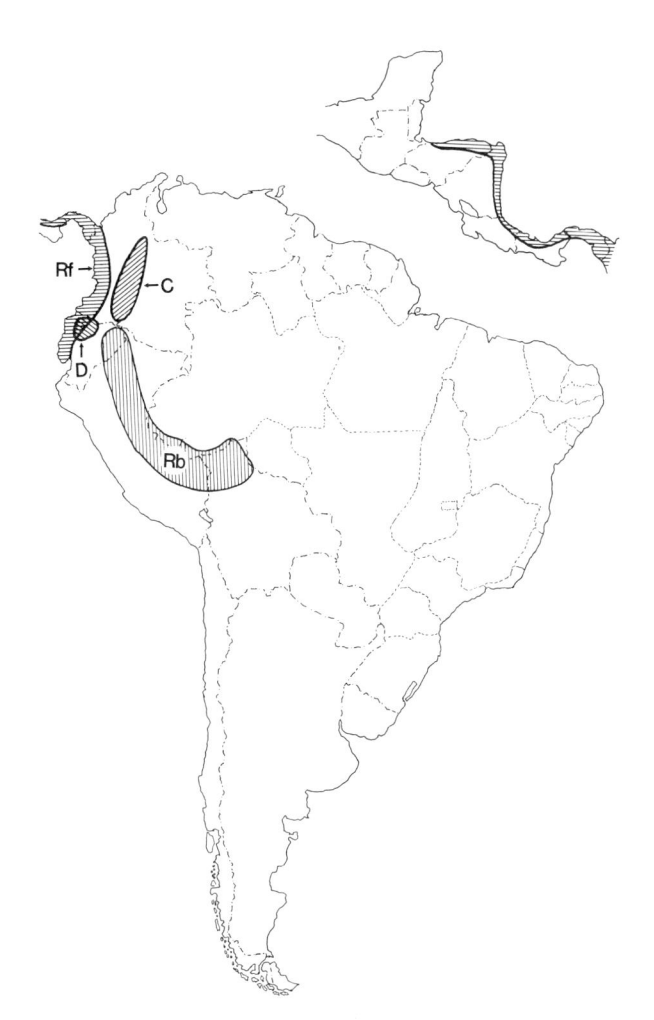

Map 9. Distribution of the chestnut (C), dark-backed (D), rufous-breasted (Rb), and rufous-fronted (Rf) wood-quails.

Distribution of species (see Map 9)

Resident of tropical forests of the Neotropics from Honduras to western Ecuador.

Distribution of subspecies

O. e. verecundus: Honduras.

　O. e. melanotis: Nicaragua, Costa Rica, and Panama. Includes *coloratus*. These first two races are sometimes considered specifically distinct, as *O. melanotis*, or black-eared wood-quail.

　O. e. parambae: Colombia and western Ecuador.
　O. e. erythrops: south-western Ecuador.

Measurements

Wings (chord), males 130–153 mm, females 131–149 mm; tail, males 43–61 mm, females 46–57 mm (Blake 1977). Estimated weight, males 340 g, females 329 g. Egg, average 37.6 × 27.9 mm, estimated weight 16.1 g.

Identification (see Fig. 13)

In the field (9.5–10 in.)

Limited mostly to the eastern slopes of Central America from Honduras southward to Panama and along the Pacific slopes of the Andes to western Ecuador. This species possibly has limited contacts with *leucolaemus* and *gujanensis*, from both of which it differs by having a black or brownish throat, although a conspicuous white gular stripe may sometimes be present below the throat. The calls include *klawcoo* and *klawcoo* sequences, frequently repeated.

In the hand

The predominantly chestnut-brown breast and sides, and usually also a similar chestnut superciliary area, serve to identify this species.

Adult female. Like the male, but lacking black on the throat, generally duller and more dingy on the head and breast, and with a coffee-brown iris and bluish black eye-ring.

Immature. Similar to the adult, but with duller crest feathers, white to buffy shaft-streaks dorsally, and heavily spotted and barred with black below (Wetmore 1965). Juvenile specimens have reddish brown (in skins) bills and dark ear-coverts that contrast with the more golden-brown sides of the head.

General biology and ecology

Wetmore (1965) noted that this species is widely distributed but little known in Panama, where it has been found at elevations of from 450 to 1600 m in the tropical and lower subtropical zones. They have been observed in forest undergrowth and reportedly when flushed fly off into surrounding trees and later disappear into the woods in twos and threes. Slud (1964) described the vocalizations as double-noted and rapidly repeated *klawooo, klawooo* or *klawuu, klawuu* sounds, which suggests that like the marbled and black-throated wood-quails these calls may well be duets between pair members.

Reproductive biology

No detailed information is available. Mendez (1979) believed that breeding probably occurs in Panama at the end of the dry period.

Evolutionary relationships

I have suggested (Johnsgard 1973, 1979) that this is part of a highly variable superspecies group that

76 *The Quails, Partridges, and Francolins of the World*

Fig. 13. Adult plumage variation in the rufous-fronted wood-quail, subspecies (A) *melanotis*, (B) *parambe*, and (C) *erythrops*. Plumage variation among the (D) dark-backed, (E) chestnut, and (F) rufous-breasted wood-quails.

extends from Honduras south to northern Bolivia, including not only *erythrops* but also the allopatric and more montane *hyperythrus* in interior Colombia and northern Ecuador, *melanonotus* of Ecuador, and *speciosus* of southern Ecuador, Peru, and Bolivia. All of these would appear to be derivatives of a single ancestral type that speciated in the northern and central Andes.

Status and conservation outlook

The fairly broad range of this species presumably favours its prospects for survival, but it is reportedly rare and locally distributed in many of these areas.

CHESTNUT WOOD-QUAIL (Plate 17)

Odontophorus hyperythrus Gould 1858
Other vernacular names: chestnut-throated

partridge; tocro marron (French); Kastanienwachtel (German); perdiz parda (Spanish).

Distribution of species (see Map 9)

Endemic to tropical forests of Colombian Andes. No subspecies recognized; possibly should be considered conspecific with *speciosus*.

Measurements

Wing, males 144–153 mm, females 140–146 mm; tail, males 55–60 mm, females 48–55 mm (Blake 1977). Estimated weight, males 392 g, females 351 g. Egg, no information.

Identification (see Fig. 13)

In the field (11 in.)

Limited to the Colombian Andes where it is perhaps in contact only with *strophium*, this species is recognized by its deep rust-red throat, sides of the head, and superciliary stripes, with contrasting white ear-coverts.

In the hand

Very similar to *melanonotus*, but not nearly so blackish on the upperparts and sides, and with the crown and upper face area rusty rather than dark brownish black.

Adult female. Differs from the male in having a blackish brown crown and the breast and rest of underparts dark grey, shading to blackish grey on the flanks. Very similar to the female of *O. s. soederstroemii* but with a bright brown supercilium rather than black and white superciliary stripes.

Immature. Wing-coverts and scapulars spotted with buff at their tips, some of the mantle feathers blotched or marked with black, and the ridge of the culmen and at least the tip of the bill orange-red. Young males are apparently rather female-like, with scattered rust-coloured feathers gradually appearing on the underparts.

General biology

No specific information.

Social behaviour

No specific information.

Reproductive biology

No specific information.

Evolutionary relationships

This species is evidently closely allopatric with *erythrops*, the latter occurring in Colombia from the Pacific coast eastward to as far as the west slope of the Central Andes of Antioquia, while the present one also has been collected in Antioquia at Santa Elena and Concordia. It is probably even more closely related to *speciosus*, as suggested by the transitional plumage of the race *soederstroemii*, and as suggested by Hellmayr and Conover (1942) might eventually prove to best be considered a subspecies of that form.

DARK-BACKED WOOD-QUAIL (Plate 18)

Odontophorus melanonotus Gould 1860
Other vernacular names: black-backed partridge; tocro à dos noir (French); Schwarzrückenwachtel (German); perdiz cuelliroja (Spanish).

Distribution of species (see Map 9)

Endemic resident of subtropical montane forests of north-western Ecuador and adjacent Colombia. No subspecies recognized; possibly should be considered conspecific with *speciosus*.

Measurements

Wing, males 130–148 mm, females 134–144 mm; tail, males 45–58 mm, females 42 mm (Blake 1977). Estimated weight, both sexes 322 g. Egg, average 39.9 × 29.9 mm (Schönwetter 1967), estimated weight 19.6 g.

Identification (see Fig. 13)

In the field (10 in.)

Limited primarily to a small area of western Ecuador where no other *Odontophorus* species is known to occur, but also recently collected in Colombia (Fitzpatrick and Willard 1982). The very dark brownish black upperparts, including the supercilium, and a cinnamon-rufous breast coloration, provide for identification.

In the hand

The combination of a deep rust-red to cinnamon underpart coloration, including the throat, and a nearly black dorsal coloration including the sides of the head, provides for distinction from other species.

Adult female. Virtually identical to the male.

Immature. A single observed juvenile specimen had a mostly reddish bill, pale greenish grey legs, and dull brown, vermiculated underparts, which became slightly brighter toward the breast.

General biology and ecology

No specific information.

Social behaviour

No specific information.

Reproductive biology

No specific information.

Evolutionary relationships

This is certainly part of a closely related assemblage of Andean allospecies, of which *hyperthrus* occurs immediately to the north, and *speciosus* to the south of the present one, and some of which might well be regarded as subspecies. Blake (1977) suggested that the three just-mentioned forms comprise a superspecies, of which *speciosus* and *melanonotus* are probably the more closely related forms.

Status and conservation outlook

The very small known range of this form, in an area where little if any conservation efforts are being made, does not bode well for its future.

RUFOUS-BREASTED WOOD-QUAIL (Plate 19)

Odontophorus speciosus Tschudi 1843
Other vernacular names: Bolivian partridge (*loricatus*); Soderstrom's partridge (*soederstroemii*); tocro à poitrine rousse (French); Rotbrustwachtel (German); codorniz de frente rojiza; perdiz pechi-castaña (Spanish).

Distribution of species (see Map 9)
Resident of tropical forests from eastern Ecuador to eastern Peru and north-eastern Bolivia.

Distribution of subspecies

O. s. soederstroemii: eastern Ecuador.
   O. s. speciosus: eastern and central Peru.
   O. s. loricatus: south-eastern Peru and eastern Bolivia.

Measurements

Wing, males 134–147 mm, females 130–142 mm; tail, males 48–68 mm, females 50–60 mm (Blake 1977). Estimated weight, males 332 g, females 302 g. Egg, no information.

Identification (see Fig. 13)

In the field (10–11 in.)
This species is in possible contact with *stellatus* and *balliviani*, but can be separated from them by its black (or brown mottled with black) throat and its predominantly rufous-brown coloration on the underparts, with no white speckling or spotting.

In the hand
Similar to both of the preceding forms, but separable from them in having dark reddish brown rather than olive-brown or brownish black on the crown and upperparts, and usually having a mottled black and white superciliary stripe.

Adult female. Differs from the male mainly in having only the chest rufous chestnut, the breast and abdomen being dark grey with a few of the feathers on the side tinged with rufous.

Immature. Inner webs of the longer scapulars and outer secondaries widely edged toward the tip with rich rufous buff; feathers of the chin throat, and sides of face chestnut with dusky margins, mantle with greatly reduced shaft-stripes, and the black and white superciliary stripe much reduced in width.

General biology and ecology

No specific information.

Social behaviour

No specific information.

Reproductive biology

No specific information.

Evolutionary relationships

This is the southernmost of the group of seemingly closely related forms that begins in Central America with *erythrops* and extends south through the ranges of *hyperythrus*, *melanonotus*, and *speciosus*. Their ranges are all extremely poorly known and their evolutionary histories can only be speculated upon at this point.

BLACK-BREASTED WOOD-QUAIL (Plate 20)

Odontophorus leucolaemus Salvin 1867
Other vernacular names: white-throated wood-quail; tocro à poitrine noire (French); Weisskehlwachtel (German); gallito del monte pechinegro (Spanish).

Distribution of species (see Map 10)

Endemic to tropical and subtropical forests of Costa Rica and western Panama. No subspecies recognized here (*smithianus* of the Dota region is usually regarded as a melanistic variant).

Measurements

Wing, males 122.5–124.5 mm, females 120–125 mm; tail, males 55.5–68 mm, females 46.5–51 mm (Ridgway and Friedmann 1946). Estimated weight, males 226 g, females 220 g. Egg, no information.

Identification

In the field (9–9.5 in.)

Limited to the subtropical forests in the Caribbean drainage of Costa Rica and Panama, probably mostly or entirely above 1000 m elevation, and the only *Odontophorus* of that area having a distinctly black breast and head colour, with a contrasting white throat patch. Its repetitive call consists of a rushing gabble of two sets of paired syllables, the first syllable accented, and often breaking out simultaneously from several individuals.

In the hand

Easily separated from other *Odontophorus* species by its extremely dark breast and head coloration, save for a more brownish crown and the sometimes nearly obsolete white throat markings.

Adult female. Virtually identical to the male.

Immature. Similar to the adult, but the breast and upper abdomen dark brown, like the sides and flanks; the malar area barred black and white. The tip of the bill is reportedly cinnamon-brown (probably reddish in life) rather than black.

General biology and ecology

In Panama this little-known species has been reported only from the subtropical zone of Chiriqui, Veraguas, and Bocas del Toro, at elevations of 1350–1600 m. In there, as in Costa Rica, it is an inhabitant of thick forests, and occurs on steep, heavily wooded slopes (Wetmore 1965). In Costa Rica it is found in scrubby, vine-rich and shaded growth adjoining taller woodlands (Slud 1964).

Social behaviour

Other than the fact that these birds apparently occur in small coveys, and have been reported as noisy during early morning hours of March and April, nothing of significance has been noted about their behaviour (Wetmore 1965).

Reproductive biology

No specific information.

Evolutionary relationships

This species is rather variable in appearance, but essentially appears to be a melanistic form of the general *Odontophorus* plumage type that is also to be seen in *atrifrons* and *columbianus*, and to a lesser degree in *dialeucos* and *strophium*, and apparently is part of this large superspecies group. It is not in sympatric contact with any of these, but may be in limited contact with *erythrops*, which is apparently generally more tropical in its distribution and is mostly found below rather than above 1000 m elevation.

Status and conservation outlook

In western Panama this species is apparently rare, and has not been collected or observed since 1933 (Wetmore 1965). It is likewise apparently rather rare but more widespread in Costa Rica (Slud 1964).

GORGETED WOOD-QUAIL (Plate 21)

Odontophorus strophium Gould 1844
Other vernacular names: gorgeted partridge; tocro à miroir (French); Kragenwachtel (German); perdiz cuelliblanca (Spanish).

Distribution of species (see Map 10)

Endemic to temperate forests of Colombian Andes from Santander south to about Cundinamarca; known from only four localities. No subspecies recognized; possibly both *dialeucos* and *columbianus* should be considered conspecific. Endangered species (King 1981).

Measurements

Wing, males 144.5–151.5 mm, females 138–146 mm; tail, males 45.6–49 mm, females 39–46.7 mm (Romero-Zambrano 1983). Estimated weight, both sexes 302 g. Egg, no information.

Identification (see Fig. 14)

In the field (10–11 in.)

This rare species is in possible contact only with *hyperythrus*, from which it can be readily dis-

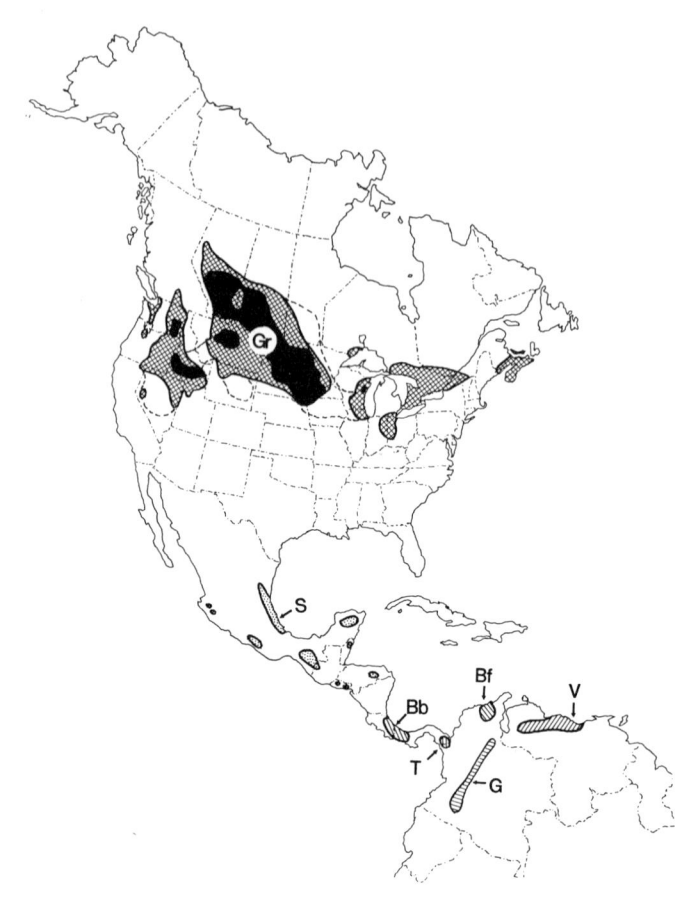

Map 10. Distribution of the black-breasted (Bb), black-fronted (Bf), gorgeted (G), Tacarcuna (T), and Venezuelan (V) wood-quails, the singing quail (S), and introduced North American distribution of the grey partridge (Gr).

tinguished by its black rather than rufous throat, and its contrasting chin and foreneck.

In the hand
Most similar to *dialeucos*, but in that species the crown is black rather than brown, and females are tawny rather than greyish below.

Adult female. Similar to the male, but dark grey below. The female also lacks the small white shaft-spots found in the adult male, and its general patterning is more obscured (Romero-Zambrano 1983).

Immature. Differs from adults in having the chest, breast, and underparts largely mixed with brownish grey, and the white spots represented by only a few narrow shaft-stripes on the sides of the chest and breast.

General biology and ecology

The very small probable range of this species has been plotted by Romero-Zambrano (1973), who noted that it is known from only seven museum specimens, which have been collected in an Andean and sub-Andean forest life zone that occurs in the departments of Santander, Boyaca, Cundinamarca, and Tolima, and is dominated by oaks (*Quercus humboldtii*) and laurels (*Nectandra* and *Persea*). A considerable number of other woody species occur in this general area, several genera of which (*Trigonobalanus, Cavendishia, Macleania, Miconia, Myrica, Rapanea, Ficus,* and *Norantea*) produce fruits that might possibly be eaten by the birds. Specimens taken so far have been found to contain various unidentified fruits and insects or larval arthropods. Three of the specimens were obtained at elevations of 1800–1970 m.

Social behaviour

No specific information.

Reproductive biology

A male adult with young was captured in the middle of May 1970, a male with well-developed gonads was collected in March 1981, a female with enlarged ovaries was obtained in late November 1979, and additionally a juvenile female was obtained in early December 1979. From this information Romero-Zambrano (1983) judged that the major reproduction period probably lasts from October to March.

Evolutionary relationships

This species is certainly a close relative of *dialeucos*, and quite possibly both should be considered as conspecific. However, too few specimens are known to judge the degree of plumage variation that might occur in *strophium*.

Status and conservation outlook

King (1981) listed this species as 'endangered', based on the fact that as of that time only four definite specimens were known, all of which had been collected over 60 years previously. The recent additional specimens and better data on this form's actual range provide some new hope for its survival, and the Colombian government has initiated action to facilitate its protection (Romero-Zambrano 1983).

TACARCUNA WOOD-QUAIL (Plate 22)

Odontophorus dialeucos Wetmore 1963
Other vernacular names: tocro de Panama (French); Tacarcunawachtel (German); gallito de monte fajaedo; perdiz oscura (Spanish).

Fig. 14. Adult plumage variation in the (A) black-breasted, (B) black-fronted, (C) Tacarcuna, (D) gorgeted, and (E) Venezuelan wood-quails.

Distribution of species (see Map 10)

Endemic resident of subtropical forests of extreme eastern Panama. No subspecies recognized; possibly should be considered conspecific with *strophium*.

Measurements

Wing (chord), males 129–131 mm, females 126–132 mm; tail, males 50–54 mm, females 46–50 mm (Blake 1977). Estimated weight, males 264 g, females 258 g. Egg, no information.

Identification (see Fig. 14)

In the field (10–11 in.)

Limited to subtropical forests of Panama, where it may be in contact with *gujanensis* and *erythrops*, but easily distinguished from both by the white superciliary stripe, anterior face, and lower throat; the rest of the plumage is relatively dark.

In the hand

Very similar to *strophium*, but with a white-spotted black crown, lacking a black collar under the white neck patch, and the underparts of males are more buffy to tawny-olive, and have less prominent white spotting.

Adult female. Very similar to the male but slightly brighter, and more tawny below.

Immature. Like the adult female, but the white of chin less extensive or replaced by black, which reaches the base of the bill.

General biology and ecology

Known only from the subtropical zone of Panama's elevations of 1200–1400 m, where the birds occur in forest undergrowth (Wetmore 1965).

Social behaviour

Wetmore (1965) found the birds in groups ranging in number from pairs to coveys of six or eight individuals during February and March. When disturbed they uttered the low, rapid calls typical of alarmed wood-quails, and one once flew up to perch in a low tree when flushed.

Reproductive biology

No specific information, except that an immature bird was collected in early June (Wetmore 1965).

Evolutionary relationships

This species is somewhat darker than *strophium* of Colombia, but the two are otherwise very similar and certainly constitute an allospecies, if they are not conspecific. They presumably have only been isolated from one another since Pleistocene times.

Status and conservation outlook

Wetmore (1965) found the birds to be 'fairly common' on the slopes of Cerro Mali and Cerro Tacarcuna, and made no comments as to any possible threatened status for this geographically highly restricted and most newly discovered species of Odontophorinae.

VENEZUELAN WOOD-QUAIL (Plate 23)

Odontophorus columbianus Gould 1850
Other vernacular names: Caracas spotted partridge; Venezuelan partridge; tocro de Venezuela (French); Venezuelawachtel (German); perdiz montañera (Spanish).

Distribution of species (see Map 10)
Endemic resident of subtropical forests of northern and western Venezuela. No subspecies recognized, but possibly should be considered conspecific with *strophium*.

Measurements

Wing, males 134–150 mm, females 132–150 mm; tail, males 45–62 mm, females 50–60 mm (Blake 1977). Estimated weight, males 343 g, females 336 g. Egg, no information.

Identification (see Fig. 14)

In the field (10–11 in.)
The very limited Venezuelan range of this species is not known to overlap with that of any other *Odontophorus*, but in any case it is the only wood-quail in which the throat and chin are both white, this area being surrounded and increasingly spotted with black laterally. The lower flanks and underparts of males have fairly large and well-marked white spots on a reddish brown background. Found in cloud forests above 1200 m, up to at least 2400 m.

In the hand
The black-rimmed and black-streaked white throat is unique to only this species of the genus, and is typical of both sexes. Otherwise, it differs from *strophium* only in its duller and less reddish underparts and its large white spots on the sides and flanks.

Adult female. Similar to male, but pale shaft-streaks of mantle and black less developed or lacking; breast and sides uniformly greyish brown, without white spots or black markings.

Immature. Mantle spotted and marked with black, the pale shafts only slightly developed, and the general tone of the underparts brownish grey, with reduced white spotting on the breast, chest, and sides. (The plate shows an apparent immature with an orange bill and reduced white spotting below.)

General biology and ecology

Gines and Aveledo (1958) stated that the species occurs in montane forests of the subtropical zone at elevations of about 2400 m. The birds occur in coveys of about a dozen individuals, consuming seeds, fruits, small insects, and worms.

Social behaviour

No specific information.

Reproductive biology

Gines and Aveledo (1958) stated that the reproductive period occurs from May to the end of July, but provided no details.

Evolutionary relationships

Except for its distinctive throat pattern, this species generally resembles *atrifrons*, and the two appear to be part of this general assemblage of forms centred in Colombia and also including *strophium*, *dialeucos*, and *leucolaemus*.

Status and conservation outlook

Like the other *Odontophorus* forms, there is no information on the status of this elusive species.

BLACK-FRONTED WOOD-QUAIL (Plate 22)

Odontophorus atrifrons Allen 1900
Other vernacular names: variegated partridge (*variegatus*); tocro à front noir (French); Schwarzstirnwachtel (German); perdiz frentinegra; perdiz cabecinegra (Spanish).

Distribution of species (see Map 10)
Resident of subtropical montane forests of northern Colombia and north-western Venezuela.

Distribution of subspecies
O. a. atrifrons: northern Colombia (Santa Marta Mountains).
  O. a. variegatus: north-eastern Colombia (eastern Andes).
  O. a. navai: north-western Venezuela (Perija Mountains).

Measurements

Wing, males 129–148 mm, females 129–142 mm; tail, males 62–91 mm, females 64–80 mm (Blake 1977). Estimated weight, males 311 g, females 298 g. Egg, no information.

Identification (see Fig. 14)

In the field (9–10 in.)
Limited to cloud forests between 4000 and 7600 ft (1200–2300 m) in a small area of northern Colombia and adjacent Venezuela, this relatively isolated species is likely to be confused only with *gujanensis*, which may occur in the same area. It differs from this and other species of wood-quails in its uniformly black forehead, chin, and throat.

In the hand
No other *Odontophorus* species is so extensively black, with white being limited to shaft-streaks and some spotting on the underparts.

Adult female. Like the male, but the median underparts extensively rufescent.

Immature. Upper mandible brownish red (in skins), and blackish throat feathers that are edged with olive-tawny, producing a scaly appearance (specimens in US National Museum).

General biology and ecology

According to Gines and Aveledo (1958), this species occurs in heavy forests of Venezuela between 1200 and 2300 m, and moves about in coveys numbering about 10 individuals except during the breeding period.

Social behaviour

The only comments that seem to have been made on this species' vocalizations are those of Gines and Aveledo (1958), who noted that it utters strong calls at dawn, and produces gabbling sounds when birds are reunited.

Reproductive biology

No information available.

Evolutionary relationships

As noted earlier (Johnsgard 1973, 1979), this species may be part of a rather large group of five currently recognized species collectively extending allopatrically from Costa Rica to Venezuela, having black or black and white throats, and mostly associated with subtropical forests of this region. All are virtually unknown except for locality data on collected specimens.

Status and conservation outlook

No specific information is available.

STRIPE-FACED WOOD-QUAIL (Plate 24)

Odontophorus balliviani Gould 1846
Other vernacular names: Ballivian's partridge; tocro de Ballivian (French); Streifengesichtwachtel (German); perdiz manchada (Spanish).

Distribution of species (see Map 11)
Resident of subtropical forests of south-eastern Peru and northern Bolivia. No subspecies recognized.

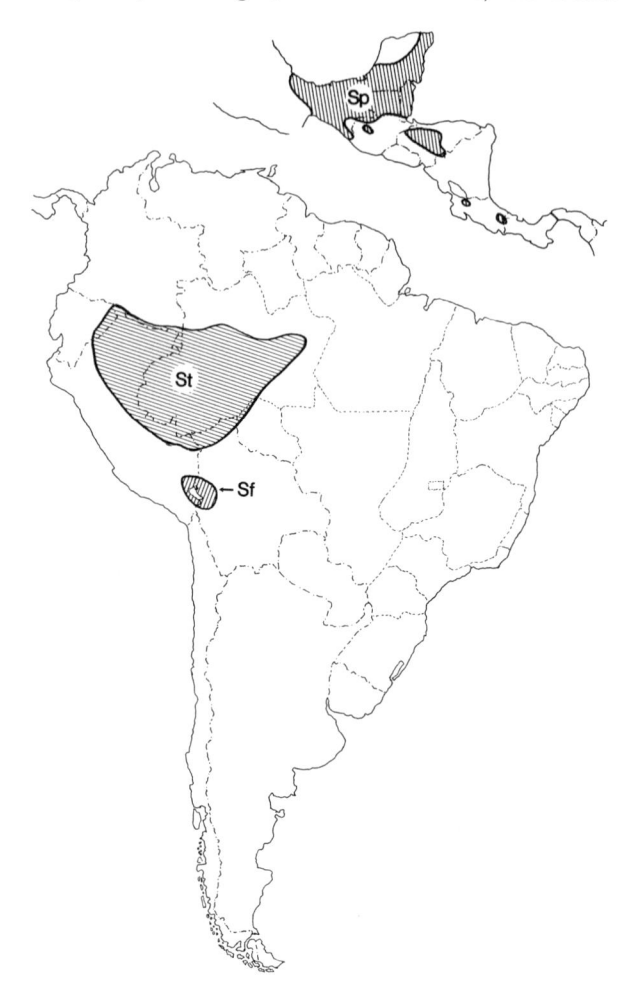

84 *The Quails, Partridges, and Francolins of the World*

Map 11. Distribution of the spotted (Sp), starred (St), and stripe-faced (SF) wood-quails.

Measurements

Wing, males 141–152 mm, females 142–156 mm; tail, males 57–71 mm, females 59–75 mm. Estimated weight, males 311 g, females 324 g. Egg, average 38 × 26.8 mm (Schönwetter 1967), estimated weight 15.1 g.

Identification

In the field (10.5 in.)

Limited to forests in the highlands of south-eastern Peru and northern Bolivia along the eastern slopes of the Andes, and probably the only *Odontophorus* species in that area. The large and conspicuous white spots on the flanks should provide a suitable field mark for visual identification.

In the hand

This species is unique among the *Odontophorus* forms in having large black-edged diamond-shaped white spots on the flanks, and a prominent black eye-stripe that is bounded above and below by cinnamon.

Adult female. Apparently identical to the male.

Immature. Undescribed.

General biology and ecology

No specific information.

Social behaviour

No specific information.

Reproductive biology

No specific information.

Evolutionary history

This seems to be a fairly isolated species, but like the more tropically adapted *stellatus* it lacks white on the throat and has chestnut underparts that are distinctly spotted with white. Thus, it seems probable that these two species are fairly close relatives. They are allopatrically distributed along the eastern Andes of Peru and Bolivia and the upper portions of the Amazonian drainage.

Status and conservation outlook

There is no information on the status of this apparently rare species, which is known from only a few specimens.

STARRED WOOD-QUAIL (Plate 25)

Odontophorus stellatus Gould 1843
Other vernacular names: starred partridge; tocro étoille (French); Sternwachtel (German); perdiz estrellada (Spanish).

Distribution of species (see Map 11)

Resident of tropical forests of the Amazonian Basin from Ecuador, Peru, and northern Bolivia to western Brazil (east to Rio Madeira, south to the northern Mato Grosso). No subspecies recognized.

Measurements

Wing, males 138–150 mm, females 134–142 mm; tail, males 56–72 mm, females 57–65 mm (Blake 1977). Estimated weight, males 358 g, females 315 g. Egg, no information.

Identification

In the field (10.5 in.)

Similar to the previous species in having diamond-shaped white spots on the sides and flanks, but these spots are smaller, and the face is not distinctly striped. Probably overlapping in range only with *gujanensis* and *speciosus*, neither of which has white flank spotting.

In the hand

Rather easily distinguished by the extensive white speckling of the breast and sides, and its grey throat, as distinct from the dusky brown throat of *balliviani*.

Adult female. Similar to the male, but the crest deep brownish black.

Immature. Very similar to the adults, but with a reddish orange to yellowish bill, and (at least in the case of a young male) with a rufous crest.

General biology and ecology

No specific information.

Social behaviour

No specific information.

Reproductive biology

No specific information.

Evolutionary relationships

Although fairly distinctive, this species is quite similar to *balliviani*, and also similar to the widely distributed and at least partially sympatric *gujanensis*. It is interesting that *gujanensis* is seemingly lacking from most of the range of *stellatus*, but overlap apparently occurs in eastern Ecuador (where both have been collected at Sarayaco, on the Rio Bobonaza, Department of Pastaza) and possibly also in eastern Peru.

Status and conservation outlook

There is no reliable information on the status of this species.

SPOTTED WOOD-QUAIL (Plate 26)

Odontophorus guttatus Gould 1838
Other vernacular names: spotted partridge; tocro tachete (French); Tropfenwachtel (German); bolonchaco (Spanish).

Distribution of species (see Map 11)

Resident of tropical and montane forests of Central America from Oaxaca and Veracruz, Mexico, south locally to extreme western Panama. No subspecies recognized by Blake (1977).

Measurements

Wing, males 134–153.5 mm, females 134.5–148.5 mm; tail, males 69.5–76.5 mm, females 61–72.5 mm (Ridgway and Friedmann 1946). Weight, both sexes 260–360 g (Leopold 1959), males average 300 g, females 288 g (Johnsgard 1973). Egg, average 40 × 29 mm (Johnsgard 1973), estimated weight 18.6 g.

Identification

In the field (10–11.5 in.)

This is the only *Odontophorus* found north of Honduras, and from there to the southern end of its range in western Panama it overlaps only with *erythrops*. It may be recognized by its black throat with white striping, and the rounded or teardrop-shaped pale spots on the underparts. Like apparently all *Odontophorus* species it is highly vocal, uttering repetitive calls such as a patterned *gahble gahble*.

In the hand

Easily separated from all other *Odontophorus* species by the black throat feathers with white shaft-streaks, producing a striped pattern.

Adult female. Differs from the male in having the head and crest browner, and the crest lacking cinnamon tones.

Immature. The rear part of the crest rust-red as in adult males, but somewhat more rufous ventrally, the white diamond-shaped spots approach the form of shaft-stripes, more black mottling is present on the neck, chest, breast, and sides, and the bill (at least in very young birds) is orange-red.

General biology and ecology

In Mexico this species is largely limited to tropical rainforest areas from central Veracruz south to the Guatemalan and Belize borders, and occurs from near sea-level up to at least 1300 m. Dense forests with an open understorey of vegetation are seemingly preferred in Mexico and in Guatemala, where in the latter area it has been recorded at elevations between 1600 and 6000 ft (500–1800 m). In Belize it occurs both in tall rainforests and in high second growth (Russell 1964), while farther south in Nicaragua and elsewhere it is especially characteristic of cloud forests, and is replaced by *erythrops* at lower elevations (Griscom 1932). In Honduras it has been found up to 2000 m, but is most common above 600 m, mainly in cloud forests or montane rainforests, although it sometimes occurs in lowland rainforests as well (Monroe 1968). In Costa Rica it has been found from the middle of the subtropical forest belt upward to the timber-line (Slud 1964), and in Panama it is mainly found between 1250 and 2100 m in subtropical forests, just below the zone of cloud forests (Wetmore 1965). Its foods in Mexico apparently consist of small bulbs, soft rootlets, and the larvae, pupae, and adults of insects, mostly dipterans and coleopterans, as well as small seeds and the meat of larger seeds or nuts. At least in captivity adults distinctly prefer large and soft fruits, such as grapes and bananas, over-dried beans and hard grains (Johnsgard 1973). Feeding is done in small groups, with the birds constantly uttering soft and low liquid sounds. There is little reason to believe that major movements are undertaken, although Slud (1964) believed that some vertical migration may occur in Costa Rica.

Social behaviour

Like the other wood-quails, these birds occur in fairly small coveys that probably represent family groups; Leopold (1959) noted that in his experience these numbered from 5 to 10 birds. However, up to as many as 20 birds have also been observed together. The call is complex, and consists of as many as six notes, which Leopold (1959) described as *wheet-o-whet-to-wheo-who*, with the last two syllables somewhat variable. The call is uttered mainly at dawn and again at dusk. Tape recordings of this call indicate that at least two call types are present, one of which is a prolonged series of uniformly spaced *to-wet!* notes. The second is initiated by a series of *whee-oh!* notes uttered at 1.8-s intervals (probably all given by a single bird) which immediately were followed by a prolonged series of repetitive and distinctly cadenced phrases, each of which lasted about 1.5 s, sounded like *whet'-o-whet, whe'oo*, and almost certainly consisted of a duet. Information from a man who had kept a pair of these birds in captivity confirmed that they regularly called in an antiphonal duet, especially at dawn and dusk, and that when the male was sick for a time the female would not 'sing', suggesting that it is the male that initiates the call and that probably utters the *whe'oh* portion of the duet (Johnsgard 1973).

Reproductive biology

There is no good information on this topic, although it is believed that in the Yucatan Peninsula the birds nest in May and June, during the first part of the wet season. In Chiapas and Guatemala young chicks have been collected in May (Leopold 1959). At the southern end of the species' range downy chicks have twice been collected in March (Wetmore 1965). No nests have been described, and although the species has occasionally been kept in captivity it has not yet bred under those conditions.

Evolutionary relationships

There are no obvious close relatives of this species, which bears some general plumage similarities to several other *Odontophorus* forms, but whose throat plumage pattern is unique.

Status and conservation outlook

Like the other wood-quails, this species is doubtlessly gradually losing out to the deforestation that is rampant through Mexico and Central America.

8 · Genus *Dactylortyx* Ogilvie-Grant 1893

The singing quail is a medium-sized and forest-adapted species of Mexico and Central America with a very short tail of 12 firm rectrices and about 40 per cent as long as the wing. The wing is rounded, with the seventh primary the longest. Some juvenal greater primary-coverts and the two outer primaries are retained in the post-juvenal moult. The tarsus is shorter than the middle toe and claw, but the latter is almost as long as the culmen, the bill being relatively small and slender. The sexes are dimorphic as adults, with males having rufous throats, but both sexes have pale superciliary stripes and predominantly grey to olive-brown body plumage. One polytypic species is recognized:

D. thoracicus Gambel: singing quail
    D. t. pettingilli Warner and Harrell 1957
    D. t. thoracicus Gambel 1848
    D. t. devius Nelson 1898
    D. t. melodus Warner and Harrell 1957
    D. t. chiapensis Nelson 1898
    D. t. moorei Warner and Harrell 1957
    D. t. dolichonyx Warner and Harrell 1957
    D. t. edwardsi Warner and Harrell 1957
    D. t. ginetensis Warner and Harrell 1957
    D. t. sharpei Nelson 1903
    D. t. paynteri Warner and Harrell 1957
    D. t. colophonus Warner and Harrell 1957
    D. t. salvadoranus Dickey and van Rossem 1928
    D. t. taylori van Rossem 1932
    D. t. fuscus Conover 1937
    D. t. rufescens Warner and Harrell 1957
    D. t. conoveri Warner and Harrell 1957

SINGING QUAIL (Plate 27)

Dactylortyx thoracicus Gambel 1848
Other vernacular names: long-toed partridge; long-toed quail; colin chanteur (French); Langkrallenwachtel; Singwachtel (German); codorniz gemidora (Spanish).

Distribution of species (see Map 10)
Resident of montane forests of Central America from Veracruz and Jalisco, Mexico, to Honduras and El Salvador.

Distribution of subspecies
D. t. pettingilli: south-western Tamaulipas and south-eastern San Luis Potosi.
    D. t. thoracicus: north-eastern Puebla and central Veracruz.
    D. t. devius: Jalisco.
    D. t. melodus: central Guerrero.
    D. t. chiapensis: central Chiapas.
    D. t. moorei: mountains of central Chiapas.
    D. t. dolichonyx: Chiapas (Sierra Madre de Chiapas).
    D. t. edwardsi: Chiapas, near Oaxaca border.
    D. t. ginetensis: Chiapas–Oaxaca border area.
    D. t. sharpei: Campeche, Yucatan, and Quintana Roo, south to northern Guatemala (Lake Peten area).
    D. t. paynteri: south-central Quintana Roo.
    D. t. colophonus: Pacific Cordillera of Guatemala.
    D. t. salvadoranus: El Salvador (Volcan de San Miguel).
    D. t. taylori: El Salvador (Mt Cacaguatique, possibly elsewhere).
    D. t. fuscus: Honduras (Tegucigalpa area).
    D. t. rufescens: Honduras (San Juancito Mountains).
    D. t. conoveri: Honduras (Department of Olancho).

Measurements

Wing, males 117–137 mm, females 113.5–133 mm; tail, males 45–60.5 mm, females 46–55.5 mm (Ridgway and Friedmann 1946). Weight, males 180–266 g, females 168–206 g (Warner and Harrell 1957). Egg, average 29 × 23.5 mm (Schönwetter 1967), estimated weight 8.4 g.

Identification

In the field (9 in.)
Associated with rather heavy brush or wooded cover, and more often heard than seen. The song (actually a duet) is a melodious series of repeated notes, preceded by low whistles and sounding like *che-va-lieu-a*. The birds are mostly greyish to olive-brown, with conspicuous whitish superciliary stripes and otherwise mostly tawny (male) or greyish (female) heads. They are relatively long-legged and typically run to heavier cover when disturbed.

In the hand
The unusually long claws, nearly crestless condition, and moderately long, firm tail of 12 rectrices serve to identify this species.

Adult female. Similar on upperparts to male, but the superciliary stripes and cheeks greyish white; the chin and throat white, and the chest and breast dull brick-red, the feathers with pale shafts, shading to buff on the abdomen.

Immature. Juvenile males have cinnamon-buff rather than tawny-orange on the head and throat, and the cheeks are mottled with blackish, while females are less ochraceous on the sides of the head, chin, and throat (Ridgway and Friedmann 1946). Older stages are increasingly similar to adults, but the two outer juvenal primaries are retained and at least the base of the bill may be relatively pale. The primary-coverts are not useful for estimating age (Warner and Harrell 1957).

General biology and ecology

This species occupies a highly discontinuous range in Mexico and Central America that is largely correlated with the distribution of cloud forests. In Mexico the species reportedly occurs in a wide variety of natural habitats including evergreen tropical forests, rainforests, humid gallery forests, semideciduous tropical forests, oak–sweet gum and beech cloud forests, mixed hardwood cloud forests, and high montane pine–oak or pine–oak–fir forests. They have also been found in coffee groves and secondgrowth cloud forests. Elevations in the Mexican range extend from as low as 1000 to as high as 10 000 ft (3000 m). In Guatemala it is associated more specifically with cloud forests, at elevations of 7000–8500 ft (2100–2600 m) (Saunders *et al.* 1950). In Honduras it is uncommon and is confined to cloud forests above 1300 m (Monroe 1968), and in El Salvador it is usually found in oak forests of the arid upper tropical zone between 2500 and 4000 ft (750–1200 m), but it sometimes also occurs in coffee groves (Dickey and van Rossem 1938). Warner and Harrell (1957) estimated a population density of about 3.5 pairs per 100 acres (11.4 ha. per pair) in a climax oak–sweet gum forest, while LeFebvre and LeFebvre (1958) noted 4–5 pairs in a 20-acre plot (or about 2 ha. per pair) of partially lumbered cloud forest. These and other surveys in similar habitats suggest that population densities of a bird per 2–5 ha. are probably frequent in favourable habitats (Johnsgard 1973). Apparently an important part of good habitat consists of moist, shady forests that have dense undergrowth and a deep and rich layer of litter favourable for foraging by scratching and exposing plant and animal materials. Although not many specimens have been examined for food consumption, coffee berries, onion-like bulbs, larvae and pupae of insects, and seeds of pokeweed (*Phytolacca*), euphorbias, and legumes have been found, suggesting that a variety of invertebrates and seeds as well as softer plant materials are probably consumed.

Social behaviour

Coveys of these birds usually range from about four to a dozen birds, of which the smaller units probably comprise families. Among observed families of young, from two to four chicks have been seen. The birds typically run faster than fly to cover, but when flushed they head for the nearest thick cover, into which they disappear and are rarely flushed again. With the approach of the breeding season the coveys break up, but even during the nesting season more than two adults may sometimes be found in a small area. The loud calls of these birds consist of repeated phrases of staccato notes that have been described as an extended series of preliminary whistled notes of increasing frequency and pitch, followed by a rapid series of three to six phrases, variously described as sounding like *cua-kaka-wak*, *che-va-lieu-a*, and *pitch-wheeler* by various authors (Warner and Harrell 1957). Whether this call is given by a single bird or a pair is unknown, but duetting has at least been observed in captive birds (Griscom 1932), and sonagraphic analysis of such a sequence suggests that the bird singing the preliminary part of the call may also be singing the final half of the more complex portion (Johnsgard 1973).

Reproductive biology

The breeding season in Mexico is probably long, extending from February to October or even later. Broods have been seen there in April, and most juveniles have been collected between May and July. Birds in juvenal plumage have also been collected as late as January. Although the nest has never been described, a female collected in early May in the Yucatan Peninsula was said to have been incubating five eggs (Paynter 1955).

Evolutionary relationships

Holman (1961) suggested that the nearest living relative of this genus is *Cyrtonyx*, a genus that has very similar skeletal adaptations for digging. I advocated (Johnsgard 1973) the position that it may actually have evolved in parallel with *Cyrtonyx* from an *Odontophorus*-like ancestor, and become even more specialized for digging and scratching than has *Cyrtonyx*.

Status and conservation outlook

Certainly the most favoured habitats of this species are the cool and moist cloud forests of middle elevations, which contain a number of economically valuable timber trees such as oaks, and which additionally are sometimes cleared for coffee plantations. However, its tolerance for other wooded habitats may make it somewhat less sensitive to deforestation than such forms as the tree-quails and perhaps the wood-quails.

9 · Genus *Cyrtonyx* Gould 1844

The Montezuma or American harlequin quails are small grassland-, woodland-, or savannah-adapted species of North and Central America with very short and soft tails of 12 rectrices that are about 40 per cent as long as the wing. The wing is rounded, with the seventh primary the longest. Some juvenal greater primary-coverts and the two outer primaries are retained in the post-juvenal moult. The tarsus is relatively short, but the anterior claws are almost as long as the culmen. The sexes are dimorphic as adults, with males having a complex black and white facial pattern, but both sexes have flat, decumbent crests. Two very similar allopatric species are recognized:

C. montezumae (Vigors): Montezuma quail
  C. m. mearnsi Nelson 1900
  C. m. montezumae (Vigors) 1830
  C. m. merriami Nelson 1897
  C. m. sallei J. Verreaux 1859
  C. m. rowleyi Phillips 1966

C. [*montezumae*] *ocellatus* (Gould) 1837: ocellated quail

KEY TO THE SPECIES OF *CYRTONYX*

A. Face and throat patterned with black and white (males)
  B. Flanks and breast extensively patterned with white to chestnut-tinted round spots: Montezuma quail (*montezumae*)
  BB. Breast without pure white spotting; flanks chestnut, with grey and black cross-markings: ocellated quail (*ocellatus*)
AA. Face and throat white to buffy, without black patterning (females)
  B. Breast darker and less pinkish; white shaft-streaks dorsally: Montezuma quail (*montezumae*)
  BB. Breast paler and more pinkish; buffy shaft-streaks dorsally: ocellated quail (*ocellatus*)

MONTEZUMA QUAIL (Plate 28)

Cyrtonyx montezumae (Vigors) 1830
Other vernacular names: harlequin quail; Massena quail; Mearns' quail (*mearnsi*); Mexican harlequin quail; Salle quail (*sallei*); colin de Montezuma (French); Massenawachtel; Montezumawachtel (German); codorniz pinta (Spanish).

Distribution of species (see Map 12)
Resident of open woodlands from Arizona, New Mexico, and western Texas south to Oaxaca, Mexico.

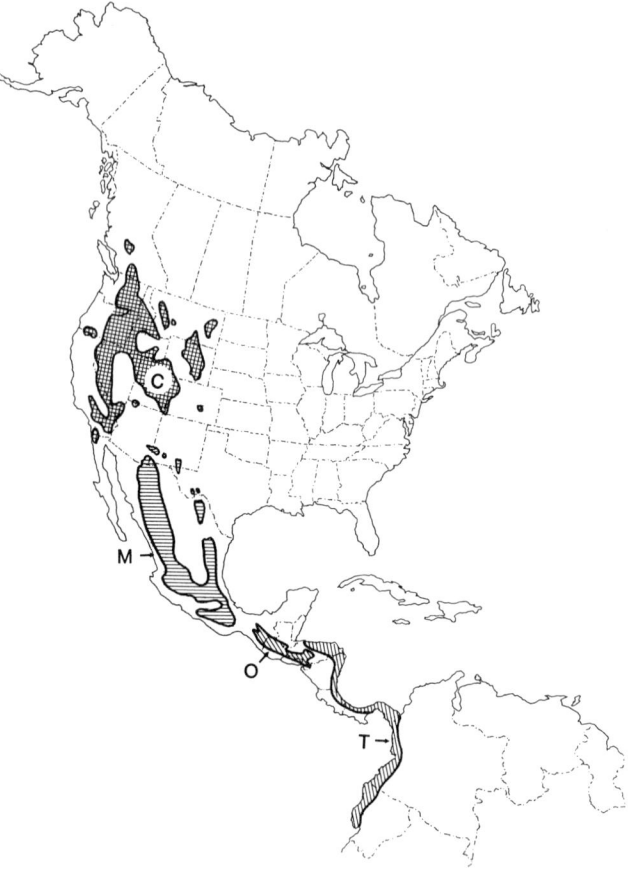

Map 12. Distribution of the Montezuma (M), ocellated (O), and tawny-faced quail (T), and introduced North American distribution of the chukar partridge (C).

Distribution of subspecies

C. m. mearnsi: west-central Texas, central New Mexico, and central Arizona south to northern Coahuila.
  C. m. montezumae: from Michoacan, Oaxaca, Distrito Federal, Hidalgo, and Puebla north and east to Nuevo Leon and west-central Tamaulipas.
  C. m. merriami: Veracruz (Mt Orizaba).
  C. m. sallei: from Michoacan south through Guerrero to east-central Oaxaca.
  C. m. rowleyi: Guerrero.

Measurements

Wing, males 113–131 mm, females 110.5–126 mm; tail, males 42.5–63 mm, females 47.5–60 mm (Ridgway and Friedmann 1946). Weight, males average 194.9 g, females 175.7 g (Leopold and McCabe 1957). Egg, average 32 × 24 mm, estimated weight 10.2 g.

Identification

In the field (8–9 in.)

At least north of the Isthmus of Tehuantepec there is no possibility of misidentifying this species if the ornately patterned male can be seen. Females are much duller, but are similarly short-tailed and plump birds that are prone to squat and 'freeze' rather than to run when disturbed. The male's advertisement call is a distinctive series of uniformly spaced whistling notes that slowly descend in pitch. South of Oaxaca this form is replaced in comparable open wooded habitats by the very similar ocellated quail.

In the hand

The combination of elongated claws, soft rectrices that are scarcely separable from their coverts, and a flat occipital crest identify the genus *Cyrtonyx*. Separation of the males from *ocellatus* is made fairly simple by the lack of chestnut-coloured flanks in *montezumae*, but females are very similar to those of *ocellatus*, and average only slightly less buffy dorsally.

Adult female. Easily separated from the male by the lack of a black and white head pattern.

Immature. Juveniles initially resemble adult females. Immature males soon acquire dark underparts and flanks, although the head remains juvenile-like for some time (Swarth 1909).

General biology and ecology

To a large degree the range of this species is coterminal with that of the pine–oak woodlands of central Mexico and similar oak woodlands of the southwestern United States. In Arizona it is primarily found in the oak–grassland habitat type, and in New Mexico it has become largely restricted to montane habitats dominated by rank grasses occurring in scattered mountain ranges. It now is only a local and rare species in parts of western Texas, the rest of its native range there having disappeared as a result of overgrazing (Johnsgard 1973). In Mexico it occurs over an altitudinal range of from about 3500 to 10 000 ft (1060–3000 m), and is essentially limited to those woodlands that have an understorey vegetation rich in bulb-bearing forbs (such as *Oxalis*) and sedges (Leopold and McCabe 1957). Probably it is able to survive without access to surface water if its supply of such succulent foods is adequate (Bishop 1964). Its density even in a fairly good range is fairly low; Leopold and McCabe (1957) estimated a density of about 10 ha. per bird in northern Chihuahua, which was very similar to an earlier estimate made in Arizona. It was believed by Leopold and McCabe that severe winters probably limit the northern range of the species and may cause considerable yearly variations in population density. The major foods of this species, at least in winter, are the bulbs of nut grass (*Cyperus*) and wood sorrel (*Oxalis*), together with tubers of lilies, fruits and seeds of various woody ericads and pines, and a variety of seeds. Animal foods are taken in quantity during summer months or whenever they are available (Leopold and McCabe 1957).

Social behaviour

Covey sizes in this species average seven to eight birds, and almost never are flocks of more than 25 reported. At night the birds assume circular roosting groups similar to those of bobwhites, but often around a rock or grass clump (Bishop 1964). Evidently the birds are highly sedentary when on their winter ranges, possibly having home ranges no greater than some 200 yd in radius, although during early autumn they may move around considerably. The separation call or assembly call of this species consists of a paced series of low, quavering whistles, during which the notes slowly descend the scale. This call is probably uttered by both sexes and birds of various ages, and is effective in attracting individual birds of both sexes. During the breeding season, lone males also uttered a thin, high-pitched buzzing note that is distinctly insect-like, and probably serves as an advertising call. Other males will vocally respond to playbacks of this call, especially during the nesting season. Various other softer 'conversational' calls have also been heard among birds in coveys.

Reproductive biology

In Arizona the breeding season is fairly late, with most pairing probably occurring from March through to May, and nesting extending from late June through to the latter part of September. The peak incubation period in Arizona is probably from late July to mid-August. This fairly late season seems to be adaptively timed so that the broods usually appear shortly after the summer rains have provided an abundance of new plant growth and insect food. The role of the male in nest-building, incubation, and nest-defence is still somewhat uncertain. Observations on captive birds (Bishop 1964) suggest that the male may help to construct the nest, as in *Odontophorus*, and the domed nests made by these two genera are quite similar. Frequently the side entrance to the nest is well hidden by a mat of grass stems that hang down like a hinged door and falls back into place whenever a bird enters or leaves the nest. The average clutch-size is fairly large, of 11.1 eggs, with a range of 6–14 (Leopold and McCabe 1957). Females in captivity have laid at a rate of about three days per egg over a two-month period, which is probably slower than the rate in the wild. Various observers have reported that males actually help incubate, but other accounts suggest that the males may only sit beside incubating females. There is no doubt that males at least help to guard and rear the young. The incubation period is the longest of any odontophorine occurring north of Mexico, of 25–26 days. Observed brood sizes have usually averaged between six and nine chicks, and as soon as they leave the nest the chicks apparently begin to eat insects, seeds, and bulbs. The growth rate of chicks is probably slightly slower than in the northern bobwhite, judging from the ages at which juvenal primaries are dropped and replaced in young birds. In captive birds the adult weight is reached in 10–11 weeks. The eighth juvenal primary is dropped at about 15 weeks and its replacement is fully grown by about the 19th week (Leopold and McCabe 1957).

Evolutionary history

The synsacrum of *Cyrtonyx* has been highly modified through evolution for scratching and digging behaviour, to an even greater degree than in *Dactylortyx* or *Rhynchortyx* (Holman 1961). It seems likely, however, that all of these forms evolved from an *Odontophorus*- or *Dendrortyx*-like ancestral type in a forested or woodland environment, and that *Cyrtonyx* became the most efficient of these digging specialists at surviving in relatively dry habitats, and thus has extended its range farther to the north and into drier habitats than have the others (Johnsgard 1973).

Status and conservation outlook

Although the situation in Mexico is less uncertain, there is no question that the range and abundance of this species has declined markedly in the United States during historical times, largely as a result of overgrazing that has diminished the species of food and cover plants that it is dependent upon.

OCELLATED QUAIL (Plate 29)

C. ocellatus (Gould) 1837
Other vernacular names: colin ocelle (French); Tranenwächtel (German); codorniz ocelada (Spanish).

Distribution of species (see Map 12)
Resident in woodlands from Mexico (Oaxaca and Chiapas) south through eastern Guatemala, El Salvador, and Honduras to northern Nicaragua. No subspecies recognized; allopatric and doubtfully distinct specifically from *montezumae*.

Measurements

Wing, males 114–130 mm, females 110.5–119.5 mm; tail, males 48–57.5 mm, females 45–55.5 mm (Ridgway and Friedmann 1946). Estimated weight, males 218 g, females 182 g. Egg, average 32.3 × 25.1 mm (Schönwetter 1967), estimated weight 11.2 g.

Identification

In the field (8–9.5 in.)
Nearly identical in general appearance to the Montezuma quail, but occurring from extreme eastern Oaxaca and Chiapas southward. The male's calls are still undescribed in detail, but apparently it has a rather melodious *colonchango* call and possibly lacks the distinctive gradually descending whistled notes of *montezumae*. Both species are usually found where bulb-bearing plants such as *Oxalis* grow in abundance and can be readily dug out of the soil.

In the hand
Adult males may be separated from *montezumae* by the absence of chestnut-coloured flanks in that form, which instead are heavily spotted.

Adult female. Very similar to females of *montezumae*, especially of *m. sallei*, but somewhat more ochraceous, generally darker above, less pinkish or vinaceous below, and with the dorsal shaft-stripes washed with buff.

Immature. Juveniles are at least initially similar to the adult female; immature males have a nearly white throat, blackish under tail-coverts, and varying amounts of adult feathering on the flanks and underparts.

General biology and ecology

Not yet described in detail, but probably much like that of *montezumae*. Areas in Chiapas that I have visited and which had been reported to have at least previously supported this species differed in no obvious way from those farther north in Mexico and occupied by *montezumae*.

Social behaviour

No specific information is available. While in the range of this species I enquired of various people who had kept it in captivity. One person informed me that the male has a beautiful whistled song sounding like *pico-de-oro*, and another said that the male has two distinct calls. One was described as a whistled *preet*, presumably comparable to the buzzing advertisement call of the male Montezuma quail, and the other a more elaborate sequence of notes sounding like *col-on-chang'-o*. This is the local vernacular name for this species, and no doubt refers to the same *pico-de-oro* call just mentioned. This is reportedly a melodious and beautiful call, and thus the ocellated quail is more highly valued in that region of Mexico as a cage bird than is the more easily obtained northern bobwhite. No call corresponding to the descending owl-like separation call of the Montezuma quail has yet been described.

Reproductive biology

No specific information.

Evolutionary relationships

General relationships of *Cyrtonyx* have already been mentioned. The affinities of *ocellatus* and *montezumae* are clearly extremely close, and the clinal aspects of male plumage variation in the latter reduce the differences between the extreme plumage types of these two populations. The allopatric distribution pattern, with the two types geographically separated by an area of unfavourable lowland habitat in the Isthmus of Tehuantepec, would certainly favour consideration of the two as subspecies, but the little so far known of the ocellated quail's vocalizations suggest that they might be quite distinctive, and thus it seems safest to regard the two forms as specifically distinct, at least for the present time.

Status and conservation outlook

I was unable to find any wild populations of this species in southern Mexico during a few days of searching in the early 1970s, and was told that the birds had become much rarer around San Cristobal in recent years, probably because of overgrazing and deforestation. Probably overgrazing is a more serious and pervasive threat to this species than is partial deforestation, because of its severe impact on the birds' subterranean food supplies.

10 · Genus *Rhynchortyx* Ogilvie-Grant 1893

The tawny-faced quail is a small woodland-adapted species of Central and South America with a tail of 10 firm rectrices and more than half as long as the wing. The wing is rounded, with the seventh primary the longest. The tarsus is longer than the middle toe, the claws are unusually short, and the bill is extremely thick. The sexes are dimorphic as adults, with males having tawny-cinnamon and females reddish brown faces, but both sexes are uncrested and have black-barred flanks. A single polytypic species is recognized:

R. cinctus (Salvin): tawny-faced quail
   R. c. pudiobundus Peters 1929
   R. c. cinctus (Salvin) 1876
   R. c. australis Chapman 1915

TAWNY-FACED QUAIL (Plate 30)

Rhynchortyx cinctus (Salvin) 1876
Other vernacular names: banded quail; long-legged colin; colin ceinture (French); Langbeinwachtel (German); gallito del monte menor; codorniz patilarga (Spanish).

Distribution of species (see Map 12)
Resident of tropical lowland forests from eastern Honduras south through north-western Colombia and north-western Ecuador.

Distribution of subspecies
R. c. pudiobundus: Honduras and Nicaragua.
   R. c. cinctus: Costa Rica and Panama.
   R. c. australis: Colombia and Ecuador.

Measurements

Wing, males 105.5–117.5 mm, females 105–112 mm; tail, males 41–48 mm, females 41–48 mm (Ridgway and Friedmann 1946). Weight, one male, 165 g (*Condor* **73**, 108). Egg, average 29.8 × 23.6 mm (Schönwetter 1967), estimated weight 9.2 g.

Identification

In the field (7.5 in.)
Found in dense tropical forests, this species is only rarely seen. However, it has a distinctive rusty brown head, a short and heavy bill, and some black banding on the rear flanks. Vocalizations are still virtually undescribed, but reportedly include piping and twittering sounds. Flights are short, infrequent, and made with loud wing noises.

In the hand
The combination of an unusually long tarsus (longer than the middle toe and claw), a very heavy bill, and a short (under 50 mm) tail of only 10 rectrices, provide for easy identification of this species.

Adult female. Upperparts similar to male, but the mantle, rump, and upper tail-coverts much browner; a fine white stripe above ear-coverts; sides of head, foreneck, and chest dull reddish brown; upper throat and posterior underparts white, the latter barred with black.

Immature. First-winter plumage apparently adult-like. The juvenile male is darker and less rufescent above than the female, which it otherwise generally resembles (Ridgway and Friedmann 1946).

General biology and ecology

Very few observations have been made on this species, which is closely associated with heavy forest undergrowth, and is the least-studied genus of all odontophorine quails. Wetmore (1965) reported the species as occurring between 500 and 1400 m in Panamanian forests. Monroe (1968) reported that in Honduras the birds are restricted to dense rainforest areas below 500 m elevation on the Caribbean slope. Karr (1971) said that the birds are generally found on the ground in thickets, making them very hard to observe for any length of time. Its foods are still undescribed, but the very stout bill of the species is seemingly well adapted for crushing hard seeds or fruits.

Social behaviour

The few observations made so far suggest these birds are usually found in pairs or very small groups. Karr

(1971) noted that a male uttered twittering calls similar to the 'chirruping' of disturbed northern bobwhites. No other specific vocalizations have been described.

Reproductive biology

No information is available on the nesting of this species. Wetmore (1965) noted that a female collected in Panama during early March had an oviducial egg, and that downy young had been collected during late March and April in Panama. Eggs have been collected in Colombia, but there is no available description of the nest.

Evolutionary relationships

Holman's data on the osteology of this species provide the only real clues as to its relationships. In addition to having an unusually heavy bill, it is the smallest representative of the *Odontophorus*-like assemblage of New World quails. Holman judged that *Rhynchortyx* is probably more closely related to *Cyrtonyx* and *Dactylortyx* than it is to *Odontophorus*, and that it is osteologically the more aberrant genus of this assemblage.

Status and conservation outlook

Nothing can be said of the current status of this highly elusive species. Undoubtedly its long-term survival will depend upon the degree to which primary lowland forests within its range can be protected from deforestation.

11 · Tribe Perdicini

The Old World quails, francolins, and partridges are a group of small to very large gallinaceous birds comprising 21 genera and 103 species that range from Africa east to the Greater Sundas, New Guinea (one genus reaches Australia and New Zealand), and north throughout the Indian subcontinent and most of Eurasia, with the largest number of species occurring in sub-Saharan Africa and tropical Asia. Successful introductions have also been made in North America and elsewhere beyond the group's native range. The Perdicini differ from the subfamily Odontophorinae in that the lower mandible edge is never serrated, the tarsus is often sharply spurred in adult males (and sometimes is also spurred in adult females), and the juvenal greater primary-coverts are regularly moulted and replaced during the post-juvenal moult when their respective remiges are lost. In most but not all species the post-juvenal moult of the juvenal primaries is incomplete, with two (rarely one or three) juvenal primaries retained in the first-basic plumage. In all species so far studied the tail moult is centrifugal, starting from the central pair or, rarely, from the immediately adjoining pair. Adult sexual dimorphism in both plumage and weight ranges from slight to substantial. Tarsal spurs are present in the majority of species and iridescent plumage occurs in adults of a few; many species have colourful eye-rings or other areas of bare skin near the eyes. In most species monogamy is the prevailing mating system, but non-monagamous matings have occasionally been reported in a few genera (*Alectoris, Coturnix*). Most species utilize diverse vocalizations for territorial advertisement, social or sexual bonding, and flock or covey coordination. Visual signals involving posturing and plumage exhibition are usually less conspicuous than in the pheasants, but the lateral 'waltzing' display typical of pheasants is well developed in at least some genera. Duetting has been observed in at least two forest-adapted genera (*Arborophila, Rhizothera*), and possible social courtship by males on a common display ground has been reported for one African genus (*Ptilopachus*). Some species (especially of *Coturnix*) undertake long seasonal migrations, and others are relatively nomadic. Most open-country species are somewhat gregarious outside the breeding season, but some forest-dwelling forms are relatively solitary. Most species are largely seed-eaters, but some forest-adapted forms may subsist primarily on invertebrates.

Nests are typically simple open scrapes, but in several forest-adapted forms (e.g. *Arborophila, Rollulus*) may be domed structures. Average clutch-sizes range from 2 to 16 eggs, and incubation periods range from 16 to 25 days. The chicks are precocial, often fledging when only about 2 weeks old, and breeding during the first year of life is regular.

KEY TO THE GENERA OF PERDICINI

A. Tail over 150 mm, of 20–22 rectrices, wing over 200 mm (snowcocks and monal partridges)
  B. Wing over 240 mm, and extensively white: *Tetraogallus* (5 spp.)
  BB. Wing under 240 mm, not extensively white: *Tetraophasis* (2 spp.)
AA. Tail under 150 mm, of 8–18 rectrices, wing usually under 200 mm
  B. Wing under 120 mm, 8–12 rectrices (typical quails and bush-quails)
    C. Tail almost as long as wing; outermost primary shorter than innermost: *Ophrysia* (1 extinct sp.)
    CC. Tail under two-thirds wing length; outermost primary longer than innermost
      D. Outermost to third primary longest; tail hidden by coverts, and less than half the length of wing: *Coturnix* (8 spp.)
      DD. Fourth primary from outside the longest; tail visible and at least half the length of wing: *Perdicula* (4 spp.)
  BB. Wing over 120 mm, 12–18 rectrices (partridges and francolins)
    C. Rectrices short and indistinct from their coverts; innermost toe with elongated nail: *Anurophasis* (1 sp.)
    CC. Rectrices longer and indistinct from their coverts; innermost toe normal
      D. Tail of 12 rectrices, usually less than half the length of the wing
        E. Hind toe clawless, or with rudimentary claw
          F. Clawless, forehead crested: *Rollulus* (1 sp.)
          FF. Claw rudimentary, uncrested: *Margaroperdix* (1 sp.)
        EE. Hind toe normally clawed
          F. Bill normal in length and depth (20–35 mm from gape to tip)
            G. Mostly sand-coloured, with a chestnut tail: *Ammoperdix* (2 spp.)
            GG. Mostly dark brown, with a rufous head: *Haematortyx* (1 sp.)
          FF. Bill unusual in length or depth
            G. Bill unusually thick and short (under 20 mm from gape to tip), both sexes unspurred: *Melanoperdix* (1 sp.)

GG. Bill unusually long (at least 35 mm from gape to tip), both sexes spurred: *Rhizothera* (1 sp.)

DD. Tail of 14–18 rectrices, usually over half the length of the wing

E. 16–18 rectrices and a dark abdominal patch: *Perdix* (3 spp.)

EE. 14 rectrices, usually no dark abdominal patch

F. Hind toe with rudimentary claw: *Caloperdix* (1 sp.)

FF. Hind toe normally clawed

G. Tail at least two-thirds length of wing; outermost primary shorter than innermost

H. Tail flattened and barred; no bare patch of skin around eye: *Bambusicola* (2 spp.)

HH. Tail vaulted and unbarred; usually with large bare patch of skin around eye

I. Adults unspurred, tail under 100 mm: *Ptilopachus* (1 sp.)

II. Adults spurred, tail over 100 mm: *Galloperdix* (3 spp.)

GG. Tail under two-thirds length of wing; outermost primary as long or longer than innermost

H. Upper tarsus feathered in front: *Lerwa* (1 sp.)

HH. Tarsus entirely naked

I. Tail no more than half the length of wing; tarsus unspurred: *Arborophila* (16 spp.)

II. Tail over half the length of wing; males usually with a tarsal knob or spur

J. With vertically barred flanks and a black collar or throat; neither sex spurred: *Alectoris* (7 spp.)

JJ. Never with combination of vertically barred flanks and a black collar or throat; males nearly always spurred: *Francolinus* (41 spp.)

12 · Genus *Lerwa* Hodgson 1837

The snow partridge is a medium-sized Himalayan alpine species having a tail of 14 rectrices that is rounded and about 60 per cent as long as the wing. The wing is relatively pointed, with the ninth primary usually the longest, and the tenth (outermost) only slightly shorter. The tail moult is apparently imperfectly perdicine (centrifugal). The tarsus is feathered for about half its length, and males have a stout and blunt spur. The sexes are similar, with strong cross-barring on most of the dorsal feathers and no bare facial skin or tarsal spurs but with bright red bill and tarsal coloration. A single weakly polytypic species is recognized:

  L. lerwa Hodgson: snow partridge
   L. l. lerwa Hodgson 1833
   L. l. major Meinertzhagen 1927

SNOW PARTRIDGE (Plate 33)

Lerwa lerwa Hodgson 1833
Other vernacular names: lerva des neiges; perdrix des neiges (French); Haldenhuhn (German).

Distribution of species (see Map 13)
Resident of alpine zone of the Himalayas, from Kashmir east through south-eastern Tibet to western Sichuan, southern Gansu, and northern Yunnan.

Distribution of subspecies
L. l. lerwa: Himalayas from Kashmir and Garhwal (Uttar Pradesh) to Assam and southern Tibet (Xizang).
  L. l. major: eastern Xizang and northern Yunnan to north-western Sichuan and southern Gansu. Doubtfully distinct from *lerwa*.

Measurements

Wing, both sexes 180–205 mm; tail, both sexes 118–138 mm. Weight, c. 454–709 g (Ali and Ripley 1978). Egg, average 54.6 × 35.4 mm (Baker 1935), estimated weight 37.8 g.

Identification

In the field (14 in.)
Associated with the more grassy alpine slopes of the Himalayas, at the same general elevations as *Tetraogallus* but perhaps in somewhat less rocky and precipitous habitats. Unlikely there to be mistaken for any other species, as no other Himalayan galliform is so barred dorsally. The birds utter a harsh whistle when alarmed, and seldom fly unless approached from above.

In the hand
The highly barred black and white upperparts, including the head, and the bright red bill and tarsal coloration, provide for easy recognition.

Adult female. Apparently almost identical to the male, but averaging slightly smaller, and differing in the intensity of red on the mandible, especially during the breeding season (Cheng *et al.* 1978).

Immature. Pale barring on the head and upperparts wider and all buff, and on the underparts of the neck it is less clearly defined, gradually merging into the chest, the feathers of which are clouded, mottled, or barred with black. Juveniles have conspicuous white shaft-stripes both dorsally and ventrally.

General biology and ecology

Throughout its range this species breeds at elevations of 10 000–15 000 ft (3000–4550 m), and usually above 12 000 ft (3600 m). During severe winter weather the birds may descend to elevations of 7000–9000 ft (2100–2700 m), but generally they are found close below the snow line at high elevations, where they are found among stunted alpine vegetation and in boulder fields. On Mount Everest they have apparently been observed at elevations in excess of 17 000 ft (5100 m) (Baker 1930). In China they are seldom seen below 4000 m, and have reportedly been observed breeding in Sichuan province at 5500 m (Cheng *et al.* 1978). Their foods primarily include vegetational materials, such as seeds and leaves, as well as mosses, but they sometimes also eat insects.

Social behaviour

According to Baker (1930), these birds are usually to be found in pairs during spring, but later on occur in groups of from 7 to 30 birds, including young. The usual call is a loud, harsh whistle, frequently repeated, and both the alarm note and that used by

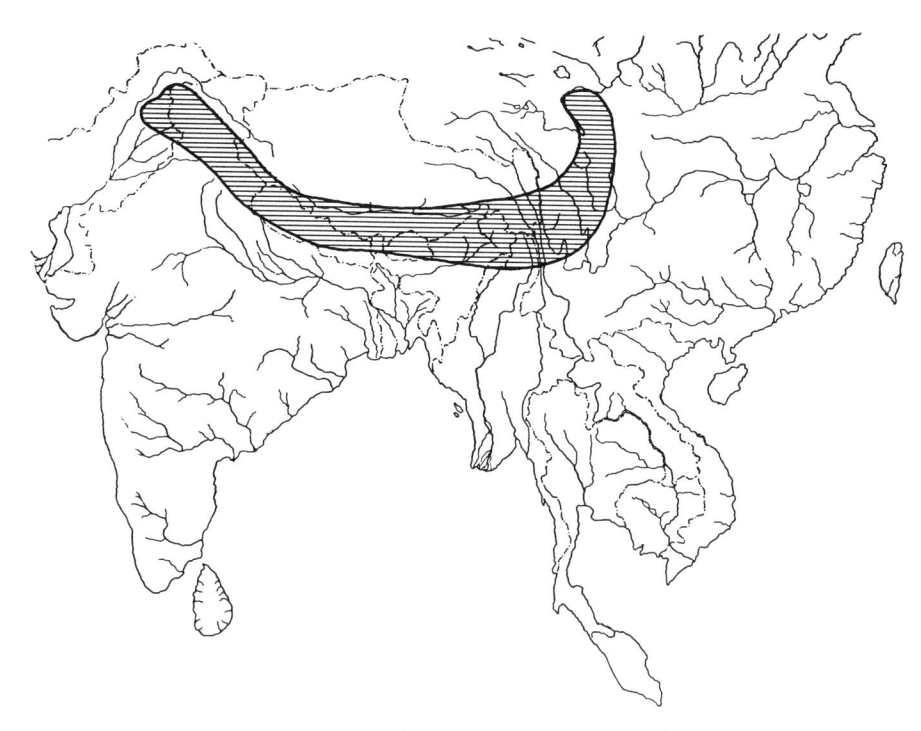

Map 13. Distribution of the snow partridge.

individuals to call to one another reportedly differ only in loudness and harshness. Softer notes are used between parents and chicks (Baker 1930).

Reproductive biology

As reported by Baker (1930), the breeding season is fairly long, with eggs being obtained from the end of May to early July, and well-grown chicks collected as early as 2 July. Reportedly the clutch numbers from three to five eggs, and perhaps rarely may be as large as seven, since families with that many chicks have been observed. Nests are well concealed, and are often hidden by projecting rocks or a clump of grass. The typical nest site is on a steep or precipitous slope with only scattered vegetation, often near the crest of a hill and always on the leeward side of it. There is no information on the incubation period or development of the young.

Evolutionary relationships

Certainly the adult plumage pattern of this species is distinctive, and its disruptive patterning is clearly an adaptation for effective camouflage in a rock- and lichen-covered environment. The distinctive downy pattern is mostly velvety black and silvery white on the head, with the upperparts of the body chestnut-brown and blackish, fading to pale buff and chestnut-buff below (Baker 1930). Neither of these plumages provides strong evidence of relationships, except perhaps to *Tetraogallus*, where the similarities may be only superficially convergent.

Status and conservation outlook

No population information exists on this species, which probably has rather few contacts with humans, and lives in an area and habitat where it is probably fairly secure for the foreseeable future.

13 · Genus *Tetraophasis* Elliot 1871

The monal-partridges are large alpine species of the Himalayan and western Chinese mountains that have wedge-shaped tails of 18 feathers that are about 75 per cent as long as the wing, and rounded wings in which the seventh primary is the longest. The post-juvenal primary moult is incomplete. The axillaries are very long, and a large grey patch of downy feathers is present below the wing. The sexes are similar, with rather heavy bills, large areas of bare skin around the eye, and white-tipped tails; males have a spur on the tarsus. Two weakly distinguished allopatric species are recognized; some authorities recognize only a single species:

T. *obscurus* (J. Verreaux) 1869: Verreaux's monal-partridge

T. [*obscurus*] *szechenyii* Madarasz, 1885: Szechenyi's monal-partridge

KEY TO THE SPECIES (OR SUBSPECIES) OF MONAL-PARTRIDGES (*TETRAOPHASIS*)

A. Chin, throat, and foreneck dark chestnut: Verreaux's monal-partridge (*obscurus*)

AA. Chin, throat, and foreneck pale fawn: Szechenyi's monal-partridge (*szechenii*)

VERREAUX'S MONAL-PARTRIDGE
(Plate 31)

Tetraophasis obscurus (J. Verreaux) 1869
Other vernacular names: moupin pheasant, Verreaux's pheasant-partridge; Verreaux's pheasant-grouse; tetraophase de Verreaux (French); Braunkehl-Keilschwanzhuhn (German).

Distribution of species (see Map 14)
Resident of the subalpine zone of China from eastern Xizang (Tibet) east to western and northern Sichuan, and north to eastern Quinghai and central Gansu. No subspecies recognized here, but probably the allopatric *szechenyii* should be considered conspecific.

Measurements

Wing, males 207–215 mm, females 197–200 mm; tail, males 153–175 mm, females 150–160 mm. Weight, males 938 g, females 720 and 840 g (Cheng *et al.* 1978). Egg, no information.

Identification

In the field (17–19 in.)
This and the following form are almost the only long-tailed and pheasant-like Old World partridges. Both have white-tipped tails that are often partially erected and spread when the birds become alarmed. Found in wooded and brushy rocky habitats at higher montane elevations near tree-line. When flushed, a loud cry usually is uttered.

In the hand
The long (at least 150 mm), wedge-shaped and white-tipped tail of 18 rectrices identifies the genus, and this form can be separated from the following one by its dark chestnut chin, throat, and foreneck coloration.

Adult female. Differs from the male in having the outer webs of the secondaries and secondary coverts more or less mottled with buff toward the margin, and in lacking tarsal spurs. The tarsus of the female is also reportedly more reddish than that of the male (Cheng *et al.* 1978).

Immature. Undescribed in detail, but probably like that of the following form.

General biology and ecology

No specific information, but apparently like that of the following form (Cheng *et al.* 1978).

Social behaviour

No specific information.

Reproductive biology

Almost nothing is known of this. There are three eggs collected in Gansu (?) by I. Berosowski in the British Museum (Natural History), and two clutches totalling eight eggs in the Berlin Museum that were collected in Gansu, at an elevation of 3150 m (Meise *et al.* 1938; Schönwetter 1967). A juvenile of this form or of *szechenyii* was collected in western Sichuan in early July (Cheng *et al.* 1978).

Map 14. Distribution of the Szechenyi's (S) and Verreaux's (V) monal-partridges.

Evolutionary relationships

The allopatric distributions of this form and *szechenyii*, together with their distinct similarities in plumage, suggest that they are very close relatives. Specimen records plotted by Cheng *et al.* (1978) indicate that their current ranges are no more than 150 km apart, and have no obvious major topographic barriers separating them.

Status and conservation outlook

This species is apparently quite rare in China; Cheng *et al.* (1978) mentioned that during studies in Sichuan this form (or *szechenyii*) was seen at the rate of about one sighting per 3 h, but that more were observed in Tsinghai Province. These authors judged that *Tetraophasis* is not now of any economic significance in China.

SZECHENYI'S MONAL-PARTRIDGE
(Plate 32)

Tetraophasis szechenyii Madarasz 1885
Other vernacular names: Szechenyi's partridge-pheasant, Szechenyi's pheasant-grouse; Tibetan pheasant; tetraophase de Szechenyi (French); Tibet-Keilschwanzhuhn; Rostkehl-Keilschwanzhuhn (German).

Distribution of species (see Map 14)

Resident of the subalpine zone of the eastern Himalayas from the Tsangpo (Brahmaputra) Valley, Xizang (Tibet), east to north-western Yunnan, China, and south to the Subansiri district of Arunachal Pradesh. Allopatric with *obscurus* and probably best regarded as conspecific with it (Cheng *et al.* 1978).

Measurements

Wing, males 216–236 mm, females 203–224 mm; tail, males 142–151 mm, females 129–133 mm (Ali and Ripley 1978; Vaurie 1965). Weight, males 1020 and 1500 g, females 880 g (Ludlow 1944; Cheng *et al.* 1978). Egg, average 54.1 × 38 mm (Schönwetter 1967), estimated weight 43.1 g.

Identification

In the field (20–25 in.)

Probably virtually indistinguishable from the previous form in the field, but separated geographically from it.

In the hand

Compared with *obscurus*, the chin, throat, and foreneck of this form are pale fawn rather than dark chestnut, the rump and tail-coverts are greyer, and the underpart feathers are chestnut rather than white on the inner web, and are tipped with buff.

Adult female. Similar to the male, but slightly smaller and lacking tarsal spurs.

Immature. A juvenile in the British Museum (Natural History) is heavily barred with black and brown on the upperparts, the feathers with white tips and shaft-streaks; on the undersides the feathers have buffy centres or wide shaft-streaks, and are edged or spotted with grey to brown outwardly.

General biology and ecology

This species occupies fir forest and rhododendron scrub, as well as rocky ravines in the Himalayan subalpine and alpine zones between about 3350 and 4600 m, according to Ali and Ripley (1978). Cheng *et al.* (1978) similarly stated that in China it occurs at 4000 m or higher, in pines, cedars, and azaleas, mostly in wooded areas, but also it is found above the tree-line in rocky alpine habitats. At night it roosts in trees, and when flushed from rocky alpine areas it quickly flies downhill to disappear in the nearest wooded cover. The crop of a specimen observed by Ludlow (1944) contained small roots, bulbs, and green leaves; of five specimens collected in China (some or all possibly of *obscurus*) four contained mosses and leaves, while a fifth had the remains of fruits present (Cheng *et al.* 1978).

Social biology

The few available reports suggest that this species moves about in small coveys of four to six birds, which probably represent family groups. Schäfer (1934) observed family groups of six to twelve individuals during August, mainly at altitudes of 3500–4000 m. Both Schäfer and Ludlow have described several vocalizations, which include softer cackles as well as a series of loud and harsh notes used as alarm calls.

Reproductive biology

In addition to the chicks that were observed by Schäfer in Tibet during August, Ludlow (1944) recorded seeing chicks in Tibet during late May, June, and July, and collected a juvenile in early September at 13 500 ft (4100 m). These very limited data suggest that a fairly extended breeding season and a relatively small clutch-size may be typical of *Tetraogallus*.

Evolutionary relationships

Beyond the obviously close and probably conspecific relationships existing between *obscurus* and *szechenyii*, the possible generic affinities of *Tetraophasis* require mention. The downy young plumage is apparently still undescribed, and the adult plumage shows no great similarities to any other species of Perdicini. Various writers have commented on the general similarities in behaviour and appearance of this genus to the koklass pheasant (*Pucrasia*), and it is even possible that the genus may eventually have to be reassigned to the Phasianini.

Status and conservation outlook

Apparently now very rare within the limits of India, but otherwise there is no information on this form's status.

14 · Genus *Tetraogallus* J. E. Gray 1832

The snowcocks are very large partridge-like species occurring in alpine habitats from the Caucasus to Tibet and Mongolia, with rounded tails that are about 60 per cent as long as the wing, and rounded wings in which the seventh to the ninth primaries are the longest. The post-juvenal primary moult is incomplete, with the three outer juvenal primaries persistent, and the tail moult is perdicine (centrifugal). There is a small area of bare skin around or behind the eye in adults of most or all species. Varying areas of white are present on the bases of the remiges, throat, and often elsewhere. The sexes are alike or slightly dimorphic, and the male has a blunt tarsal spur. Five allopatric species are recognized:

T. caucasicus (Pallas) 1811: Caucasian snowcock
T. caspius (S. G. Gmelin): Caspian snowcock
  T. c. tauricus Dresser 1876
  T. c. caspius (S. G. Gmelin) 1874
  T. c. semenowtianschankii Zarudny 1908
T. himalayensis G. R. Gray: Himalayan snowcock
  T. h. sewerzowi Zarudny 1910
  T. h. incognitus Zarudny 1911
  T. h. himalayensis G. R. Gray 1843
  T. h. koslowi Bianchi 1898
  T. h. grombczewskii Bianchi 1898
T. tibetanus Gould: Tibetan snowcock
  T. t. tibetanus Gould 1854
  T. t. aquilonifer R. and A. Meinertzhagen 1926
  T. t. henrici Oustalet 1891
  T. t. przewalskii Bianchi 1907
T. altaicus (Gebler) 1836: Altai snowcock

KEY TO THE SPECIES OF SNOWCOCKS (*TETRAOGALLUS*)

A. Primaries about half white, with dark tips; underparts medium grey to brownish
  B. Throat-band pale rufous to greyish; secondaries white for about half their lengths
    C. Breast with narrow black barring; flanks more chestnut-toned: Caucasian snowcock (*caucasicus*)
    CC. Breast with slight black spotting; flanks less chestnut-toned: Caspian snowcock (*caspius*)
  BB. Throat-band chestnut; secondaries whitish only at bases: Himalayan snowcock (*himalayensis*)
AA. White on primaries absent or limited to their bases only; underparts greyish white
  B. Feathers of flanks with distinct black edges, no white on primaries: Tibetan snowcock (*tibetanus*)
  BB. Feathers of flanks lacking distinct black edges, primaries white basally: Altai snowcock (*altaicus*)

CAUCASIAN SNOWCOCK (Plate 34)

Tetraogallus caucasicus (Pallas) 1811
Other vernacular names: tetraogalle du Caucase (French); Kaukasisches Konigshuhn (German); Kavkazky ular (Russian).

Distribution of species (see Map 15)

Resident of alpine and subalpine zones of Main Range of Caucasus. No subspecies recognized.

Measurements

Wing, males 273–285 mm, females 245 male 250 mm (Dementiev and Gladkov 1952); tail, males 146 mm, females 152–161 mm. Weight, males average 1.93 kg, females 1.73 kg (Cramp and Simmons 1980). Egg, average 67 × 46 mm (Cramp and Simmons 1980), estimated weight 78.2 g.

Identification (see Fig. 15)

In the field (21–22 in.)

This is the only snowcock of the northern Caucasus, so any very large grouse-like bird of that region with white on the sides of the head and wings is of this species. It is found in steep, rocky habitats, usually in pairs or small flocks, and has a loud repeated *ooy* call, a prolonged whistled note that is uttered by males, and other loud calls.

In the hand

Differs from the other species of *Tetraogallus* in having dark grey underparts that are strongly barred and scalloped with black and buff on the chest and breast. The nape is also distinctly rust-toned, and a chocolate-brown band extends down each side of the throat.

Adult female. Duller than male, with mantle and chest paler due to wider feather-margins and larger

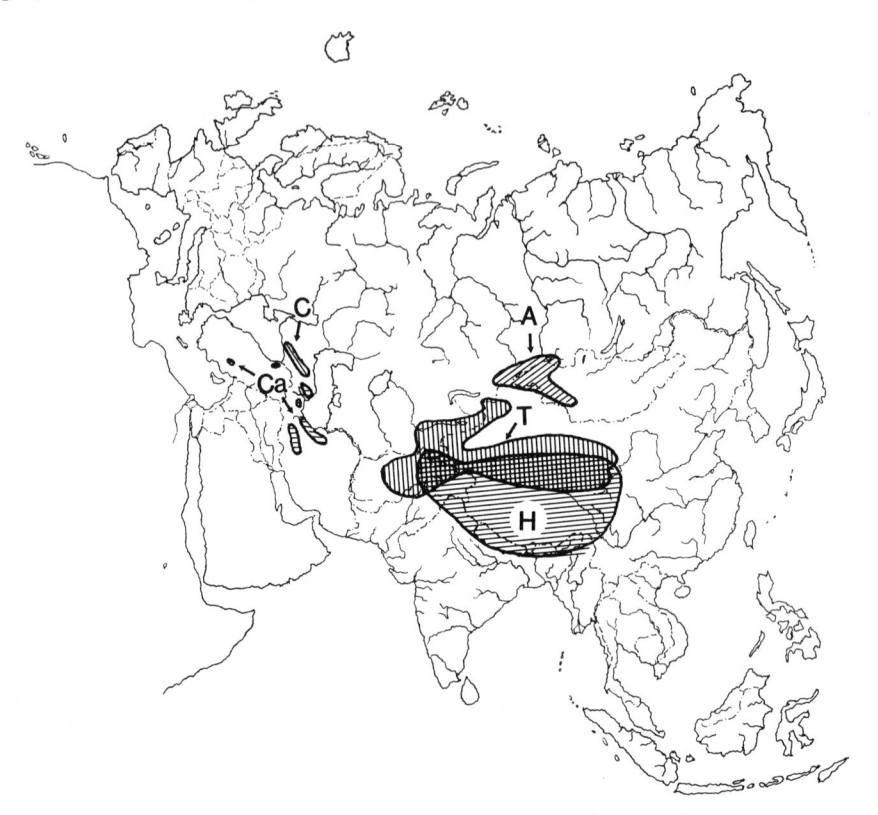

Map 15. Distribution of the Altai (A), Caspian (Ca), Caucasian (C), Himalayan (H), and Tibetan (T) snowcocks.

spots; crown, hindneck, and neck patch all duskier or browner; chestnut flank streaks less obvious; tarsi unspurred.

Immature. First-year birds have less well-developed plumage patterning and retain the three outer juvenal primaries, which are relatively pointed and worn (Cramp and Simmons 1980).

General biology and ecology

This species is found over an altitudinal range that centres around 1800–4000 m between alpine tree-line and snow line, and there it occurs on steep slopes with rocky outcrops, talus and scree slopes, and alpine meadows. It avoids tall montane forests and also rhododendron scrub, and is usually to be found in the vicinity of snow patches. It is almost exclusively vegetarian, and eats a variety of plant parts including leaves, stolons, fruits, seeds, and stems, as well as underground tubers and bulbs. Young birds eat a higher proportion of legumes than do adults, apparently using these plants to supplement their protein intake, inasmuch as insects are scarce at alpine elevations (Dementiev and Gladkov 1952; Baziev 1968).

Social behaviour

During much of the year this species is gregarious, forming coveys of three to nine individuals, which probably consist of up to as many as three family units. By March pair-bonding begins to become evident; first courtship displays are seen in early April, and these displays peak in late April. During that period the males chase females persistently, sometimes flying over them, while both birds whistle frequently. At times the male will circle the female with his plumage ruffled and his tail fanned and lifted, in what perhaps corresponds to the lateral waltzing display. The advertising call of the male is prolonged, lasting about 6 s, and consists of repeated *ooy* syllables that terminate in a *oooooeeey-yeeo-yeeoo*, audible for up to a kilometre. Males also utter shorter but similar whistled calls, and both sexes vocalize prior to flight, on taking flight, and while in the air. Females also utter notes to summon their young, and chicks also produce cheeping calls (Baziev, in Cramp and Simmons 1980).

Reproductive biology

Eggs are laid from late April through May, but with renesting attempts occurring as late as July.

Fig. 15. Adult plumage variation among species of *Tetraogallus*: (A) Caucasian, (B) Caspian, (C) Altai, (D) Himalayan, and (E) Tibetan snowcocks. Broken line indicates condition of probable limited sympatry between species.

Although at any one level the laying period is quite short, birds laying at the highest elevations may do so up to 1.5 months later than those at the lowest elevations. The average clutch is of 6.5 eggs, with a range of five to eight, but sometimes is larger when two females lay in the same nest. The nest is located in the open on the ground, or under partial rock shelter. The egg-laying interval is two, or occasionally three days, and the incubation period is 28 days. Incubation is by the female alone and some observations suggest that only she tends the young, although it has also been suggested and seems probable that both parents normally remain with their brood. The age of independence is not known, but initial breeding reportedly occurs at a year (Baziev 1968; and in Cramp and Simmons 1980).

Evolutionary relationships

Baziev (1972) suggested that *Tetraogallus* was probably initially adapted to steppe-like habitats during the late Tertiary period. With the building of mountain systems since that time the genus has become disruptively distributed on various mountain ranges as the birds have become adapted to progressively more alpine-line habitats and separated from other populations by intervening unsuitable habitats. He believed that all five of the currently extant species of *Tetraogallus* differentiated from a single ancestral population independently of one another, and thus presumably regards them all as representing approximately the same degree of genetic isolation from each other. The close plumage similarities of *caucasicus* and *caspius*, as well as their closely allopatric distributions, suggest to me that these two forms have probably been isolated only since Pleistocene times and represent an allospecies pair that is also clearly related to *himalayensis*. Marien (1961) suggested that *caspius* and *himalayensis* might well be considered subspecies, with *caucasicus* a part of the same species group but apparently slightly less closely related.

Status and conservation outlook

This species is included in the list of rare and threatened animals of the USSR (Bannikov 1978). Its population has recently been estimated at about 410 000 birds, with fairly stable numbers (Baziev, in Cramp and Simmons 1980).

CASPIAN SNOWCOCK (Plate 35)

Tetraogallus caspius (S. G. Gmelin) 1874
Other vernacular names: tetraogalle de Perse (French); Kaspi-Konigshuhn (German); Kaspiysky ular (Russian).

Distribution of species (see Map 15)

Resident of alpine and subalpine zones of mountains of eastern Turkey east through Armenia to southern Transcaspia.

Distribution of subspecies

T. c. tauricus: southern and eastern Turkey and Armenia, probably intergrading with *caspius* in south-western Azerbaydzhan, eastern Armenia, and north-western Iran. Not recognized as distinct from *caspius* by Vaurie (1965).

    T. c. caspius: northern Iran east to southern Transcaspia.

    T. c. semenowtianschankii: western Zagros to near Shiraz, Iran.

Measurements

Wing (*tauricus*), males 298–307 mm, females 281–295 mm; tail, males 186–191 mm, females 176–177 mm. Weight, males average 2.68 kg, females 2.34 kg (Cramp and Simmons 1980). Egg, average 66 × 46 mm, estimated weight 77.1 g.

Identification (see Fig. 15)

In the field (21–22 in.)

Limited to the southern Caucasus and the mountains of Transcaucasia, where it is the only species of *Tetraogallus*. Found in rocky alpine areas, and with white on the bases of the flight feathers and head. Its calls are like those of *caucasicus*, differing only in timbre.

In the hand

Similar to the Caucasian snowcock, but with a grey chest that is lightly spotted with dark blotches rather than closely barred, less chestnut streaking on the flanks, a less cinnamon-tinted tail, a generally greyer overall cast, and the nape tending toward pale bluish rather than chestnut.

Adult female. Like the male, but smaller and duller; crown, neck-stripes, and chest buff-grey, faintly marked with black and white, and unspurred.

Immature. Like the adults but with a more indistinct head and neck pattern, and black streaking less well developed. The outer three primaries are also pointed and worn (Cramp and Simmons 1980).

General biology and ecology

This species is primarily found at subalpine and alpine elevations centred at around 1800–3000 m, mainly between the tree-line and the zone of permanent snow, but it occasionally enters the zone of rhododendron scrub and even the tree zone below, at least in summer months. Like the other snowcocks it favours rocky outcrops and steep slopes with talus and scree, especially those having scattered snow-fields. There are probably slight altitudinal changes with the seasons, with the lowest elevations apparently reached in late summer and autumn. It is a vegetarian, probably showing no significant difference from *caucasicus* in its diet, although it occupies somewhat lower elevations and occurs in a considerably drier climate (Baziev 1968; and in Cramp and Simmons 1980).

Social behaviour

The birds are gregarious, gathering in groups of up to about eight in late summer (presumably family units); these later merge to form larger flocks in November. Courtship reportedly begins in April and extends until late May, with males greatly outnumbering females. During April the loud whistling call of males can be frequently heard, and it apparently closely resembles the corresponding call of the Caucasian snowcock. Various alarm calls, calls uttered by birds among foraging flocks, and notes uttered before taking flight have also been described (Dementiev and Gladkov 1952; Cramp and Simmons 1980).

Reproductive biology

Egg records are for late April and May, with the nest site typically on the ground in the open, or under a sheltering rock. The clutch is said to range from six to nine eggs, and although single-brooded the birds are said to lay replacement clutches. The incubation period is unreported but probably close to the 28-day period reported for the Caucasian snowcock. The young reportedly are cared for only by the female, and the age of breeding is probably one year (Dementiev and Gladkov 1952; Cramp and Simmons 1980).

Evolutionary relationships

See the preceding account for a discussion of this.

Status and conservation outlook

Included in the list of threatened and endangered species of the USSR (Bannikov 1978), the population in Transcaucasia was estimated at no more than 850 birds in 1964, with some population isolates near extinction. The Turkish population's status is uncertain but it may be fairly common in some remote mountains (Cramp and Simmons 1980).

HIMALAYAN SNOWCOCK (Plate 36)

Tetraogallus himalayensis G. R. Gray 1843
Other vernacular names: tetraogalle de l'Himalaya (French); Himalaja-Konigshuhn (German); Gimalayskya ular (Russian).

Distribution of species (see Map 15)

Resident of alpine and subalpine zones of mountains from northern and eastern Afghanistan east to Ladakh, Nepal, and from eastern Kazakhstan and Xinjiang north to the Zaysan Basin and east to the borders of Qinghai and Gansu.

Distribution of subspecies

T. h. sewerzowi: Tian Shan range, north to Zaysan Basin and east to eastern Xinjiang. Intergrades in Pamirs with *incognitus*.

T. h. incognitus: mountains of southern Tadzhikistan and northern Afghanistan. Includes *bendi* (Vaurie 1965).

T. h. himalayensis: mountains from eastern Afghanistan east to Ladakh and western Nepal.

T. h. koslowi: mountains of Nan Shan and south Ching Hai Hu (Koko Nor) ranges, Qinghai, and southern Gansu.

T. h. grombczewskii: western Kunlun range of northern Tibet and southern Xinjiang.

Measurements

Wing, males 320–340 mm, females 275–315 mm (Dementiev and Gladkov 1952); tail, both sexes 173–193 mm (Ali and Ripley 1978). Weight, males 2.2–3.1 kg, females 2–2.57 kg (Dementiev and Gladkov 1952). Weight, males c. 1.9–3 kg, females 1.36–1.8 kg (Ali and Ripley 1978); game-farm males average 2.5 and females 1.95 kg (Stiver 1984). Egg (*himalayensis*), average 68.5 × 45.4 mm (Dementiev and Gladkov 1952), estimated weight 77.9 g.

Identification (see Fig. 15)

In the field (20–22 in.)

Geographically isolated from all other species of *Tetraogallus* except *tibetanus*, which lacks this species' dark chest and neck markings. Both are found in similar rocky alpine habitats, and utter similar melodic whistles as well as harsher and more croaking calls.

In the hand

Most like *caspius*, but the chest paler and more horizontally streaked, the flanks more streaked with blackish, the head generally lighter, and with a distinct brown collar at the base of the white throat.

Adult female. Similar to the male but with no tarsal spur, and with the forehead and area about the eye usually buff, barred with grey.

Immature. The chestnut patches on both sides of the nape more or less united, and the mantle usually buffy. The outer three (juvenal) primaries are pointed and mottled with rufous buff toward their tips.

General biology and ecology

Baker (1930) reported that these birds occur from 12 000 to 17 000 ft (3600–5100 m) during summer, and sometimes in winter as low as 7000 ft (2100 m), and prefer rocky, precipitous slopes having little vegetation. Mountain meadows are a favourite habitat, and in some areas it may feed on small grass-like herbs such as *Gager lutea*, according to Baker. Evidently grass is a major source of food, and they also eat seeds and moss roots, but probably very little animal materials. Introduced birds in Nevada favour well-vegetated alpine turf and alpine tundra habitats, usually in glacial cirques. There most birds have remained above 10 000 ft (3000 m) during winter months (Stiver 1984).

Social behaviour

These are gregarious birds, and form fairly large groups of 10–20 individuals, and rarely may number up to as many as 50. Like the other *Tetraogallus* species, various loud whistled notes are uttered, one of which certainly must serve as a male advertisement display. When courting, the male reportedly spreads his wings slightly, depresses his tail, and slightly ruffles his feathers. In this posture he runs back and forth in front of the hen or circles around her, presumably in what corresponds to the waltzing display (Baker 1930).

Reproductive biology

Nests in the wild have often been found at the extreme crest of a hill or just beyond on the leeward side, sheltered by scrubby grass or rocks but never in bushes or dense grassy vegetation. The clutch reportedly numbers four to five, but sometimes up to seven, and rarely more. Only the female incubates, but the male remains close to the nest and acts as a lookout (Baker 1930). In captivity, breeding behaviour was found to begin during February and probably peaked in April, with egg-laying starting in mid- to late March, and usually peaking in early May. Game-farm females usually did not begin breeding until two years old, and had an average annual egg production of 11.3. The incubation period is 25 days. Based on a small sample of seven broods, the average size in Nevada was found to be 5.7 chicks (Stiver 1984).

Evolutionary relationships

See the Caucasian snowcock account for some comments on this.

Status and conservation outlook

This species has a relatively broad range and occurs in several countries (USSR, Afghanistan, Pakistan, India, and China). It is also now well established in Nevada, where over 1700 wild and game-farm birds from Pakistan stock have been released. They now are found in the Ruby and East Humboldt ranges of Elko County and probably numbered 250–500 individuals in the early 1980s (Stiver 1984).

TIBETAN SNOWCOCK (Plate 37)

Tetraogallus tibetanus Gould 1854
Other vernacular names: tetraogalle du Tibet (French); Tibet-Konigshuhn (German); Tibetsky ular (Russian).

Distribution of species (see Map 15)
Resident of alpine and subalpine zones of Pamir and Himalayan mountains, from Tadzhikistan and Kashmir east to eastern Xizang (Tibet), north-western Sichuan, and the Mishmi Hills of north-eastern India.

Distribution of subspecies
T. t. tibetanus: Pamirs and Himalayas of western Xizang, Ladakh, and adjacent Tadzhikistan. Includes *tschimenensis*.
  T. t. aquilonifer: Himalayas from western Nepal east to Bhutan.
  T. t. henrici: mountains of eastern Xizang, intergrading with next race in north-western Sichuan.
  T. t. przewalskii: mountains of Qinghai, south-western Gansu, and north-western Sichuan, south to the Arbor and Mishmi Hills of north-eastern India. Includes *centralis*; sometimes considered part of *henrici*.

Measurements

Wing, males 270–280 mm, females 260–270 mm (Ali and Ripley 1978); tail, males 163–174 mm, females 160–176 mm. Weight, males 1.5–1.75 kg, females 1.17–1.6 kg (Cheng *et al*. 1979). Egg (*tibetanus*), average 63.8 × 44.1 mm (Baker 1935), estimated weight 68.5 g.

Identification (see Fig. 15)

In the field (19–20 in.)
Geographically isolated from the other *Tetraogallus* species except in Tsinghai and southern Gansu, where *himalayensis* may also occur parapatrically if not sympatrically in the Nan Shan and associated Koko Nor ranges (Marien 1961; Cheng *et al*. 1978). Both species are found in similar rocky alpine habitats, and apparently both have similar sharp whistled calls. This species lacks the dark brown markings on the sides and front of the neck, and on

Plate 33. Snow partridge, adults. Painting by H. Jones.

Plate 34. Caucasian snowcock, pair. Painting by H. Jones.

Plate 35. Caspian snowcock, pair. Painting by H. Jones.

Plate 36. Himalayan snowcock, pair. Painting by H. Jones.

Plate 37. Tibetan snowcock, pair. Painting by H. Jones.

Plate 38. Altai snowcock, pair. Painting by H. Jones.

Plate 39. Arabian red-legged partridge, pair. Painting by H. Jones.

Plate 40. Przhevalski's rock partridge, pair. Painting by H. Jones.

Plate 41. Rock partridge (*saxatilis*), pair. Painting by H. Jones.

Plate 42. Chukar partridge, pair. Painting by H. Jones.

Plate 43. Philby's rock partridge, adult. Painting by Mark Marcuson.

Plate 44. Barbary partridge (*barbara* and *spatzi*), males. Painting by H. Jones.

Plate 45. Red-legged partridge, pair. Painting by H. Jones.

Plate 46. See-see partridge, pair. Painting by H. Jones.

Plate 47. Sand partridge, pair. Painting by H. Jones.

Plate 48. Black francolin, pair. Painting by H. Jones.

Plate 49. Painted francolin, pair. Painting by H. Jones.

Plate 50. Chinese francolin, pair. Painting by H. Jones.

Plate 51. Red-necked francolin, pair. Painting by H. Jones.

Plate 52. Swainson's francolin, pair. Painting by H. Jones.

Plate 53. Grey-breasted francolin, pair. Painting by H. Jones.

Plate 54. Yellow-necked francolin, pair. Painting by H. Jones.

Plate 55. Erckel's francolin, pair. Painting by H. Jones.

Plate 56. Chestnut-naped francolin, pair. Painting by H. Jones.

Plate 57. Jackson's francolin, pair. Painting by H. Jones.

Plate 58. Cameroon Mountain francolin, adult and juvenile males. Painting by H. Jones.

Plate 59. Ahanta francolin, pair. Painting by H. Jones.

Plate 60. Scaly francolin, pair. Painting by H. Jones.

Plate 61. Grey-striped francolin, male. Painting by H. Jones.

Plate 62. Double-spurred francolin (*bicalcaratus* and *thornei*), pair. Painting by H. Jones.

Plate 63. Pale-bellied (upper left), handsome (upper right), Hartlaub's (middle, male on left), Swierstra's (lower left), and Nahan's (lower right) francolins, adults. Painting by Timothy Greenwood.

the chest typical of *himalayensis*; instead it has narrow black flank stripes.

In the hand

Separated from the other *Tetraogallus* species by the extensively white underparts, which are streaked with grey and black on the sides and flanks. The chest is greyish, with indefinite whitish horizontal blotches, and there is no white on the basal halves of the outer primaries.

Adult female. Like the male, but smaller, and with no tarsal spur.

Immature. Paler above, with a more conspicuous white superciliums; chin and throat white below, with no pectoral band; breast grey, mottled with brown and buff; flanks and abdomen white, lacking the black streaks (Ali and Ripley 1978). Presumably the outer three juvenal primaries are retained into the first-winter plumage, as in the other species of *Tetraogallus*.

General biology and ecology

This species occurs up to 5800 m elevation during summer months, and down to between 3000 and 4000 m during winter, or possibly even lower under conditions of heavy snowfall. Like the other snowcocks, it is associated with alpine meadows, rockfields, sparsely vegetated ridges, and the edges of snowfields above the tree-line. It tends to occur at even higher elevations than does the Himalayan snowcock. It may also assemble in somewhat larger groups during the autumn migration to lower elevations; reportedly at times these number in the thousands (Baker 1930). The species' foods are not well known, but include grasses and a variety of herbs such as *Stellaria, Saxifraga, Oxytropis, Potentilla,* and *Primula* (Dementiev and Gladkov 1952).

Social behaviour

Little specifically known, but at least as gregarious as the other snowcocks, with flocks often numbering 30–50. The vocalizations are likewise similar, and include a clear whistle, a call similar to that of a curlew (*Numenius arquatus*), and a subdued chuckling that gradually becomes louder until it reaches a climax (Ali and Ripley 1968).

Reproductive biology

In Ladakh and north-western Xinzang (Tibet) breeding mainly occurs from the end of May until early July, although one late egg record is for 25 August. The clutch-size has been generally reported as from four to seven eggs, and usually only four or five (Baker 1930). There is no information on the incubation period or development of the young, although fledged broods of four to six young have been seen in mid-July (Dementiev and Gladkov 1952). It is also unreported as to whether males regularly accompany broods or whether they instead gather into flocks following the nesting season.

Evolutionary relationships

This is the smallest member of the entire genus *Tetraogallus*, which Marien (1961) has suggested might facilitate ecological compatibility with *himalayensis* where the two are sympatric. In plumage it exhibits some similarities with *altaicus*, and I believe the two are derived from a common ancestor more recently than that giving rise to the assemblage of *caspius, caucasicus,* and *himalayensis*.

Status and conservation status

No specific information exists on this topic.

ALTAI SNOWCOCK (Plate 38)

Tetraogallus altaicus (Gebler) 1836
Other vernacular names: tetraogalle de l'Altai (French); Altai-Konigshuhn (German); Altaysky ular (Russian).

Distribution of species (see Map 15)

Resident of alpine and subalpine zones of Altai and Sayan mountains, Tuva ASSR, east to western Mongolia. Includes *orientalis* (Vaurie 1965). (There are no subspecies.)

Measurements

Wing, males 290–325 mm, females 280–332 mm (Dementiev and Gladkov 1952); tail (sex?), 173 mm (Ogilvie-Grant 1893). Estimated weight, males 3.0 kg, females 2.54 kg. Egg, average 70 × 47.1 mm, estimated weight 85.7 g.

Identification (see Fig. 15)

In the field (20–22 in.)

Limited to the Altai and Sayan mountains, where it is isolated from all other *Tetraogallus* species, but occupies comparable rocky alpine habitats. Its soft whistled notes are said to be more melodious than those of the Himalayan snowcock, but both probably are very similar.

In the hand

Most similar to the Tibetan snowcock, but the white underparts are not streaked with black, and the grey chest is lightly spotted with black rather than blotched with white. The head is almost entirely brown, with no chestnut or pale greyish blue tones, and white is confined to the chin and throat. The basal parts of the remiges also lack white.

Adult female. Like the male, but without a tarsal spur and somewhat smaller.

Immature. Similar to adults, but the outer three primaries (retained from the juvenal plumage) relatively pointed and frayed.

General biology and ecology

Like the other snowcocks, this species occurs in alpine and subalpine zones at altitudes of 2000 m or higher, on open stony slopes, rockfields among valley meadows, and on generally precipitous slopes where broken rubble-strewn ground and alpine meadows alternate with rock outcrops. In winter the birds descend to somewhat lower levels, sometimes even to steppelands of foothills. Their foods reportedly consist of shoots, berries, seeds, buds, tubers, bulbs, and insects, including the berries of *Berberis* and the shoots and buds of dwarf birches (*Betula*) (Dementiev and Gladkov 1952).

Social behaviour

Evidently these birds move about in small groups of from two to twelve birds, which probably represent family groups. They forage in scattered flocks, and often call during morning and evening hours from crags or rocks. The call is a loud whistle, as in the other *Tetraogallus* forms, and probably serves similar diverse functions. Mating activity reported to occur in April, although loud calling can be heard into late summer (Dementiev and Gladkov 1952).

Reproductive biology

Eggs have been found in May, although there is still no detailed information on clutch-sizes, nest sites, or other aspects of the reproductive biology of this rare species. One early observer found a brood tended only by a female, while at the same time he saw flocks of adults, of which one captured specimen proved to be a male, suggesting that males may not participate in brood care (Dementiev and Gladkov 1952).

Evolutionary relationships

This is the most northerly and perhaps the most isolated of the *Tetraogallus* forms, with its greatest similarities to *tibetanus*.

Status and conservation outlook

This species is included in the list of rare and endangered species of the USSR (Bannikov 1978), but there is no detailed information on its status.

15 · Genus *Alectoris* Kaup 1829

The red-legged partridges are medium to large arid-adapted Palaearctic and Arabian Peninsula species that have rounded tails of 14 grey or chestnut-coloured rectrices that are about 60 per cent as long as the wing. The wing is rounded, with the eighth primary the longest. The post-juvenal primary moult is incomplete, and the tail moult is perdicine (centrifugal). The sexes are alike or very similar, with uniformly greyish upperparts, strongly barred sides and flanks, and red to orange-red eye-rings, bills, and tarsi. A blunt tarsal knob is present in one or both sexes. Seven largely allopatric species are recognized here; some authorities recognize fewer:

A. melanocephala (Rüppell) 1835: Arabian red-legged partridge
    A. m. quichardi Meinertzhagen 1951
A. magna (Przhevalski) 1876: Przhevalski's rock partridge
A. graeca (Meisner): rock partridge
    A. g. saxatilis (Bechstein) 1805
    A. g. graeca (Meisner) 1804
    A. g. whitakeri Schiebel 1934
A. chukar J. E. Gray: chukar partridge
    A. c. cypriotes Hartert 1917
    A. c. sinaica Bonaparte 1858
    A. c. kurdestanica Meinertzhagen 1923
    A. c. werae Zarudny and Loudon 1904
    A. c. koroviakovi Zarudny 1914
    A. c. subpallida Zarudny 1914
    A. c. dzungarica Sushkin 1927
    A. c. falki Hartert 1917
    A. c. pallescens Hume 1873
    A. c. pallida Hume 1873
    A. c. fallax Sushkin 1927
    A. c. chukar J. E. Gray 1830
    A. c. pubescens Swinhoe 1871
    A. c. potanini Sushkin 1876
A. [chukar] philbyi Lowe 1934: Philby's rock partridge
A. barbara Bonnaterre: Barbary partridge
    A. b. koenigi Reichenow 1899
    A. b. barbara Bonnaterre 1790
    A. b. spatzi Reichenow 1895
    A. b. barbata Reichenow 1896
A. rufa (L.): red-legged partridge
    A. r. rufa (L.) 1758
    A. r. hispanica Seone 1894
    A. r. intercedens A. E. Brehm 1858

KEY TO THE SPECIES OF ROCK PARTRIDGES (*ALECTORIS*)

A. Crown and nape black, outer tail feathers grey: red-legged Arabian partridge (*melanocephala*)
AA. Crown and nape brown or grey, outer tail feathers chestnut
  B. Throat entirely black: Philby's rock partridge (*philbyi*)
  BB. Throat white or grey, enclosed by a dark band
    C. Throat set off by an unspotted, continuous black to brownish band
      D. Band surrounding throat mostly rusty brown and very narrow; eyes yellowish grey: Przevalski's rock partridge (*magna*)
      DD. Band surrounding throat mostly black and quite broad; eyes brown
        E. Black loral stripe broader, reaching base of bill; auriculars black and buff; throat white and usually ending at an angle below: chukar partridge (*chukar*)
        EE. Black loral stripe narrower, ending at top of bill; auriculars chestnut; throat more buffy and terminating roundly below: rock partridge (*graeca*)
    CC. Throat-band spotted with white or having irregular lower border
      D. Throat-band brown, throat and cheeks grey: Barbary partridge (*barbara*)
      DD. Throat-band black, throat and cheeks white: red-legged partridge (*rufa*)

ARABIAN RED-LEGGED PARTRIDGE
(Plate 39)

Alectoris melanocephala (Rüppell) 1835
Other vernacular names: Rüppell's partridge; perdrix à tête noire (French; Arabisches Steinhuhn; Schwartzkopf-Felsenhuhn (German).

Distribution of species (see Map 16)

Resident of desert habitats of southern Saudi Arabia from Jiddah east to Muscat in Oman, and south to the Aden Protectorate.

Distribution of subspecies

A. m. melanocephala: range as indicated above.
    A. m. quichardi: known only from El Hajar, eastern Hadramaut, Saudi Arabia, but probably intergrading with *melanocephala* in western Hadramaut (Meinertzhagen 1954).

112 *The Quails, Partridges, and Francolins of the World*

Map 16. Distribution of the Arabian red-legged (A), Barbary (B), chukar (C), Philby's (P), Przhevalski's (Pr), red-legged (R), and rock (Ro) partridges. See also map 12 for introduced North American distribution of chukar partridge.

Measurements

Wing, males 177–210 mm, females 166–181 mm (Meinertzhagen 1954); tail, males 146–149 mm, females 140 mm (Hartert 1903). Estimated weight, males 724 g, females 522 g. Egg, no information.

Identification (see Fig. 16)

In the field (16–17 in.)

Easily separated from the other *Alectoris* species by its considerably larger size, grey tail feathers (visible in flight), and black crown. Calls are similar to those of other *Alectoris* forms, including a repeated *cook* as a rally call.

In the hand

The large size (wing over 165 mm) and grey outer rectrices immediately distinguish this species from the other forms of *Alectoris*.

Adult female. Like the male, but slightly smaller, and with no tarsal knob.

Immature. Birds in first-winter plumage are similar to adults, but have a shorter crest, a blackish bill, and are perhaps in general not so brightly coloured. A juvenile bird is illustrated by Gallagher and Woodcock (1980).

General biology and ecology

This largest species of *Alectoris* occurs in mountains, wadis, and vegetated upland plains, especially favouring rocky hillsides, where escape from danger is attained by running uphill. The birds forage on vegetable materials, seeds, and insects, regularly visiting sources of water each morning and evening (Gallagher and Woodcock 1980). They occur from sea-level to about 4500 ft (1350 m), and occasionally to 7000 ft (2100 m) (Watson 1962a). Most birds examined by Meinertzhagen (1954) had eaten almost exclusively a small grass (*Schimus barbatus*) and a dwarf cudweed (*Gnaphalium pulvinatum*), although some had eaten a higher percentage of cultivated grain and about 15 per cent insect materials.

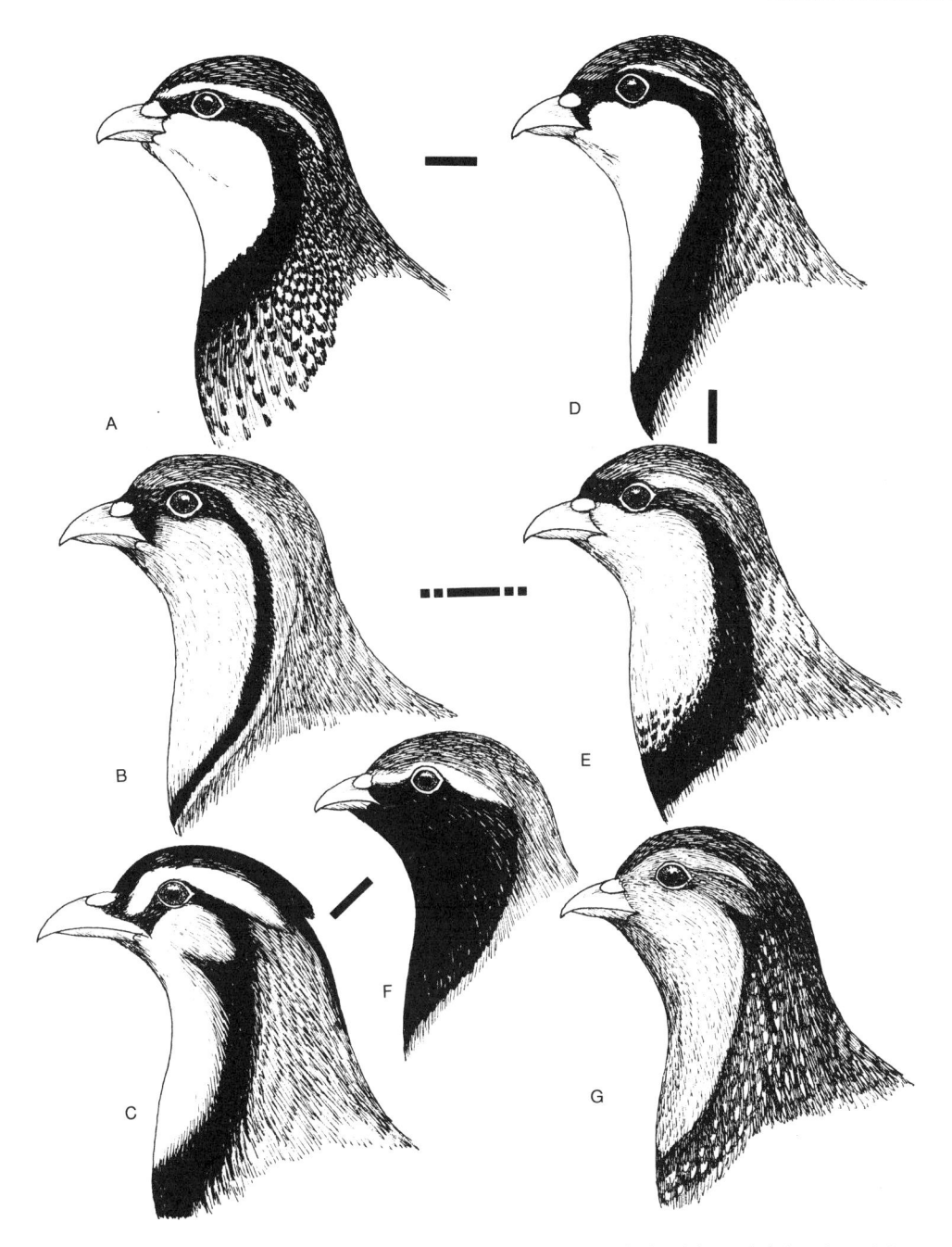

Fig. 16. Adult plumage variation among species of *Alectoris*: (A) red-legged, (B) Przhevalski's rock, (C) Arabian red-legged, (D) rock, (E) chukar, (F) Philby's rock, and (G) Barbary partridges. Lines indicate known (solid) or probable (broken) conditions of limited sympatry between species.

Social behaviour

Like the other rock partridges this is a gregarious species, typically occurring in groups of five to eight birds that probably represent family groups. Their calls include a few preliminary *cucks* that gradually increase in tempo and loudness to a *owk-owk-owk* or *crowk, crowk, crowk*. There is also a rally or contact call of quickly repeated *cook* notes, and the flight call uttered in alarm is a repeated *kerowk*. There are also various soft conversational notes that are punctuated by high-pitched *mew* notes (Gallagher and Woodcock 1980). Their calls are louder and lower-pitched than those of the smaller *Alectoris* species such as *chukar*, but do include notes sounding like *chukor* or *caccaba* (Meinertzhagen 1954).

Reproductive biology

Breeding occurs in Arabia during March, and was recorded during July and August on the Eritrean coast, where it now no longer occurs. The nest is usually found at elevations above 600 m, and is placed under cover of a rock or bush. The usual clutch is apparently rather small, of five to eight eggs, but up to eleven or more have also been reported (Meinertzhagen 1954; Gallagher and Woodcock 1980). There is no information on incubation or development of the young.

Evolutionary relationships

Watson (1962a) believed that this species is sufficiently distinctive to warrant its placement in a separate monotypic subgenus, but it is probably closer to the *graeca* superspecies than to the *chukar* assemblage. He also believed that its morphological and size differences from *philbyi* represented a prime example of character displacement associated with reducing interspecific competition and avoiding possible disadvantageous hybridization.

PRZEVALSKI'S ROCK PARTRIDGE
(Plate 40)

Alectoris magna (Przevalski) 1876
Other vernacular names: great partridge; perdrix de Przevalski (French); Chinesisches Steinhuhn (German).

Distribution of species (see Map 16)
Resident of rocky habitats of China from Qinghai to south-eastern Gansu. Sometimes considered conspecific with *chukar*, but regarded as distinct and as part of the *graeca* superspecies by Watson (1962a).

Measurements

Wing, males 169–180 mm, females 160–172 mm; tail, males 105–122 mm, females 90–120 mm. Weight, males 445–710 g, females 442–615 g (Cheng *et al.* 1978). Egg, average 39.5 × 30.6 mm (Cheng *et al.* 1978), estimated weight 20.4 g.

Identification (see Fig. 16)

In the field (15 in.)
Apparently geographically isolated from *chukar*, but very similar to it in appearance and behaviour. It is said to utter a peculiar hollow note sounding something like *cuta-cuta* when taking flight, but is probably not generally separable from *chukar* under field conditions unless the distinctive bicoloured collar can be observed.

In the hand
Closely resembling *chukar*, but larger, with black lores (as in *graeca*), a narrow double neck collar (fuscous to black inside, reddish brown outwardly), and indistinctly marked under the throat. The black flank stripes are as narrow as in *graeca*, and the black gorget is similarly rounded, but the dorsal body colour is almost identical to the Chinese forms of *chukar*. Iris yellowish grey.

Adult female. Nearly identical to the male, but slightly smaller.

Immature. Undescribed.

General biology and ecology

Apparently the habitats of this species are very similar to those of *chukar* in the same region, namely rocky hillsides and gorges covered with grasses and small bushes. In Tsinghai and Gansu it has been recorded only at about 7800 ft (2400 m), and *chukar* below 7000 ft (2100 m), so some slight altitudinal differences might exist. Nothing is known of its foods.

Social behaviour

No specific information, but unlikely to differ from the other species. Reportedly more silent than *chukar*.

Reproductive biology

No specific information.

Evolutionary relationships

Watson (1962a) judged that *magna* is a member of the *graeca* superspecies that as a single population once had a far more extensive range, but which is now represented by two highly disjunctive high-montane relict forms. There is probably only a very limited amount of current sympatry with *chukar* in western Gansu, and no evidence of hybridization or character displacement in this area.

ROCK PARTRIDGE (Plate 41)

Alectoris graeca (Meisner) 1804
Other vernacular names: Greek partridge; perdrix

bartavelle (French); Steinhuhn (German); perdiz moruna (Spanish).

Distribution of species (see Map 16)
Resident of European Alps from France and Italy to Austria, Italy south to Sicily, and east to Greece and Bulgaria.

Distribution of subspecies
A. g. saxatilis: France to Austria, the Apennines, and western Yugoslavia. The Apennine population has recently been separated as *orlandoi* (Priolo 1984).
  A. g. graeca: south-eastern Yugoslavia, Greece, and Bulgaria.
  A. g. whitakeri: Sicily.

Measurements

Wing (various races), males 162–174 mm, females 155–167 mm; tail, males 82–95 mm, females 76–90 mm. Weight (various races), males 550–850 g, females 410–650 g (Cramp and Simmons 1980). Egg, average 41 × 30 mm, estimated weight 20.3 g.

Identification (see Fig. 16)

In the field (13.5–15 in.)
Not always separable visually from *chukar* unless the minor points mentioned in the species description can be seen. The calls are often distinct from those of *chukar*, and the male's advertisement or rally call is a highly variable staccato, grating series of usually quadrisyllabic notes often sounding something like *tchertsi-ritt-chi* or *tchertsivitchi* but occasionally consisting of more rasping two- or three-syllable sounds such as *pair-chuck* or *chukara* similar to those of the following species. These call differences are most likely to assist in species separation in those few areas where either or both might occur, as in Bulgaria, where narrow geographical overlap and local hybridization has been reported.

In the hand
Plumage almost exactly as in *chukar*, but with more extensively black lores, a more nearly white throat and face, and predominantly black ear-coverts. The inner black bands on the flank feathers also tend to be narrower, the black eye-stripe is broader and more often continuous above the eye, and the black gorget is usually more sharply defined along its anterior and posterior edges. The birds often are more generally bluish grey and less brownish on the upperparts, but this distinction may not apply to all races.

Adult female. Like the male, but slightly smaller, usually with no tarsal knob, and a duller head pattern. The black forehead band typically is also narrower and less blackish in females than males.

Immature. Juveniles are smaller and much duller than adults, with little flank barring or facial patterning; juveniles generally have whiter throats and darker brown ear-coverts than those of *chukar*. The two outer primaries from the juvenal plumage are retained in the otherwise adult-like first-winter plumage, and are narrow, pointed, frayed, and relatively short (Cramp and Simmons 1980).

General biology and ecology

This species is primarily associated with dry, rocky habitats, mainly between 900 and 1500 m, but rarely reaching as high as 2700 m in Italy, and using habitats ranging from dwarf heaths, pastures, areas of grass, and low scrub, high vineyards, to rather barren stony ground, scree, and rocky escarpments or craggy areas. Open woodlands of coniferous or deciduous trees are sometimes utilized, but not closed forests, and access to drinking water is apparently required. There are usually altitudinal movements within the seasons, although some birds may remain at their breeding elevations even through midwinter if sufficient snow-free habitat persists. Foods predominantly consist of plant materials such as seeds, fruits, and green portions, plus such invertebrates as insects and other terrestrial arthropods, the latter taken mainly by breeding females and chicks (Glutz von Blotzheim 1973; Cramp and Simmons 1980).

Social behaviour

The social behaviour of this species has been extensively studied in a series of German papers by A. Menzdorf, which recently have been summarized in English by him (1982) and also by Cramp and Simmons (1980). These birds alternate seasonally between flocking and pair-dispersion behaviour during the breeding season. Additionally unpaired males may form summer flocks. However, most flocking occurs in early autumn, as family-based coveys begin to gather into larger groups that usually reach a maximum of 50 but rarely may reach 100 birds. Pair-bonding is monogamous and may be long-term, although there have been recorded instances of successive bigamy by males. Additionally some mate-switching occurs between breeding seasons, and re-mating by widowed birds occurs. Displays associated with pair formation in red-legged and chukar partridges have been described and illustrated in Chapter 5, and there is little apparent difference in *graeca* from this general pattern of dominance-establishment by males and submissive responses by females that serve to avoid attack and eventually stimulate sexual responses. The flocks begin to dissolve in late winter, often in February, and territorial advertisement occurs during March and April. These individual male territories, which are advertised by rally-calling and defended by both members of the pair, are initially often non-contiguous. They are held only until about the time of hatching, when males begin to gather into small unisexual groups.

Reproductive biology

Nesting in the Alps lasts from mid-May to June, and in Greece from late April to June. The nest site is apparently chosen by the male (Menzdorf 1975c), but the nest scrape is made by the female. The eggs are laid at a rate of about 1.3 days per egg, and a typical clutch is from 8 to 14 eggs, averaging about 11. Larger clutches of up to 21 have been recorded but possibly these are artefacts caused by the efforts of more than one female. Incubation is by the female alone, and lasts 24–26 days. The young are led from the nest within hours of hatching, and within a few days the family may move several hundred metres from the nest site. A female regularly lays replacement clutches after loss of a nest, and reportedly may sometimes produce two separate clutches, incubating the first herself and turning the second over to the male. However, Menzdorf (1975b) did not find any evidence of such double-brooding among the birds he studied, and considered it unlikely that this is of normal occurrence. The young are capable of precocious flight when only 7–10 days old, their last-growing (eighth) first-winter primaries are fully grown by about 90 days, and they reach their adult weight by about 120 days (Menzdorf 1975e).

Evolutionary relationships

The analysis by Watson (1962b) strongly indicates that *graeca* is specifically distinct from *chukar*, and deserves specific recognition from it. Watson determined that a zone of sympatric contact between these two species occurs in Thrace, where he found no intermediate specimens or signs of character introgression. He also judged that there the chukar probably generally occurs at lower elevations than does the rock partridge, thereby reducing the degree of interspecific interactions. However, it has since been determined that a narrow zone of hybridization does occur through Greece and Bulgaria (Dragoev 1974), and that limited hybridization between *graeca* and *rufa* also reportedly has occurred in the Roya Valley of the Maritime Alps in Italy (Spano 1978). The latter hybrids are probably fertile, and the two forms might best be regarded as 'semispecies'

(Bernard-Laurent 1984). Watson (1962a) considered *graeca* to be derived from an ancestral *Alectoris* assemblage that also includes *magna* and that eventually gave rise to the *chukar* superspecies group, probably in the Himalayas or elsewhere in the Middle East.

Status and conservation outlook

The population of *graeca* is declining in most areas of its range (at least in France, Switzerland, Austria, and Italy), probably through the effects of overhunting, and possibly also as a result of long-term climatic changes (Cramp and Simmons 1980).

CHUKAR PARTRIDGE (Plate 42)

Alectoris chukar J. E. Gray 1830
Other vernacular names: chukar; perdrix chukar (French); Chukarsteinhuhn (German); perdiz chukar (Spanish).

Distribution of species (see Map 16)

Resident of rocky habitats from the south-eastern Balkan Peninsula and adjacent Mediterranean islands (Aegean Islands, Crete, Cyprus) east through Turkey, the Caucasus, Transcaucasia, and Iranian region to Sind, Nepal, Xinzang (Tibet), north-eastern China (Gansu, Shandong, and southern Manchuria); also north to Kazakhstan, Xinjiang, and Mongolia. Also locally introduced in the USSR, England, western North America (British Columbia to Baja, east to Montana and Colorado), the Hawaiian Islands (all larger islands), New Zealand (South Island), and on St Helena (Long 1981).

Distribution of subspecies (excluding introductions)

A. c. cypriotes: Bulgaria, the Aegean Islands, Crete, Rhodes, Cyprus, and Asia Minor, intergrading with *sinaica* in southern Syria. Includes *kleini* and *scotti*.

A. c. sinaica: Syrian desert south to the Sinai Peninsula.

A. c. kurdestanica: Caucasus and Transcaucasia south to Iran and Kurdistan. Includes *caucasica*.

A. c. werae: eastern Iraq and south-western Iran.

A. c. koroviakovi: eastern Iran east to Sind and western Afghanistan. Includes *shestoperovi*, *laptevi*, and *dementievi*.

A. c. subpallida: Kyzyl Kum and Kara Kum mountains of Tadzhikistan.

A. c. falki: north-central Afghanistan south to the Pamirs, north to Kazakhstan, and east to western Xinjiang in Tian Shan range.

A. c. dzungarica: north-western Mongolia, Russian Altai, and western Xinjiang. Includes *obscurata*.

A. c. pallescens: north-eastern Afghanistan east to Ladakh and western Tibet (Xinzang).

A. c. pallida: Tarim Basin of western and southern Xinjiang.

A. c. fallax: eastern and southern Tian Shan mountains, Xinjiang.

A. c. chukar: eastern Afghanistan east to eastern Nepal (Jiri).

A. c. pubescens: Liaoning and Shantung west through Nei Monggol (Inner Mongolia) to north-western Sichuan and eastern Qinghai.

A. c. potanini: western Nei Monggol and Mongolia.

Measurements

Wing (*chukar*), both sexes 146–180 mm; tail, both sexes 78–105 mm (Ali and Ripley 1978). Wing (*kleini* and *sinaica*), males 162–172 mm, females 148–160 mm; tail, males 76–87 mm, females 74–86 mm. Weight (*cypriotes*), males 460–595 g, females 365–545 g (Cramp and Simmons 1980). Weight (*chukar*), males c. 540–765 g, females c. 370–540 g (Ali and Ripley 1969). Egg (*chukar*), average 43 × 31.4 mm (Baker 1935), estimated weight 23.4 g.

Identification (see Fig. 16)

In the field (13.5–15 in.)

Very similar to *graeca* and probably not unquestionably separable from it under most field conditions, either visually or by calls, but usually identifiable by its hoarse *chuck, chukar,* or *chukara* notes (see *graeca* account).

In the hand

Best separated visually from *graeca* by its more extensively white lores, the greater width of the inner black bar on the flank feathers, its more buffy throat, more black spotting near the base of the gorget, obscuring its borders, and chestnut rather than black on the upper ear-coverts. Hybridization may prevent certain identification of some specimens.

Adult female. Like the male, but slightly smaller and usually lacking a tarsal knob. Additionally the colour of the bill and tarsus tend to be duller than in males, and the head patterning is often less contrasting (the buffy superciliary area is usually lighter, and the black forehead band is usually darker and wider in males).

Immature. Birds in first-winter plumage usually retain one or two outermost juvenal primaries, which tend to be faded and pointed, and often are yellowish near the tip. First-year birds also tend to have narrower black forehead markings than adults, and young males usually have smaller tarsal knobs (Watson 1962b).

General biology and ecology

Although very similar in its ecology to *graeca*, this species is found at generally lower altitudes, and is perhaps somewhat better adapted to aridity, at least in some races. Like *graeca*, the birds avoid both wet and heavily wooded habitats, and favour rocky slopes with scattered grasses and bushes, although in some cases they are found in fairly flat desert situations. Introduced birds in western North America are most abundant where up to half of the surface area is comprised of talus slopes, rocky outcrops, cliffs, and bluffs, the other half being covered with sagebrush (*Artemisia*) and grass (especially the annual *Bromus tectorum*, but including perennials such as *Agropyron* and *Poa*), with small amounts of brushy creek-bottom areas (Galbreath and Moreland 1953). In some parts of northern North America persistent snow cover during winter may be a significant limiting factor, and in some southern areas the availability of water during the summer may also be important. Precipitation during late winter and spring may be a significant factor affecting reproductive success, at least in California, where it influences food supplies and availability of nesting cover (Harper *et al.* 1958). Studies in North America suggest that grasses are the most important food source, including both leafy materials and seeds, and the seeds of various weedy forbs. Although young birds consume the usual array of insects and other invertebrates, this is a minor food source for adults. Most foraging is done during early morning and late afternoon hours, with the hottest part of the day spent in shady cover, often near a source of water. The birds are quite mobile, with released birds sometimes moving as far as 33 miles in about 2 years (Harper *et al.* 1958), and probably able to move as far as 2 or 3 miles in a day in order to reach water supplies. Seasonal altitudinal or horizontal movements related to variable water supplies have also been documented (Johnsgard 1973).

Social behaviour

Covey sizes in chukars range from as few as about five to as many as 40 or more birds, and probably average about 20. These birds remain in coveys until late winter, when pair-formation activities cause their dissolution. The birds are basically monogamous, although observations in western North America indicate that a small proportion of birds may form bigamous matings (Mackie and Buechner 1963). Probably few differences exist between the pair-forming behaviour of this species and other *Alectoris* forms, and Stokes (1961) has described these activities in detail, primarily for captive birds. He believed that when uttered by paired birds the rally call tends to repel other males, and thus may have some spacing effects. Although other writers such as Mackie and Buechner have questioned whether true territoriality occurs in *Alectoris*, Stokes was of the opinion that well-defined territories are established by wild birds. Postures and calls associated with pair formation in this species were summarized in Chapter 5.

Reproductive biology

Before building a nest the male typically performs a 'nest ceremony' that involves entering a clump of vegetation and performing nest-scraping movements. This may serve to strengthen the pair-bond and also attach the male to the nest site in the event that he takes over incubation later on. Females deposit their eggs at the rate of about 1–2 days per egg, with the interval shortest early in the laying season. In western North America the clutch averages about 15.5 eggs (Mackie and Buechner 1963), which is probably fairly typical of the species. Renesting following the loss of a clutch is normal, and there is at least one case of a captive male incubating one of two clutches laid by his mate (Portal 1924). There is no good evidence that this is typical of wild birds, however. Incubation lasts about 24 days. Fairly shortly after hatching there is a tendency for brood amalgamation into larger assemblages of adults and young, and even prior to hatching the males may begin to assemble into flocks. Age ratios in autumn flocks based on hunter harvests in North America indicate that about 80–85 per cent of these flocks are comprised of young birds, suggesting that a relatively high annual adult mortality rate may be typical of the species (Johnsgard 1973).

Evolutionary relationships

Watson (1962a) suggested that *chukar*, *philbyi*, *barbara*, and *rufa* comprise a superspecies, which has fairly recently become broken into these four isolate forms, all now well removed from their probable geographic origin in the Middle East or the Himalayas.

Status and conservation outlook

Within its native range in the western Palaearctic the chukar population has been locally reduced in areas such as Greece and Lebanon, but elsewhere has been maintaining or even increasing its numbers (Cramp and Simmons 1980). It has attained a remarkable population status in North America, where it has become well established and is hunted in nine states and one Canadian province, and where during the past few decades over half a million birds have been harvested annually (Johnsgard 1973). It has also become well established on the South Island of New Zealand (Long 1981).

PHILBY'S ROCK PARTRIDGE (Plate 43)

Alectoris philbyi Lowe 1934
Other vernacular names: perdrix de Philby (French); Philbysteinhuhn (German).

Distribution of species (see Map 16)
Resident of rocky desert habitats of western Arabia from south-western Saudi Arabia to northern Yemen. Allopatric and sometimes considered conspecific with *graeca* or *chukar*.

Measurements

Wing, 156 mm (females) to 172 mm (males) (Meinertzhagen 1954); tail, both sexes 95–114 mm. Estimated weight, both sexes 441 g. Egg, not separable from those of *Alectoris c. sinaica* (which average 40.2 × 29.7 mm), according to Meinertzhagen.

Identification (see Fig. 16)

In the field (13–14 in.)
Easily separated from all the other Arabian partridges by the black throat and chin, but otherwise almost identical to *chukar*. Calls are not well known, but probably are not significantly different from those of *chukar*.

In the hand
Comparable in most respects to *chukar* save for the black chin, cheeks, and throat.

Adult female. Like the male, but slightly smaller, and often (always?) lacking a tarsal knob.

Immature. Closely resembling adults after the post-juvenal moult, but probably with a smaller tarsal knob in the case of males, and presumably both sexes retaining juvenal outer primaries as in other *Alectoris*.

General biology and ecology

According to Meinertzhagen (1954), who considered this to be a race of *graeca*, the birds are usually found above 4500 ft (1350 m), and occur up to as high as 9000 ft (2700 m), but in general occur slightly lower than and sometimes in the same areas of *melanocephala*. He said that they closely resemble this species in their habits, foods, and calls.

Social behaviour

No specific information is available.

Reproductive biology

Eggs are reportedly laid from the end of March onward, and the clutch-size is said to be quite small, of five to eight eggs (Meinertzhagen 1954). This would seem to be a surprisingly small clutch, and more detailed observations are needed. It is possible that this form breeds about a month earlier than does *melanocephala*, and natural hybrids are unknown, although they have been produced in captivity (Watson 1962a).

Evolutionary relationships

This isolated form was considered by Watson (1962b) to comprise part of the *chukar* superspecies and a very recent offshoot of *chukar* that was a late arrival to the Arabian Peninsula, perhaps being separated from *chukar* by desiccation of the northern and central Arabian desert after pluvial periods of the late Pleistocene.

BARBARY PARTRIDGE (Plate 44)

Alectoris barbara Bonnaterre 1790
Other vernacular names: perdrix gambra (French); Felsenhuhn (German); perdiz moruna (Spanish).

Distribution of species (see Map 16)
Resident of rocky habitats of North Africa from Morocco to north-western Egypt and on Sardinia, where possibly introduced. Also introduced on Canary Islands and Gibraltar (where very rare). Many unsuccessful introductions elsewhere (Long 1981).

Distribution of subspecies
A. b. koenigi: north-west Morocco and Canary Islands (Tenerife, Gomera, Lanzarote, and Fuerteventura).
  A. b. barbara: northern Morocco east to northern Algeria; also Sardinia. Includes *theresae*.
  A. b. spatzi: southern Morocco, central Algeria, and southern Tunisia. Includes *duprezi*.
  A. b. barbata: Libya east to north-western Egypt.

Measurements

Wing (*barbara*), males 162–171 mm, females 149–162 mm; tail, males 89–106 mm, females 75–91 mm (Cramp and Simmons 1980). Estimated weight, males 461 g, females 376 g. Egg, average 40.5 × 30.4 mm, estimated weight 20.4 g.

Identification (see Fig. 16)

In the field (13.5–15 in.)
Geographically separated from but parapatric with *rufa*, from which it can be readily separated visually by its white-spotted brown collar and mostly grey face and throat. Vocalizations are similar to those of the other *Alectoris* forms (see below) and probably are not useful for field recognition.

In the hand
Easily distinguished from all other *Alectoris* partridges by the chestnut-brown crown and broad neck-band, and by the grey face and throat.

Adult female. Apparently almost identical to the male, but slightly smaller, and probably with a smaller tarsal knob on average.

Immature. Distinctly yellowish throughout in the juvenal plumage, and similar to *rufa*, but more brownish on the sides and back of the neck. Like adult after attaining first-winter plumage, but the outermost two primaries narrower, more pointed, and more worn (Cramp and Simmons 1980).

General biology and ecology

Unlike the other *Alectoris* species, this one is found not only on rocky and arid hillsides but also extending into brushy and even wooded habitats, at altitudes ranging as high as 3300 m. Besides occurring in sandy dune areas and rocky hills, the birds are found in shrubby stands along dry river beds, in citrus groves, stands of *Thuja*, *Eucalyptus*, or *Euphorbia*, and in thorny *maquis* habitats. Like its habitats, its foods are apparently quite variable, but primarily consist of leaves, shoots, fruits, and seeds of a wide variety of grasses, forbs, and shrubs, with the leaves of succulent plants probably of importance in some arid regions (Cramp and Simmons 1980).

Social behaviour

Evidently similar behaviourally to the other *Alectoris* species, few studies have been performed on this one. The birds probably have monogamous pair-bonding, with a dispersal of pairs occurring during the breeding season and possible non-breeding occurring during very dry years. The advertising call of the territorial male is a harsh, grating, and deliberate *krrraiik*. This is apparently distinct from the rally call that is uttered by dispersed adults and is climaxed by a series of repeated *kutchuk* or *kekelik* notes that may be interspersed with louder squawking notes and are often preceded by softer *tic* sounds. The call's climax may be followed by a series of slower *tchouk . . . tchoukor* notes. Courtship display by the male consists of an uttering of the advertisement call while moving in a semicircular manner that very probably corresponds to the waltzing behaviour typical of other *Alectoris* species. Advertising calls may result in countercalling responses from neighbouring birds (Cramp and Simmons 1980).

Reproductive biology

Although little specific information is available, eggs have been found in Africa from the end of March to the end of May, and clutches of from 8 to 18 eggs have been recorded, with the normal range being about 10–14 (Etchécopar and Hüe 1957; Cramp and Simmons 1980). Like the other *Alectoris* species, the nest is located on the ground, usually under the shelter of a bush, and the incubation period is probably of comparable length, or around 25 days. Although no information exists with respect to replacement clutches or double-brooding among wild birds, double-clutching has been reported in captivity. The relative roles of the sexes as to providing parental care are also still to be determined (Cramp and Simmons 1980).

Evolutionary relationships

Watson (1962a) believed that the somewhat aberrant plumage characteristics of *barbara* might have been the result of introgressive hybridization of an early *chukar* form with *melanocephala*, perhaps during a Pleistocene meeting of these two gene pools in northern Arabia. Presumably this hybrid gene pool continued to move westward across northern Africa and eventually became stabilized there as *barbara*. Later, a European isolate of this population may in turn have evolved into the present-day *rufa* in Watson's view.

Status and conservation outlook

Although little specific data exists, there have apparently been population declines in Egypt, Morocco, and the Canary Islands (excepting Lanzarote), while the species' numbers are being maintained in Sardinia and Tunisia in spite of hunting pressures (Cramp and Simmons 1980).

RED-LEGGED PARTRIDGE (Plate 45)

Alectoris rufa (L.) 1758
Other vernacular names: French partridge; perdrix rouge (French); Rothuhn (German); perdiz roja (Spanish).

Distribution of species (see Map 16)
Resident of rocky habitats from France south to the Iberian Peninsula, Corsica, and east to north-western Italy. Introduced locally in England, the Azores, Gran Canaria, and Madeira; also temporarily elsewhere (Long 1981). Allopatric with *barbara*; both are probably part of the *chukar* species group and form a closely related species pair (Watson 1962a).

Distribution of subspecies (excluding introductions)

A. r. rufa: France to north-western Italy, Elba, and Corsica. Includes *corsa*.

A. r. hispanica: northern and western Iberian Peninsula.

A. r. intercedens: eastern and southern Spain.

Measurements

Wing (*rufa*), males 161–169 mm, females 152–161 mm; tail, males 87–97 mm, females 77–92 mm. Weight (England), males 480–547 g, females 391–514 g (Cramp and Simmons 1980). Egg (*rufa*), average 40 × 31 mm, estimated weight 21.2 g.

Identification (see Fig. 16)

In the field (13.5–15 in.)
Rather easily distinguished in the field from *graeca* (with which it is largely parapatric but locally sympatric in western Italy) by the more conspicuous rufous barring on the flanks, the discontinuous lower border of the black gorget, and the more tawny underpart coloration. Calls are probably not separable under field conditions from those of *graeca*.

In the hand

Separable from *barbara* by its white throat and forehead, and from the other black-gorgeted species of *Alectoris* by its white forehead and the disrupted lower edge of the black gorget. It is also more tawny below than are these latter species, and less greyish on the rump and upper tail-coverts.

Adult female. Like the male, but slightly smaller and with a duller head and throat colour. According to Pepin (1985), females can be distinguished from adult males by their smaller body size (total bill-to-tail length no more than 356 mm) and their lack of tarsal spurs.

Immature. Juveniles closely resemble those of *barbara* and are generally more yellowish throughout than juveniles of *chukar* and *graeca*. Similar to the adult in the first-winter plumage, but with the two outermost juvenal primaries retained; these being narrower, more pointed, and with tips sometimes slightly mottled with buff (Cramp and Simmons 1980).

General biology and ecology

Like the other *Alectoris* species, this one prefers dry and usually hilly land, varying from sandy to rocky substrates, and from level steppes to montane foothills. Areas with low or open vegetation, allowing unobstructed vision and running, and with sloping grades, seem to be particularly favoured. It is usually found at fairly low elevations, rarely occurring as high as 2000 m and frequently almost reaching sea-level. Foods consist primarily of seeds, leaves, and roots, and include a considerable number of grasses and legumes, especially in winter. In mountainous areas of Spain some altitudinal movements downward during the colder months have been recorded, but generally the birds seem to be quite sedentary (Cramp and Simmons 1980).

Social behaviour

Coveys or flocks comprise the typical social unit of this species for most of the year; assemblages of up to 70 are fairly common, and exceptionally groups of up to 200–300 have been recorded during cold weather. There is also some evidence that unisexual groups of males may be formed at times (Bump 1958). The usual mating system is one of long-term monogamy, but males sometimes acquire two mates, and females have also been known to associate with two males (Jenkins 1957). Vocalizations and display postures have been described by Goodwin (1953, 1954), and the major displays have been summarized earlier in Chapter 5. Vocalizations are quite similar to those of *graeca* (Menzdorf 1984), and include a rally call that varies but usually consists of a grating series of *chuk*, *chukar*, or similar-sounding notes that increase in loudness. The advertising or 'steam-engine' call is a similar series of *go-chak'*, *geg-geg'*, or *go-gog'* notes that increase in volume and sometimes terminate in a hoarse crowing (von Frisch 1962). The same or a very similar call is used as a territorial call and by birds in a thwarted or frustrated aggression situation, with fear the slightly dominating tendency (Goodwin 1953).

Reproductive biology

In Europe the breeding season is variable, extending from late April to May in England, during late April and early May in Portugal, and from May to mid-June in France. The male selects the nest site, and attracts the female to it by uttering a special call (Goodwin 1953). The eggs are laid at the rate of about two per three days, and the clutch-size in England averages 11.2–12.7 eggs (Cramp and Simmons 1980; Jenkins 1957). In England as well as on the continent there have been numerous instances of double-brooding reported in both wild and captive birds (Ricci 1983; Green 1984; Podor 1984). In these cases the second clutch is laid only a few days after the first, and each member of the pair tends a clutch separately, with the male normally incubating the first. In addition, the female typically lays replacement clutches after losing the first nest. The incubation period requires 23–24 days, and the chicks are initially capable of short flights at about 10 days. When double-brooding has occurred the young are cared for by a single parent, but otherwise are tended by both members of the pair.

Evolutionary relationships

Watson (1962a) regarded this species as a fairly late offshoot of early *barbara* stock that had crossed from Africa into south-eastern Europe and become isolated there. *Barbara* in turn was considered part of the *chukar* superspecies group by Watson. There is a limited amount of sympatry between *rufa* and *graeca* in western Italy, and natural hybridization has been reported there (Spano 1978; Bernard-Laurent 1984).

Status and conservation outlook

In England the introduced population is fairly healthy, numbering about 100 000–200 000 pairs, and perhaps at least locally is increasing. However, in its native range the species is much reduced in Italy, is possibly decreasing in France, and the situations in Spain, Portugal, and Corsica are unknown. The introduced island populaties (Canaries, Madeira, Azores) are generally small and except on Madeira are probably declining (Cramp and Simmons 1980).

16 · Genus *Ammoperdix* Gould 1851

The sand or see-see partridges are medium-sized and arid-adapted Palaearctic species having somewhat rounded tails of 12 rectrices that are about 50 per cent as long as the wing. The wing is rounded, with the eighth primary usually the longest. The post-juvenal primary moult is incomplete. The sexes are somewhat dimorphic, with males having white post-ocular patches and longitudinally striped sides and flanks; both sexes are otherwise strongly isabelline in general coloration except for chestnut outer rectrices. The claws are relatively short, tarsal spurs are absent, and no bare facial skin occurs in either sex. Two allopatric species are recognized:

A. [heyi] griseogularis Brandt 1843: see-see partridge

A. heyi Temminck: sand partridge
  A. h. heyi Temminck 1825
  A. h. intermedia Hartert 1917
  A. h. nicolli Hartert 1919
  A. h. chomleyi Ogilvie-Grant 1897

KEY TO THE SPECIES OF SAND PARTRIDGE (*AMMOPERDIX*)

A. Throat grey, or head with whitish superciliary stripe and pale spotting on sides of neck: see-see partridge (*griseogularis*)

AA. Throat reddish brown, or head lacking distinct superciliary stripe and pale spotting on sides of neck: sand partridge (*heyi*)

SEE-SEE PARTRIDGE (Plate 46)

Ammoperdix griseogularis Brandt 1843
Other vernacular names: Bonham's partridge; desert partridge; see-see; perdrix see-see (French); Persisches Wüstenhuhn; Sandhuhn; Sisi (German); pustynnaya kuropatka (Russian); perdiz gorgigris (Spanish).

Distribution of species (see Map 17)
Resident of arid and stony habitats from southeastern Turkey and Iraq east through Iran to Pakistan (Baluchistan and North West Frontier Province to the Salt Range and Kirthar Hills), and north to the USSR (south-eastern Uzbekistan and western Tadzhikistan). No subspecies accepted by Vaurie (1965).

Measurements
Wing, males 132–141 mm, females 126–134 mm; tail, males 51–61 mm, females 52–59 mm (Cramp and Simmons 1980). Weight, both sexes 182–205 g (Dementiev and Gladkov 1952). Egg (*griseogularis*), average 34.8 × 25.5 mm (Baker 1935), estimated weight 12.5 g.

Identification

In the field (9.5 in.)
Associated with desert and steppe areas, especially low, rocky hills with sparse growth. Larger than *Coturnix*, but smaller than *Alectoris*, and similar to the latter in exhibiting chestnut-coloured outer tail feathers in flight. Otherwise distinctly yellowish in appearance, and often heard uttering a distinctive *who-it* advertising call, which is endlessly repeated. Also utters a soft contact or alarm whistle, *sil-sil*, the basis of the common name. Not in contact with the next species, with which confusion might otherwise be most likely.

In the hand
Distinction from *heyi* is fairly easy in males, since *griseogularis* has a grey throat, a black forehead band, more brownish dorsal coloration, and a paler chestnut colour on the flanks and tail. Females are much more difficult to distinguish from *heyi*, but the mottling on the sides of the neck and lower head contains small white spots rather than pinkish buff bars.

Adult female. Differs from the male in having the upperparts and crown uniform darker isabelline, irregularly barred with faint rufous buff bars; the wings more coarsely marked, and the inner webs of the scapulars blotched with black; the forehead, rest of head, throat, sides, and foreneck whitish, marked and barred with dusky grey; and the rest of the underparts darker isabelline, clouded all over with dusky and vermiculated with blackish on the sides and flanks.

Immature. Juveniles are unusual in (at least in most cases) having male-like head markings rather than resembling the female. Resembles the adult after completing the post-juvenal moult, but with the

Map 17. Distribution of the see-see (Se) and sand (Sa) partridges.

outer two juvenal primaries retained, which are slightly shorter, narrower, and more pointed and frayed (Cramp and Simmons 1980).

General biology and ecology

This lowland and middle altitude-adapted species is generally found below 100 m on arid foothills and semi-desert to desert climatic situations, but occasionally has been reported as high as about 2000 m on plateaux. The favoured habitats are similar to those of *Alectoris*, consisting of rocky hillsides, rough and broken ground, and sandy or gravelly to loess or clay-substrate areas, which vary from level to hilly or dune-like in topography. Trees and well-vegetated areas are avoided, but sources of surface water such as springs or streams are frequented, and apparently regular access to such direct water supplies is essential. Sources of water are usually visited twice a day, and the hottest parts of the day are spent in shady areas. Foods consist of a mixture of plant and insect materials, including seeds and green parts of grasses in particular (Dementiev and Gladkov 1952; Cramp and Simmons 1980).

Social behaviour

This is a fairly gregarious species, usually occurring in pairs, families, or flocks of up to about 20 birds, or even more when groups collect near water. Family groups occur from summer onward, but there is some separation into pairs as early as the winter months (Baker 1930). From about April the birds are found in dispersed pairs, and the mating system is evidently monogamous, although the duration of the pair-bond and the extent of territorial spacing behaviour are still unknown. The male's advertising call is an endlessly repeated note sounding like *ooak* or *who-wit*, uttered at the rate of slightly more than one per second, and while perched on a low rock or mound, or at times on the roof of a ruined building. The distinctive squeaky *sil-sil* or *chil-chil* note is uttered under various circumstances, as for example when the birds are feeding, when surprised, or when alarmed (Cramp and Simmons 1980). Details of sexual display are lacking, but a lateral waltzing display, with one wing drooping, is known to be present (Ali and Ripley 1978).

Reproductive biology

Nesting has been observed to occur from the lowest foothills all the way up to about 7000 ft (2100 m), and the site is usually sheltered by a tuft of coarse grass or a boulder on exposed ground, or at times in a ravine in a protected crevice or under a bush (Baker 1930). In the USSR eggs have been reported from mid-April to July (Dementiev and Gladkov 1952), while in the Indian region the reported breeding months are April to June. The clutch is usually of five to fourteen eggs, most often six to nine. Dump-nests, containing up to 28 eggs dropped by several females, have also been observed. Incubation is by the female alone, although the male remains nearby, and is of uncertain length but at least 21 days (Dementiev and Gladkov 1952). Although some have believed that only the female cares for the chicks, others have reported that the male assists in brood care (Baker 1930; Cramp and Simmons 1980), which would seem more probable. There is no information on the development of the young, but broods eventually merge to form the autumn flocks.

Evolutionary relationships

The rather strong similarities in ecology, behaviour, and juvenal plumages that exist between *Alectoris* and *Ammoperdix*, as well as similarities in flank and tail coloration in adults, suggest that these two genera are very closely related. The plumages of *Ammoperdix* are more isabelline and less greyish throughout, and they blend into a stony, desert-like background even more effectively than do those of *Alectoris*, which genus seems in general to be adapted to a more upland and montane-like environment. In a recent phrenetic study (Crowe and Crowe 1985), *Ammoperdix* and *Alectoris* both were closely associated with *Francolinus*.

Status and conservation outlook

Although it has a generally very low population density, this species is quite widely distributed, and occurs in areas that are unlikely to be developed for agricultural or other purposes. It is possible that limited human activities which result in increased surface water availability may actually benefit the species.

SAND PARTRIDGE (Plate 47)

Ammoperdix heyi Temminck 1825
Other vernacular names: Arabian sand partridge; Arabian see-see; Hey's sand partridge; perdrix de Hey (French); Arabisches Wüstenhuhn; Ost-Sandhuhn (German); perdiz arenicola (Spanish).

Distribution of species (see Map 17)
Resident of arid and stony habitats from the Jordan Valley of Israel south to southern and south-eastern Saudi Arabia, and west to Egypt and Sudan east of the Nile.

Distribution of subspecies

A. h. heyi: Jordan Valley south to the Sinai Peninsula and north-western Saudi Arabia, where it intergrades with *intermedia*, and locally to Oman and Hadramaut.

A. h. intermedia: western Saudi Arabia, south of *heyi*, south to Aden and Muscat.

A. h. nicolli: Egypt east of the Nile, south to about 27° N latitude.

A. h. chomleyi: Egypt, south of *nicolli*, south to northern Sudan and east to the Red Sea.

Measurements

Wing (*heyi*), males 126–135 mm, females 123–132 mm; tail, males 56–65 mm, females 56–63 mm (Cramp and Simmons 1980). Average weight, both sexes 181 g (Pinshow *et al.* 1984). Egg (*heyi*), average 37 × 27 mm, estimated weight 13.8 g.

Identification

In the field (9.5 in.)

Associated with arid, often hilly and rocky or sandy habitats, where the species' overall sandy yellow coloration makes it difficult to see. In flight, the bright chestnut outer tail feathers are visible, but the birds often run rather than fly from danger. Their calls include repeated yelping-like *quay*, *kew*, or *proop* notes, uttered especially at dawn and dusk, and which sound something like two stones being struck together. Allopatric with the very similar *griseogularis*, so confusion in the field is unlikely, but sometimes found in the same areas as chukar partridges.

In the hand

Males are easily identified by the combination of having chestnut outer rectrices, linear flank stripes, and chestnut throat coloration. Females and immatures are extremely similar to those of *griseogularis*, but at least adult females tend to have a less definite superciliary stripe and lack distinct white spotting on the sides of the head, where instead they have short and pale pinkish buff barring.

Adult female. Like the male, but sandier and greyer, lacking a pale patch on the ear-coverts and chestnut on the throat, neck, and flanks. Bill a duller yellow to yellow horn.

Immature. Juveniles are female-like. Probably inseparable from adults after the post-juvenal moult, but possibly (as in *griseogularis*) retaining two outer juvenal primaries (Cramp and Simmons 1980).

General biology and ecology

Like the preceding species, this one is associated with desert environments that offer local sources of surface water. Typically the birds occur on fairly steep, rocky slopes having only sparse vegetation, where shady areas can be found during the middle of the day, but they also are sometimes seen on more sandy substrates. Their altitudinal distribution is from about 400 m below sea-level to about 2000 m above (Meinertzhagen 1954). They apparently eat a variety of seeds, berries, and insects (Etchécopar and Hüe 1967). In the summer, when their principal food consists of seeds, the birds probably get about 70 per cent of their water directly from surface sources. However, like chukars in the same area, they can survive for at least three days without water. Captives of these two species lost weight at about the same rate when deprived of water, although the sand partridges inhabit somewhat drier habitats than do the chukars (Pinshow *et al.* 1984).

Social behaviour

This is a little-studied topic, but the species is probably similar to the see-see partridge. Generally the birds are to be found in pairs, or sometimes occur in groups of up to as many as about 12 during late autumn and winter months. Like the see-see partridge, the birds prefer to run for cover among rocks when frightened rather than to fly, and when taking flight they remain very close to the ground, invariably heading downslope. Their calls include a variety of rapidly repeated notes that have a yelping or ringing quality, and which probably variously serve as contact, separation, and advertisement functions, but which still remain to be described in detail.

Reproductive biology

On the Arabian Peninsula nesting is said to occur in April (Meinertzhagen 1954), and in northern Africa laying in April has also been reported (Etchécopar and Hüe 1967). The clutch has been reported to range from four to seven eggs, and the nest is a scrape that is often placed under an overhanging rock or a bush. No information on incubation period or on the role of the sexes in parental care is available.

Evolutionary relationships

The two species of *Ammoperdix* form an obvious superspecies whose components are apparently allopatrically separated by the interior of the Arabian Peninsula.

Status and conservation outlook

No information exists on this, but the birds are little exploited by humans, owing to the inhospitable locations in which they occur.

17 · Genus *Francolinus* Stephens 1819 (including *Pternistis* Wagler 1832)

The francolins are a group of small to large species adapted to varied but primarily tropical unforested habitats, centred in sub-Saharan Africa but extending north to the Caspian Sea and east to central Asia and southern China. The tail is of 14 rectrices, and is about half as long as the wing. The wing is rounded, with the fifth to the seventh primaries the longest. In some but evidently not all species the post-juvenal primary moult is incomplete, and the tail moult is perdicine (centrifugal). In four African species ('*Pternistis*') the throat is bare of feathers and yellow to crimson in adults; in these and several other species the legs are also brightly coloured. The sexes are alike, similar or rarely dimorphic. Males of most species have singly or doubly spurred tarsi, and females are sometimes also spurred. Following the taxonomy of Hall (1963), 41 species are recognized here. A more recent review of the genus by Crowe and Crowe (1985) was received too late to incorporate their suggested taxonomic modifications, which apart from a shift of the ring-necked francolin from the striated group to the red-winged group, and associations of the Latham's and Nahan's forest francolins with the red-tailed and scaly groups, respectively, were mostly of minor sequential nature:

1. Spotted group
   F. francolinus (L.): black francolin
      F. f. francolinus (L.) 1766
      F. f. arabistanicus Zarudny and Härms 1913
      F. f. bogdanovi Zarudny 1906
      F. f. henrici Bonaparte 1856
      F. f. asiae Bonaparte 1856
      F. f. melanonotus Hume 1888
   F. pictus (Jardine and Selby): painted francolin
      F. p. pallidus (J. E. Gray) 1831
      F. p. pictus (Jardine and Selby) 1828
      F. p. watsoni Legge 1880
   F. pintadeanus (Scopoli): Chinese francolin
      F. p. phayrei (Blyth) 1843
      F. p. pintadeanus (Scopoli) 1786
2. Bare-throated group (= *Pternistis* Wagler 1832)
   F. afer (Müller): red-necked francolin
      F. a. harterti (Reichenow) 1909
      F. a. cranchii (Leach) 1818
      F. a. leucoparaeus Fischer and Reichenow 1884
      F. a. melanogaster (Neumann) 1898
      F. a. itigi (Bowen) 1930
      F. a. bohmi Reichenow 1885
      F. a. intercedens (Reichenow) 1909
      F. a. loangwae (Ogilvie-Grant and Praed) 1934
      F. a. benquellensis (Bocage) 1893
      F. a. punctulatus (J. E. Gray) 1834
      F. a. afer (P. L. S. Müller) 1776
      F. a. humboldtii Peters 1854
      F. a. swynnertoni (W. L. Sclater) 1921
      F. a. lehmanni (Roberts) 1931
      F. a. castaneiventer (Gunning and Roberts) 1911
      F. a. notatus (Roberts) 1924
   F. swainsoni (A. Smith): Swainson's francolin
      F. s. gilli (Roberts) 1932
      F. s. damarensis (Roberts) 1931
      F. s. lundazi White 1947 (= *chobiensis* Roberts 1935, preoccupied)
      F. s. swainsoni (A. Smith) 1836
   F. rufopictus (Reichenow) 1887: grey-breasted francolin
   F. leucoscepus G. R. Gray: yellow-necked francolin
      F. l. leucoscepus G. R. Gray 1867
      F. l. infuscatus (Cabanis) 1868
3. Montane group
   F. erckelii (Rüppell): Erckel's francolin
      F. e. erckelii (Rüppell) 1835
      F. e. pentoni Praed 1920
   F. ochropectus Dorst and Jouanin 1952: pale-bellied (Djibouti) francolin
   F. castaneicollis Salvadori: chestnut-naped francolin
      F. c. castaneicollis Salvadori 1888
      F. c. ogoensis Praed 1920
      F. c. kaffanus Ogilvie-Grant and Praed 1934
      F. c. atrifrons Conover 1930
   F. jacksoni Ogilvie-Grant: Jackson's francolin
      F. j. jacksoni Ogilvie-Grant 1891
      F. j. pollenorum Meinertzhagen 1936
   F. nobilis Reichenow: handsome francolin

F. n. nobilis Reichenow 1908
F. n. chapini Ogilvie-Grant and Praed 1934
F. camerunensis Alexander 1909: Cameroon Mountain francolin
F. swierstrai Roberts 1929: Swierstra's francolin
4. Scaly group
F. ahantensis Temminck: Ahanta francolin
F. a. hopkinsoni Bannerman 1934
F. a. ahantensis Temminck 1854
F. squamatus Cassin: scaly francolin
F. s. squamatus Cassin 1857
F. s. schuetti Cabanis 1881
F. s. maranensis Mearns 1910
F. s. usambarae Conover 1928
F. s. uzungwensis Bangs and Loveridge 1931
F. s. doni Benson 1939
F. griseostriatus Ogilvie-Grant 1890: grey-striped francolin
5. Vermiculated group
F. bicalcaratus (L.): double-spurred francolin
F. b. ayesha Hartert 1917
F. b. bicalcaratus (L.) 1766
F. b. thornei Ogilvie-Grant 1902
F. b. adamauae Neumann 1915
F. b. ogilviegranti Bannerman 1922
F. icterorhynchus Ogilvie-Grant: yellow-billed (Heuglin) francolin
F. i. icterorhynchus Ogilvie-Grant 1890
F. i. dybowskii Oustalet 1892
F. clappertoni Children: Clapperton's francolin
F. c. clappertoni Children 1826
F. c. heuglini Neumann 1907
F. c. gedgii Ogilvie-Grant 1891
F. c. sharpii Ogilvie-Grant 1892
F. c. nigrosquamatus Neumann 1902
F. hildebrandti Cabanus: Hildebrandt's francolin
F. h. altumi Fischer and Reichenow 1884
F. h. hildebrandti Cabanus 1878
F. h. fischeri Reichenow 1919
F. h. johnstoni Shelley 1894
F. natalensis Smith: Natal francolin
F. n. neavei Praed 1920
F. n. natalensis Smith 1834
F. hartlaubi Bocage: Hartlaub's francolin
F. h. hartlaubi Bocage 1869
F. h. bradfieldi Roberts 1928
F. h. crypticus Stresemann 1939
F. harwoodi Blundell and Lovat 1899: Harwood's francolin
F. adspersus Waterhouse 1838: red-billed francolin
F. capensis (Gmelin) 1789: Cape francolin
6. Striated group
F. sephaena Smith: crested francolin
F. s. spilogaster Salvadori 1888
F. s. somaliensis Ogilvie-Grant and Praed 1934
F. s. grantii Hartlaub 1865
F. s. rovuma G. R. Gray 1867
F. s. sephaena Smith 1836
F. s. zambesiae Praed 1920
F. streptophorus Ogilvie-Grant 1891: ring-necked francolin
7. Red-winged group
F. psilolaemus G. R. Gray: montane red-winged (moorland) francolin
F. p. psilolaemus G. R. Gray 1867
F. p. ellenbecki Erlanger 1905
F. p. elgonensis Ogilvie-Grant 1891
F. p. theresae Meinertzhagen 1936
F. shelleyi Ogilvie-Grant: Shelley's francolin
F. s. uluensis Ogilvie-Grant 1892
F. s. macarthuri Van Someren 1938
F. s. shelleyi Ogilvie-Grant 1890
F. s. whytei Neumann 1908
F. africanus Stephens 1819 [= *F. afer* (Latham)]: grey-winged francolin
F. levaillantoides Smith (= *F. garipensis* Smith 1843, in part): acacia (Orange River) francolin
F. l. gutturalis Rüppell 1835
F. l. archeri Sclater 1927
F. l. lorti Sharpe 1897
F. l. jugularis Büttikofer 1889
F. l. pallidior Neumann 1920
F. l. kalaharica Roberts 1932
F. l. levaillantoides Smith 1836
F. levaillantii Valenciennes: Levaillant's (red-winged) francolin
F. l. kikuyuensis Ogilvie-Grant 1897
F. l. crawshayi Ogilvie-Grant 1896
F. l. levaillantii Valenciennes 1825
F. finschi Bocage 1881: Finsch's francolin
8. Red-tailed group
F. coqui Smith: coqui francolin
F. c. spinetorum Bates 1928
F. c. buckleyi Ogilvie-Grant 1892
F. c. maharao Ogilvie-Grant 1892
F. c. angolensis Rothschild 1902
F. c. ruahdae Van Someren 1926
F. c. hubbardi Ogilvie-Grant 1895
F. c. thikae Ogilvie-Grant and Praed 1934
F. c. coqui Smith 1836
F. c. kasaicus White 1945
F. c. vernayi Roberts 1932
F. c. hoeschianus Stresemann 1937
F. [coqui] schlegelii Heuglin 1863: Schlegel's banded francolin
F. albogularis Hartlaub: white-throated francolin
F. a. albogularis Hartlaub 1854
F. a. buckleyi Ogilvie-Grant 1892
F. a. dewittei Chapin 1937
F. a. meinertzhageni White 1944
9. Not assigned to any group by Hall (1963)
F. lathami Hartlaub: Latham's forest francolin

F. l. lathami Hartlaub 1854
F. l. schubotzi Reichenow 1912
F. nahani Dubois 1905: Nahan's forest francolin
F. pondicerianus Gmelin: grey francolin
  F. p. mecranensis Zarudny and Härms 1786
  F. p. interpositus Hartert 1917
  F. p. pondicerianus Gmelin 1789
F. gularis (Temminck) 1815: swamp francolin

KEY TO THE SPECIES OF FRANCOLINS (*FRANCOLINUS*)

A. Non-African species (occurring from Cyprus east to southern China)
  B. Outer tail feathers not chestnut, flight feathers barred or spotted ('spotted' group)
    C. Whitish throat and cheeks separated by black malar stripe; no submarginal buff band on scapulars: Chinese francolin (*pintadeanus*)
    CC. Scapulars with buffy submarginal band, throat, and cheeks not as described
      D. Chestnut on sides and back of neck; throat black (males) or white (females): black francolin (*francolinus*)
      DD. No chestnut collar or patch; throat buffy yellow: painted francolin (*pictus*)
  BB. Outer tail feathers chestnut; flight feathers without bars or spots
    C. Throat buffy, fringed with black; flanks with narrow cross-bars: grey francolin (*pondicerianus*)
    CC. Throat chestnut; flanks with white stripes: swamp francolin (*gularis*)
AA. African species
  B. A conspicuous area of red to yellow skin on the throat; and a large bare eye-patch ('bare-throated' group)
    C. Throat skin red
      D. Legs black, underparts earth-brown: Swainson's francolin (*swainsoni*)
      DD. Legs red, underparts grey to black: red-necked francolin (*afer*)
    CC. Throat skin yellow to orange-pink, legs blackish
      D. Flight feathers barred to tips, chest greyish: grey-breasted francolin (*rufopictus*)
      DD. Flight feathers forming a clear buffy patch, chest streaked with buffy and brown: yellow-necked francolin (*leucoscepus*)
  BB. No bare skin present on the throat
    C. Almost entirely blackish or dark brown throughout body, with contrasting white spots on flanks and underparts (forest francolins)
      D. Legs red, no white shaft-streaks above, red skin around eye: Nahan's forest francolin (*nahani*)
      DD. Legs yellow, white shaft-streaks above, no red skin around eye: Latham's forest francolin (*lathami*)
    CC. Not blackish, contrastingly spotted with white on flanks and underparts
      D. Smaller species (wing under 150 mm) having ochre on sides of face, chestnut to pinkish outer tail feathers, yellow legs, and a variably yellow bill ('red-tailed' group)
        E. Wings tawny-chestnut, no barring on flanks or underparts; underparts with buffy ground colour: white-throated francolin (*albogularis*)
        EE. Wings greyer; barring present at least on the upper flanks; ground colour of underparts white
          F. Upperparts distinctly rufous; often with rufous extending to the flanks: Schlegel's banded francolin (*schlegelii*)
          FF. Upperparts browner and with more *Coturnix*-like markings; no rufous present on flanks: coqui francolin (*coqui*)
      DD. Mostly larger species (wings usually over 150 mm), not coloured as described
        E. Dorsal pattern *Coturnix*-like (white or buff bars and streaks on a darker background), under wing-coverts and primaries usually bright chestnut; legs yellow, bill mostly blackish ('red-winged' group)
          F. Dark barring on tips of primaries; breast chestnut to tawny with small black spots: montane red-winged (moorland) francolin (*psilolaemus*)
          FF. No barring on primary tips; breast feathers not tipped with black
            G. Primaries bright chestnut for most of their length; exposed culmen 30–33 mm
              H. Head and neck patterned with black, tawny, and white, chest bright brown: Levaillant's red-winged francolin (*levaillantii*)
              HH. Head and neck mostly grey and brown, chest grey to olive-brown: Finsch's francolin (*finschi*)
            GG. Primaries chestnut-coloured for no more than about half of their total length; exposed culmen 25–30 mm
              H. Primaries somewhat reddish only at their bases; white throat not bounded by a definite black necklace; finely barred below: grey-winged francolin (*africanus*)
              HH. Primaries reddish brown for about their basal half; white throat bounded with a narrow black necklace; not finely barred below
                I. Upper breast and sides with reddish brown blotches, becoming more buff-coloured below: acacia red-winged francolin (*levaillantoides*)
                II. Upper breast feathers with chestnut blotches, lower breast and abdomen feathers with coarse black barring: Shelley's red-winged francolin (*shelleyi*)
        EE. Dorsal plumage not quail-like; wings, bill, and leg colour variable
          F. A white superciliary stripe and rufous eye-stripe, a white throat edged with chestnut spots, a definite barred or striped breast-band above a pale abdomen, and dark upperparts with white shaft-stripes; generally smaller species (wing 120–170 mm) ('striated group')
            G. Legs yellow; bill yellow at base; black

flank stripes; breast barring extends around back of neck: ring-necked francolin (*streptophorus*)

GG. Legs red; bill all black; no flank stripes; breast and neck striped: crested francolin (*sephaena*)

FF. Plumage not as just described, nearly all (but one) larger species (wing usually over 160 mm)

G. With a 'scaly' breast pattern; upperparts brown with darker vermiculations and paler edges, throat whitish, rest of underparts brown to buffy, legs and bill mostly orange to reddish ('scaly' group)

H. Feathers of mantle chestnut, broadly edged with grey, wings and tail also bright chestnut: grey-striped francolin (*griseostriatus*)

HH. Feathers of mantle narrowly edged with grey; general pattern dull brownish, with no bright tones on wings or tail

I. Upperparts with some white patterning; underparts streaked with white: Ahanta francolin (*ahantensis*)

II. Upperparts with faint greyish patterning; underparts dull brownish: scaly francolin (*squamatus*)

GG. Lacking scaly breast pattern and other described features

H. Males with crown, lower back, primaries, and tail plain brown to chestnut-brown; females similar but with primaries, lower back, and tail vermiculated; generally quite large species (wing usually over 190 mm) ('montane' group)

I. With black forehead and lores, forming a dark eye-stripe

J. Feet and bill red

K. With a chestnut nape and no breast-band: chestnut-naped francolin (*castaneicollis*)

KK. With a brown nape and a black breast-band: Swierstra's francolin (*swierstrai*)

JJ. Feet yellow, bill black to yellowish

K. Brighter, with black bill and chestnut flank streaks: Erckel's francolin (*erckelii*)

KK. Duller, bill yellowish below and ochre flank streaks: pale-bellied (Djibouti) francolin (*ochropectus*)

II. Without black forehead and lores

J. With scarlet eyelid but no bare eye-patch: Jackson's francolin (*jacksoni*)

JJ. With a bare red eye-patch

K. Wing-coverts chestnut-maroon, head greyish blue: handsome francolin (*nobilis*)

KK. Wing-coverts dark brown, head brown: Cameroon Mountain francolin (*camerunensis*)

H. With uniformly brown or greyish brown head, back, wings, and tail, the body feathers vermiculated or marked with V- and U-patterning; lores blackish, and often with a white superciliary stripe ('vermiculated' group)

I. With a bare eye-ring but no white superciliary stripe

J. Eye-ring red, dark scalloping below: Harwood's francolin (*harwoodi*)

JJ. Eye-ring yellow, dark narrow barring below: red-billed francolin (*adspersus*)

II. Usually without a bare eye-ring; if so a pale superciliary stripe is also present

J. Underparts streaked with white; wing usually over 200 mm: Cape francolin (*capensis*)

JJ. Underparts spotted or unmarked; wing under 200 mm

K. Bill and legs not bright yellow

L. Bill and legs greenish, no eye-ring: double-spurred francolin (*bicalcaratus*)

LL. Legs red, bill red or partially red

M. With a red eye-ring, lower mandible red basally: Clapperton's francolin (*clappertoni*)

MM. No eye-ring, upper and lower mandible variably red

N. Culmen blackish, the underparts not finely patterned with blackish: Hildebrandt's francolin (*hildebrandti*)

NN. Bill all red, underparts intricately patterned with black: Natal francolin (*natalensis*)

KK. Bill largely or entirely yellow, legs also yellow

L. Wing 155–175 mm, neck and mantle with crescentic markings: yellow-billed francolin (*icterorhynchus*)

LL. Wing 135–150 mm, neck and mantle not with crescentic markings: Hartlaub's francolin (*hartlaubi*)

BLACK FRANCOLIN (Plate 48)

Francolinus francolinus (L.) 1766

Other vernacular names: black partridge; francolin noir (French); Halsbandfrankolin (German); turach (Russian); francolin de collar (Spanish).

Distribution of species (see Map 18)

Resident from Cyprus, Asia Minor, and the southern Caspian south-eastward through Iraq, Iran, and southern Afghanistan to Pakistan, and across India

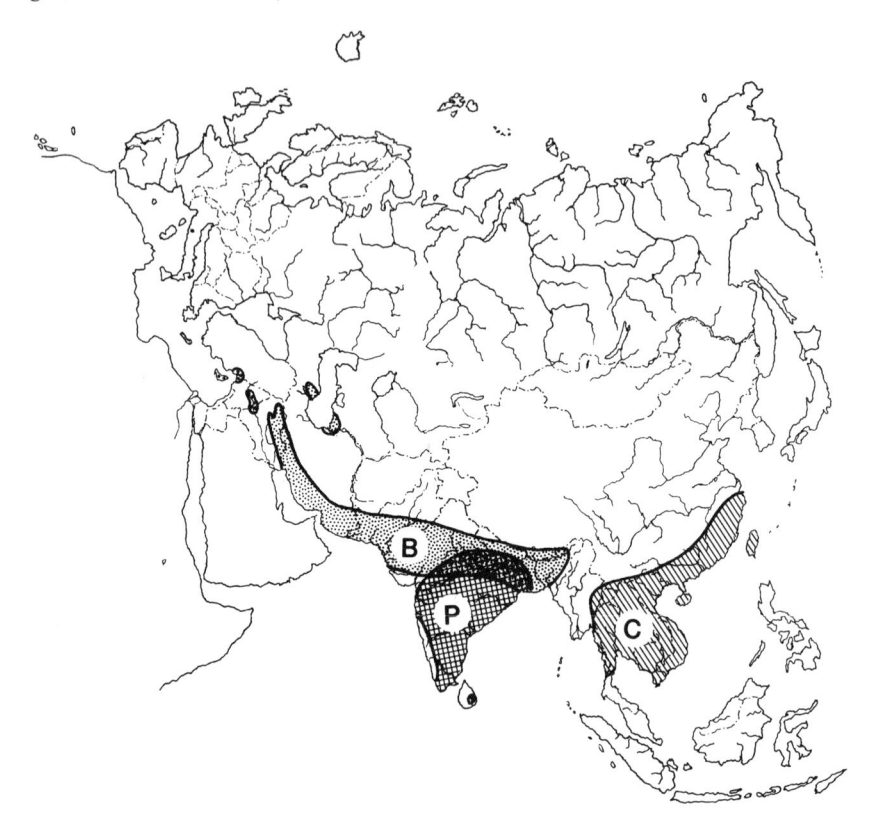

Map 18. Distribution of the spotted group of *Francolinus*, including the black (B), Chinese (C), and painted (P) francolins.

east to Assam and south to the Deccan plateau. Previously much more widespread in Europe, west to Spain. Introduced in south-eastern USA (Louisiana and southern Florida), Hawaii (Kauai, Molokai, Maui, and Hawaii), and Guam; locally re-established in central Italy, from where it had been extirpated.

Distribution of subspecies (excluding introductions)

F. f. francolinus: Cyprus, Asia Minor, and eastern Transcaucasia from the southern Caspian Sea south to Syria and Israel, and east to Iraq and Iran. Listed in the *USSR red data book* (Bannikov 1978). Includes *billypayni*. Previously west to Spain; now locally re-established in Italy (Tuscany).

F. f. arabistanicus: southern Iraq and western Iran.

F. f. bogdanovi: southern Iran and Afghanistan, east to Baluchistan in southern Pakistan.

F. f. henrici: Baluchistan east to Sind and Kutch.

F. f. asiae: western India except for Kutch, east to about western Nepal, western Bihar, and south to northern parts of Gujarat, Madjya Pradesh, and Orissa.

F. f. melanonotus: India east of *asiae* east to Sikkim, Assam, Bengal, and Manipur.

Measurements

Wing (*asiae*), males 145–168 mm, females 138–167 mm; tail, males 77–110 mm, females no data. Weight, males c. 283–566 g, females c. 227–482 g (Ali and Ripley 1978). Wing (*francolinus*), males 168–181 mm, females 160–172 mm. Weight (*francolinus*), males 425–550 g, females 400–450 g (Dementiev and Gladkov 1952); *asiae*, average of 19 males, 482 g, 18 females, 425 g (Bump and Bump 1964). Egg (*asiae*), average 37.8 × 31.3 mm (Baker 1935), estimated weight 20.4 g.

Identification

In the field (13 in.)

Associated with rather dense vegetation, often near water, and tending to move into heavy cover when disturbed. Males have an unmistakable combination of black underparts, a chestnut collar, and a white ear-patch. Females are much more nondescript, but have somewhat 'scaly' flank patterning, white throats, and a much less conspicuous chestnut-tinged nape pattern. The male's advertising call is a preliminary guttural note followed by a series of clicking or cicada-like notes that are primarily uttered in early morning and evening, sometimes sounding like 'be quick, pay your debts' or 'fixed bayonets, straight ahead', and uttered with the neck outstretched.

In the hand

Both sexes are best identified by the chestnut on the sides of the neck and lower hindneck, and by the bold flank patterning (black edged with white in males; buffy with black chevrons in females). Females are most similar to those of *pintadeanus*, which latter tend to have a rust-tinted superciliary stripe but no definite rust tones on the sides and back of the neck.

Adult female. Upperparts like the male, but the upper back resembling the middle back and wing-coverts, and the general tone of the lower back, rump, and upper tail-coverts browner. Sides of the face buff, dotted with black; the ear-coverts blackish brown; chin and throat white; the chestnut collar confined to the nape; front and sides of the throat and rest of underparts white mixed with buff, with black limited to Y-shaped bars that vary greatly in intensity.

Immature. Young males resemble adults after the post-juvenal moult. First-year birds only occasionally retain the outer two juvenal primaries (the trait possibly varying geographically), but tend to show slightly more white on feather-bases of cheeks and chin than do adults (Cramp and Simmons 1980).

General biology and ecology

This species occurs over a variety of generally fairly well-vegetated habitats that range to as high as 2500 m, and typically is found fairly near water, such as along canals or rivers, where fairly dense natural woody and tall grassy jungle vegetation occurs in close proximity to grain fields (such as millet or barley), sugar-cane fields, or vegetable plantings. Scrub-covered river bottoms and thickets of low trees such as tamarisk (*Tamarix*) or such salt-tolerant bushes as sagebrush (*Artemisia*) are also utilized in some areas. Foods include a very wide array of plant and animal materials, but probably mainly consist of seeds and insects, the latter eaten throughout the year but probably especially important during summer (Faruqi et al. 1960; Ali and Ripley 1978; Bump and Bump 1964).

Social behaviour

Evidently rather little flocking occurs in this species; the birds usually are to be found as singles, pairs, or trios, and at times as groups of up to five or so birds outside of the breeding season (Ali and Ripley 1978).

Bump and Bump (1964) considered the species to be non-flocking, but Muller (1966) reported that captive birds were gregarious when they were not breeding and formed territories during the breeding period. They are evidently monogamous. The length of the pair-bond is still unreported except for observations that apparently paired birds can be seen outside of the breeding season and that both sexes protect the young, suggesting a fairly permanent pair-bonding tendency. The male's advertising call is uttered at intervals of about 10–15 s from elevated locations throughout the spring months, and is most frequently heard during the early morning and evening hours. Different observers have described this call quite variably, but it seems to consist of a soft preliminary clicking sound that is immediately followed by a variably emphasized series of five to seven *cheek*, *kek*, or *zzzeee* notes (Cramp and Simmons 1980). This is mainly uttered during spring, but there is also a resurgence of calling in late summer (Bump and Bump 1964).

Reproductive biology

The breeding season extends collectively from March to October in the Indian region, with an April peak in the south, compared with June and July breeding in the central and western areas, and nesting about September (after the monsoon rains) in the drier areas of Bihar. In southern Turkey and northern Iraq nesting occurs from March to May, during the spring rains. It has been suggested but not established that two broods might be raised, at least at the southern end of its range (Baker 1930; Bump and Bump 1964). At the northern end of the range in the USSR eggs are laid over a much shorter period, from the second half of April to June (Dementiev and Gladkov 1952). The nest is usually well hidden in grasses, tamarisk, or scrub jungle, or at times may be placed in cultivated fields of grain or sugar-cane. The clutch in the USSR is evidently usually of 8–12 eggs, rarely to 14 or 15, but estimates from India are generally lower and six to eight seems to represent a normal clutch there (Baker 1930). The incubation period lasts 18–19 days, and is performed by the female alone.

Evolutionary relationships

Hall (1963) considered this form and *pictus* as comprising two separate species, inasmuch as hybrids in the narrow zone of geographic contact are scarce and adult plumages of the two forms have diverged appreciably, especially in males. These two populations are now largely parapatric, and Hall suggested that the ancestral *francolinus* evolved in Palestine and possibly Arabia, whereas the early *pictus* was probably isolated in southern India. A third member of this closely related species group is *pintadeanus*, which in Hall's opinion probably originated in southeast Asia. She believed that all three probably were derived from a common ancestral species.

Status and conservation outlook

At one time this species' European range extended west to Spain and the Balearic Islands (in part as a result of early introductions by the Moors and Saracens), as well as including Sardinia, Sicily, some of the Italian mainland, and parts of Greece and Lebanon. It is now gone from all these areas except for Italy, where it has recently been re-established in Tuscany (Cramp and Simmons 1980). In the USSR this species is at the northern edge of its range, and has been included in the list of rare and endangered species there (Bannikov 1978). In Cyprus it has also been threatened with extinction as a result of hunting, and is still quite rare there (Bell and Summers 1982). Its situation elsewhere within its native range is not known, but it has done well in various areas of introduction. These include Guam, where it is apparently still expanding its range, and Hawaii, where it now occurs on all the main islands (Long 1981). In the continental United States it is locally established in Louisiana (Craft 1966; Emfinger 1966), and is less well established in southern Florida (Genung and Lewis 1982).

PAINTED FRANCOLIN (Plate 49)

Francolinus pictus (Jardine and Selby) 1828
Other vernacular names: painted partridge; Burmese francolin (*phayrei*); che-kua (Chinese); francolin peint (French); Tropfenfrankolin (German).

Distribution of species (see Map 18)
Widespread resident in grassland and edge habitats of central and southern India from Gujarat, Uttar Pradesh, and Madjya Pradesh southward except in humid areas of Karnataka and Kerala; Sri Lanka.

Distribution of subspecies
F. p. pallidus: north-central India as indicated above, southward to about 20° N latitude, intergrading there with *pictus*.
 F. p. pictus: central and southern India north to about 20° N latitude.
 F. p. watsoni: eastern Sri Lanka (Uva Province).

Measurements

Wing (*pallidus*), males 140–149 mm, females 140–151 mm; tail, males 65–68 mm, females 64–69 mm. Weight, both sexes (*pictus*) c. 242–340 g (Ali and Ripley 1978). Egg (*pallidus*), average 35.7 × 29.5 mm (Baker 1935), estimated weight 17.1 g.

Identification

In the field (12.5 in.)
Associated with semi-moist grasslands with some scrub and cultivation, in generally drier habitats than the previous species, which is also more northerly in distribution, and the two only slightly overlap in range. Both sexes are distinctive in having a pale chestnut supercilium and face, and being heavily spotted with black and white on the chest and underparts. Usually skulks and runs into cover rather than flies. In flight the birds show black outer tail feathers (as in *francolinus*) and rufous-tinted wings. The advertising call is a series of repeated, high-pitched clicking notes very similar to those of *francolinus*.

In the hand
The distinctive tawny to chestnut tones of the face, with a paler superciliary stripe and throat, are distinctive, as is the heavily streaked breast pattern. The general pattern of *pondicerianus* is similar, but that species has a barred breast pattern and chestnut outer tail feathers.

Adult female. Resembles the male, but the chin and throat are usually paler and not spotted with black. There is also a greater tendency for barred patterning on the abdomen, and the white of the underparts and rump area tends to be replaced with buffy tones. The legs are duller than in males, and the bill is more brownish and paler at the base.

Immature. Like the female, but with black arrow-shaped marks on the pale buffy flanks and lower breast (Ali and Ripley 1978).

General biology and ecology

This is largely a tropical lowland and foothills species that locally (in Sri Lanka) reaches elevations of up to about 1200 m. A combination of undulating grasslands and cultivated lands that are bordered and interspersed with patches of scrub jungle and shrubs, and with well-vegetated waterways between the higher areas, provide the optimum habitat of this species, which is somewhat more xeric-adapted than *francolinus* (Ali and Ripley 1978). Baker (1930) stated that the species' favourite habitat consists of evergreen vegetation having rather thin grass about two or three feet high, and interspersed with bushes, and that it avoids extensive areas of uninterrutped grasslands. It also avoids heavy jungle, but instead likes very dry jungle, and short grass on broken and stony plateaux and plains where some trees for perching are available. The birds forage during morning and evening hours on grain, seeds, insects, green shoots, and the like—termites are a favourite food—and birds typically drink every evening.

Social behaviour

These reportedly are not gregarious birds, and are usually to be seen in pairs or small family groups of adults and up to four young. It is believed that males form strong, possibly permanent, pair-bonds. They are evidently quite territorial, and during the breeding season males utter advertisement calls from elevated sites such as tree stumps or earthen mounds every 20 s or so for extended periods of 15 min or more. These calls are a series of about five high-pitched notes quite similar to those of the black francolin, and have been reported as sounding like *chee-kee-kerray, chee-kee-kerray* (Baker 1930) or *click . . . cheek-cheek-keray* (Ali and Ripley 1978).

Reproductive biology

The nesting season is evidently closely tied to the rainy season in India, beginning with the start of the south-west monsoon rains in June and continuing until September in most areas. In Sri Lanka it has been reported to breed during March–June, with newly fledged young having been seen in May, and in central India nesting during the April–June period has also been reported. The nest is often placed in cropland fields or in a patch of grass and scrub jungle, usually under the shelter of a bush. The clutch-size ranges from four to eight eggs, with six probably being a normal number. Incubation is by the female, but is of undetermined length, and there is also no information on the role of the male in caring for the brood. The chicks are highly precocial, and able to fly for short distances when little larger than sparrows (Baker 1930; Ali and Ripley 1978).

Evolutionary relationships

Hall (1963) considers this form to be an incompletely genetically isolated relative of *francolinus* that was probably separated from it and from *pintadeanus* in Pleistocene times, a hypothesis that seems quite believable judging from available information.

Status and conservation outlook

Little specific information on the current status of this species is available for India, but in Sri Lanka the endemic race has been reported as becoming very rare as a result of uncontrolled hunting (Wikramanayake 1969; Henry 1971).

CHINESE FRANCOLIN (Plate 50)

Francolinus pintadeanus (Scopoli) 1786
Other vernacular names: Chinese partridge; perdrix de Chine; francolin perle (French); Perlhuhnfrankolin (German).

Distribution of species (see Map 18)

Resident in grassy and woodland habitats of Manipur, Burma (Arrakan, Pegu, and east and south to northern Tenasserim and Shan State), eastern Indochina from Tonkin to Cambodia, and southeastern China (Zhejiang to Guangdong, Guangxi, and Yunnan; also on Hainan Island).

Distribution of subspecies

F. p. phayrei: range as indicated above except for China.
  F. p. pintadeanus: China as indicated above.

Measurements

Wing, males 141–150 mm, females 138 mm; tail, males 78–87 mm, females 78 mm. Weight, males 347–388 g, females 310 g (Cheng *et al.* 1978). Egg (*phayrei*), average 35.3 × 28.7 mm (Ali and Ripley 1978), estimated weight 16.1 g.

Identification

In the field (12.5 in.)

Associated with low oak scrub, grassy openings in dipterocarp forests, and similar relatively dry habitats. Males have a strongly black and white spotted breast and flank pattern, a black and white face and throat pattern, and chestnut scapulars and superciliary stripes. Females show a duller chestnut superciliary stripe, and have a comparably subdued black and rufous buff to whitish head pattern, but have no chestnut on the wings. (See black francolin account for distinction from the very similar females of that species. Both of these species exhibit black outer tail feathers in flight, but are more prone to skulk than to fly.) The advertisement call is a five-noted sequence of notes sounding something like 'Do be quick papa'.

In the hand

The combination of a blackish tail, scapulars that are chestnut (males) or black, margined with brownish (females), a white throat, and a rufous superciliary stripe should serve to distinguish this species.

Females are similar to those of *francolinus*, but the barring is narrower and more sharply contrasting.

Adult female. Differs from the male in having the upper back, scapulars, and wing-coverts black, margined with brownish and with short transverse bars and spots of buff; the scapulars not conspicuously mixed with chestnut, although most of them have a chestnut stripe down the outer half of the shaft; the lower back, rump, and upper tail-coverts black, tipped with brownish buff and barred with pale buff; the sides of the head washed with rufous buff; and the underparts whitish buff, barred with black. The soft-parts are duller than in the male, and the tarsus usually lacks spurs.

Immature. Immature males resemble adult females, but the chest and breast feathers are partly ocellated, and the scapulars are mixed with chestnut.

General biology and ecology

Of the three species of francolins in the 'spotted' group, this one is adapted to the most xeric conditions. It is generally found in dry, open forest and scrub jungle, and is more often present in hilly country than on the low and flat plains. Heavy, humid forests are avoided, but the birds do occur where low trees and brushy habitats are found adjacent to small cultivated areas (Baker 1930). In Thailand it occurs up to about 5000 ft (1500 m), and extends from dry, deciduous forests of the plains through open hill-forests to areas on the mountains where the evergreen forests have been destroyed and replaced by *Eupatorium* and lalang grasses (Deignan 1945). Its foods are not well studied, but apparently consist of seeds, vegetative shoots, and insects.

Social behaviour

Apparently rather non-gregarious, these birds are typically found singly or in pairs, and reportedly never assemble in coveys. However, in favoured sites groups sometimes accumulate. During the breeding season the males are highly vocal, uttering their advertisement calls from elevated sites. They are also said to call at other times during the year, but certainly do so mainly during the spring and summer months. The harsh, grating calls are usually five-noted, and have been variously rendered as '*More beer hah*', '*Do-be-quick-papa*', and '*Come to the peak, ha-ha*'. This call clearly also functions as a territorial challenge, inasmuch as caged birds that utter the call will effectively attract wild males and thus facilitate their snaring, by placing the captive 'decoys' in the territories of these wild males.

Reproductive biology

Breeding in the India and Burma region occurs from March to September (Ali and Ripley 1978), and this general breeding period (coinciding with the wet season) probably applies to the rest of the range as well. In lower Burma the species has been reported breeding in March and April, and nesting has been observed as late as September and even October in the Chin and Kachin Hills. It has even been suggested (without apparent proof) that double-brooding might occur, the first clutch being laid at the onset of the rains and the second at their end. Nests are frequently placed on the ground in bamboo jungle, and less often in grassy areas or scrub jungle. The clutch-size varies from three to seven or rarely up to eight, but usually is of four or five (Baker 1930). There is no information on the incubation period, which is likely to be similar to the 18–19-day period reported for *francolinus*.

Evolutionary relationships

Hall (1963) regarded this species, *pictus*, and *francolinus* as comprising an essentially allopatric species group evolving from a common ancestral type, with *pintadeanus* the easternmost representative. It is interesting that this species, the most arid-adapted of the three, also has the most white patterning of the male's plumage, whereas the most mesic-adapted species, *francolinus*, likewise has the most blackish males. All seem to have remarkably similar male advertisement/territorial calls of about five or six grating notes.

Status and conservation outlook

No specific information is available, but like the two previous species this one seems to survive well near human habitations and associated cultivation on marginal lands, and so should not be of any special concern to conservationists. It has also become locally established in some extralimital areas through introductions, including Mauritius, Oman (vicinity of Muscat), and perhaps a few other areas (Long 1981).

RED-NECKED FRANCOLIN (Plate 51)

Francolinus afer (P. L. S. Müller) 1776
Other vernacular names: bare-throated francolin; Boehm's francolin (*bohmi*); Cranch's francolin (*cranchii*); Humboldt's francolin (*humboldtii*); red-necked spurfowl; Sclater's francolin; francolin à gorge rouge (French); Nachkehlfrankolin; Rotkehlfrankolin (German).

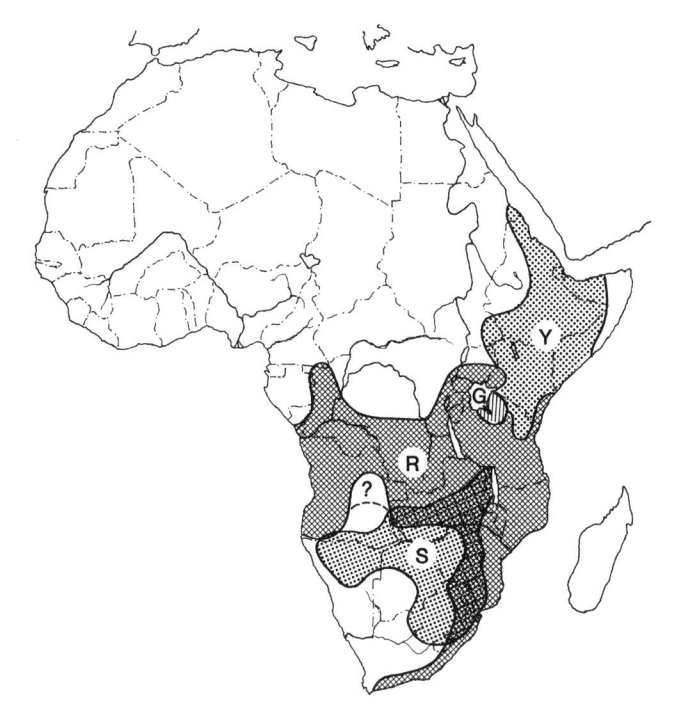

Map 19. Distribution of the bare-throated group of *Francolinus*, including the grey-breasted (G), red-necked (R), Swainson's (S), and yellow-necked (Y) francolins.

Distribution of species (see Map 19)

Resident in brushy habitats from Angola, Zaire, Uganda, and Kenya south to the Cape, excepting Namibia and eastern South Africa.

Distribution of subspecies

F. a. harterti: Ruzizi valley of Burundi.

F. a. cranchii: northern Angola, northern Zambia, Uganda, western Kenya, and western Tanzania. Includes *nyanzae*. Urban *et al.* (1986) also include *intercedens*.

F. a. intercedens: south-western Tanzania, south-eastern Zaire, and northern Zambia. Included in *cranchii* by Urban *et al.* (1986).

F. a. leucoparaeus: eastern Kenya, northern Tanzania.

F. a. loangwae: north-eastern Zambia, locally hybridizing and intergrading with *cranchii*.

F. a. melanogaster: eastern Tanzania, northern Mozambique. Urban *et al.* (1986) include *loangwae*.

F. a. itigi: central Tanzania; an unstable hybrid population (*intercedens* × *melanogaster*) approaching *melanogaster*, not accepted by Urban *et al.* (1986).

F. a. bohmi: western Tanzania; an unstable hybrid population (*intercedens* × *melanogaster*) approaching *intercedens*, not accepted by Urban *et al.* (1986).

F. a. humboldtii: Zambesi valley of western Mozambique: an unstable hybrid population involving *melanogaster* and *swynnertoni*, not accepted by Urban *et al.* (1986).

F. a. swynnertoni: Zimbabwe; southern Mozambique.

F. a. benquellensis: western Angola, included in *afer* by Urban *et al.* (1986).

F. a. punctulatus: central Angola, included in *afer* by Urban *et al.* (1986).

F. a. afer: southern Angola, northern Namibia (includes *cunenensis*).

F. a. castaneiventer: eastern and south-eastern Cape Province, southern Natal. Includes *krebsi*; Urban *et al.* (1986) also include *notatus* and *lehmanni*.

F. a. lehmanni: eastern Transvaal.

F. a. notatus: southern Cape Province.

Measurements

Wing, males 151–215 mm, females 147–188 mm; tail, males 90–110 mm, females 79–105 mm (Clancey 1967; Urban *et al.* 1986). Weight (various subspecies, with northern forms averaging smaller than southern ones), males 480–907 g, females 370–652 g (Newman 1983; Urban *et al.* 1986). Egg, average 40–45 × 30–35 mm, estimated weight 19–29 g.

Identification (see Fig. 17)

In the field (12–16 in.)

Associated with forest clearings, forest edges in low country, weedy roadsides, riverine woodlands, and other fairly heavy vegetation. Highly variable in appearance geographically, but always with red legs as well as red facial and throat skin, and with varying amounts of black and white streaking on the sides and flanks. The call is a harsh, repeated *choorr* that tends to fade at the end, and a repeated *kek* is uttered in flight. Sometimes found with *swainsoni*, and probably hybridizes occasionally with it, as in Zambia, but usually found in moister areas than is that species.

In the hand

The combination of a bare scarlet throat, red legs, variably white-streaked flanks, and underparts ranging in colour from black to grey, with varying amounts of white present, should identify adults of this extremely variable form. The feathers of the sides of the face are white, black and white, or black in various races; the sides, flanks, and underparts are correspondingly variable, ranging from heavily vermiculated, with sparse chestnut streaks (vermiculated/rufous-striped 'type'), to strongly patterned in black, white and grey (bland and white 'type'), heavily streaked below with chestnut (as in

Fig. 17. Adult plumage variation in the red-necked francolin, subspecies (A) *cranchii*, (B) *afer*, (C) *humboldtii*, and (D) *castaneiventer*.

castaneiventer), or even entirely black on the abdomen, with boldly streaked black and white flanks (as in *swynnertoni*).

Adult female. Differs from the male in having the tarsi unspurred and also sometimes in having broader dark shaft-streaks in the feathers.

Immature. Birds in first-basic plumage usually retain some feathers on the lores, chin, and throat. Some juvenal remiges are also retained in this plumage, which are paler than those of adults and are heavily marked with blackish vermiculations (Clancey 1967). The bill is dark brown, and the legs are yellow (Urban *et al.* 1986).

General biology and ecology

Habitats of this species are usually thickets in moister areas, but also include wooded gorges, evergreen forest fringes, rank grasses, *Brachystegia* woodlands, grassy plains with forest or woodland patches, and similar diverse types, especially where it is the only francolin species. Where it occurs sympatrically with the Swainson's or yellow-necked francolins it is likely to occur in moister areas, such as in dense cover near stream beds. In such areas it may associate with the Swainson's francolin, forming mixed groups that presumably compete to some degree. As with other francolins, plant materials predominate among its foods, especially tubers, bulbs, shoots, berries, roots, seeds, as well as grain or legume crops, supplemented by some invertebrate materials (Clancey 1967; Urban *et al.* 1986). There is no information on the mobility or movements of this species, which are probably rather limited except in areas of extensive dense screening cover.

Social behaviour

This is a monogamous, territorial species, with pair-bonds that probably in at least some areas (where breeding is nearly year-round) are likely to be permanent. Mated pairs tend to remain within their territorial areas outside the breeding season, which during that period are advertised and defended by the male. Calling by the male, a harsh, repeated crowing that tends to fade or trail toward the end of the sequence, occurs mainly in the very early morning hours and again in late afternoon. While calling, the male stands with his plumage sleeked, the neck stretched, the flank feathers fluffed, and the whitish cheek patches (in *melanogaster*) conspicuous. Male courtship (and probably aggressive display) is similar to that of *Gallus*, with the wings lowered to the ground while in a posture that tends to be more frontally oriented than in *Gallus*, and courtship-feeding of the female by the male has also been observed (Peek, cited in Urban *et al.* 1986).

Reproductive biology

The breeding season of this species occurs during the cool and dry season in Natal and the eastern Cape, in Zambia it occurs late in the rainy season or just after the rains, in interior Tanzania nearly throughout the year but peaking in December–January, and in Zaire peaking at the end of the rains or the start of the dry season. Generally breeding seems to be timed to occur so that the chicks hatch when there is dense cover available early in the dry season. From three to nine (usually four to seven) eggs are laid in a scrape under a shrub, rank grass, or herbage, the eggs being deposited on alternate days. Incubation is evidently by the female alone, and records from captivity indicate a 23-day incubation period. The young hatch simultaneously, and flying is possible in birds as young as 10 days. The birds are virtually fully grown at 3–4 months, and remain for some time with their adults in groups of four to six, perhaps until the onset of the next breeding season (Urban *et al.* 1986).

Evolutionary relationships

This is part of a superspecies with the following three species, and seems especially close to *rufopictus*, with which it is closely allopatric.

Status and conservation outlook

Because of its affinities for rather heavy cover, this species is less hunted than the other bare-throated francolins, and as such has probably suffered less. In some areas of Zimbabwe it has decreased, perhaps as a result of competition with the Swainson's francolin, which has colonized these areas (Urban *et al.* 1986).

SWAINSON'S FRANCOLIN (Plate 52)

Francolinus swainsonii (A. Smith) 1836
Other vernacular names: Swainson's spurfowl; francolin de Swainson (French); Swainsonfrankolin (German).

Distribution of species (see Map 19)

Resident of grasslands and brush from northern Namibia east to Botswana, Zambia, Transvaal, and Mozambique.

Distribution of subspecies

F. s. gilli: northern Namibia, northern Botswana, and western Zambia.

F. s. damarensis: endemic to northern interior Namibia.

F. s. lundazi: north-eastern Botswana, Zimbabwe, and western Mozambique.

F. s. swainsoni: southern Botswana, Transvaal, and southern Mozambique.

Measurements

Wing, males 194–207 mm, females 170–175 mm; tail, males 80–89 mm, females 66–75 mm (Clancey 1967). Weight, males 400–875 g, average (of 90) 706 g, females 340–750 g, average (of 100) 505 g (Newman 1983). Egg, average 45 × 35 mm, estimated weight 30.4 g.

Identification

In the field (14 in.)

Associated with habitats ranging from dry grasslands to bushveld, often but not always near water, and sometimes near cultivated lands. Similar to the last species, but with blackish legs and the body generally uniformly brownish, the sides and underparts lacking definite white or grey patterning; instead the lighter areas are more mottled and speckled with black. The advertising call is a repeated *gok* or *kwahli* that gradually becomes a more prolonged *kowarrr* and diminishes in volume.

In the hand
Most like *afer*, but with sides, flanks, and underparts that are earth-brown, the feathers with black edges and shaft-stripes, and with dark blackish legs and toes.

Adult female. Similar to the male, but most of the feathers of the abdomen resemble those of the breast and chest, and lack chestnut margins. No tarsal spurs.

Immature. Similar to the adult female, but retaining juvenal outer primaries, which are mottled basally with cinnamon (Clancey 1967). Younger birds have the chin and throat partially feathered, and broad blackish brown markings on the scapulars and innermost secondaries (Mackworth-Praed and Grant 1952–73). The legs are yellowish brown, and the bill is dark, with a yellow base (Urban *et al.* 1986).

General biology and ecology

Dense grasslands with nearby water and cultivated areas provide optimum habitat in South Africa, while in Zimbabwe and elsewhere the birds occur in riparian forests, woodland edges, thornbush with varying amounts of cover, bushveld, and similar partially wooded areas. Like the red-necked, this francolin is dependent on water for drinking, which is usually done twice a day. Roosting is in low trees and bushes, and sometimes the birds also perch in trees on wet mornings. They often flock with red-necked, red-billed, and Natal francolins, and probably feed on much the same materials, including a variety of plant foods as well as some animal materials such as arthropods and other invertebrates. In a Transvaal analysis agricultural crops (maize, wheat, beans) comprised 30 per cent of crop weights, followed by 14 per cent indigenous seeds, 14 per cent roots and corms, 7 per cent arthropods, and 2 per cent green leaves. During summer the arthropod proportion increased to 20 per cent, and agricultural materials predominated during winter and spring (Kruger, cited in Urban *et al.* 1986). No information is available on mobility and movements.

Social behaviour

This species is monogamous and territorial, usually occurring in pairs or small coveys of up to about eight birds, which probably represent family groups. Territorial calling, a deep, rasping croak that is repeated with diminishing volume and descending pitch, is performed by males and is sometimes responded to by females, with a much weaker call reminiscent of the cry of a human infant. When calling, the male stands upright and raises his bill, while moderately inflating his bright red throat skin. Calling is usually performed from a small promontory, especially around the time of roosting. Courtship displays include a lateral display with drooping wings, with the head held upright, as well as a more crouched lateral posturing, with the bill pointed downward and the tail depressed, as the mantle feathers are partially erected (van Niekirk 1983; Urban *et al.* 1986).

Reproductive biology

Breeding in South Africa usually occurs between December and May, peaking in February–March, but the species may breed in every month, possibly even twice within one year. In Zimbabwe the records extend from November to August, also peaking in February and March, and in Mozambique and Zambia the records are scattered from December to August. The nest is a well-hidden scrape, and the clutch-size is normally four to eight eggs, averaging 5.5. Incubation is probably by the female alone, and requires about 21 days. Little is known of the post-hatching development or survival of the chicks, but major predators include lizards, mongooses, snakes, baboons, and the ground hornbill (*Bucorvus leadbeateri*) (Urban *et al.* 1986).

Evolutionary relationships

Part of a superspecies with *afer*, *leucoscepus*, and *rufopictus*, this species is sympatric with *afer*, with which it has hybridized, and hybrids have also been reported with the Natal francolin, both cases occurring in Zimbabwe (Urban *et al.* 1986).

Status and conservation outlook

Although actively hunted in many areas, this species has locally increased with the spread of agriculture, and is generally common to abundant throughout its range.

GREY-BREASTED FRANCOLIN (Plate 53)

Francolinus rufopictus (Reichenow) 1887
Other vernacular names: grey-breasted spurfowl;

Reichenow's francolin; francolin à poitrine grise (French); Graubrustfrankolin (German).

Distribution of species (see Map 19)

Endemic resident of grasslands in western Tanzania south of Lake Victoria (Serengeti Plains and around Lake Eyasi) and on that lake's Ukerewe Island. No subspecies recognized.

Measurements

Wing, males 193–222 mm, females 180–190 mm; tail, males 90–102 mm, females 81–94 mm (Urban *et al.* 1986). Weight, males 779–964 g, average 848 g, females 439–666 g, average 588 g. Egg, average 44 × 37 mm (Mackworth-Praed and Grant 1952–73), estimated weight 33.2 g.

Identification

In the field (15 in.)

Associated with dry grasslands with scattered acacia trees in a very limited geographic area of northwestern Tanzania. Identified by the yellow bare chin and throat, and the orange area of skin around the eyes. The underparts and chest are mostly grey to white, with some black streaking in the upper chest. Vocalizations are much like those of the yellow-necked francolin, a loud, grating, and repeated *ka-waaaark* or *koarrrk*, descending terminally.

In the hand

Most like the following species in having yellowish throat skin, but with a greyish to white breast, more rufous on the flanks, and the primaries barred for their entire length.

Adult female. Like the male, but without a tarsal spur.

Immature. Like the adult, but the upperparts greyish black, with white central shaft-streaks and barring, and grey-buffy margins, and the underparts broadly barred with black and white (Urban *et al.* 1986).

General biology and ecology

Associated with the grassland–acacia woodland ecotone and with thickets of thornbush or other dense vegetation along waterways, this species occurs in areas of somewhat greater rainfall than does the following species, with which it is in peripheral contact and has been reported to hybridize in Serengeti National Park. It is apparently not sympatric with any of the other bare-throated francolins, but probably is in contact with Hildebrandt's and Shelley's francolins. It is essentially a vegetarian, feeding primarily on tubers of sedges (*Cyperus*), obtained by digging. These materials are supplemented by weed and grass seeds, and some insects, as well as fallen grain in cultivated fields. It is apparently not very mobile, moving from cover out into grasslands only in early morning and late afternoon hours (Urban *et al.* 1986).

Social behaviour

Monogamous and territorial, calling occurs throughout the year from elevated points during morning and evening hours. Calling is especially frequent after the onset of rains, suggesting a timing of breeding with the rains and associated male advertisement. Apparently the courtship displays are similar to those of the yellow-necked francolin, as are the advertisement calls, but no details are available. Generally the birds are found as singles, paired, small (presumably family) groups, and occasionally in larger assemblages at favoured foraging sites (Urban *et al.* 1986).

Reproductive biology

Breeding records extend from February to April, plus June and July, the breeding occurring late during the main rainy season, and in the subsequent dry period. As usual, the nest is a scrape hidden in tall grass, and the usual clutch is of four to five eggs. The incubation period is unreported, but both parents are said to tend the brood, forming family parties that remain intact until the start of the next breeding season (Urban *et al.* 1986).

Evolutionary relationships

The distribution of this species (parapatric with *afer*) suggests that it a recent offshoot of red-necked francolin stock, and it might almost be regarded as a subspecies of *afer*. However, it is also closely related to *leucoscepus*, and as noted above has hybridized with it.

Status and conservation outlook

This species is still common locally, but has suffered recently from the effects of cultivation and pastoralism (Urban *et al.* 1986).

YELLOW-NECKED FRANCOLIN (Plate 54)

Francolinus leucoscepus G. R. Gray 1867

Other vernacular names: Gray's francolin; Cabanis' francolin (*infuscatus*); yellow-necked spurfowl;

francolin à cou jaune (French); Gelbkehlfrankolin (German).

Distribution of species (see Map 19)
Resident of brush and forest edge habitats in Ethiopia, Somalia, Kenya, Uganda, and northern Tanzania. Introduced, apparently unsuccessfully, to the Hawaiian Islands between 1958 and 1961.

Distribution of subspecies
F. l. leucoscepus: eastern Ethiopia and northern Somalia.
F. l. infuscatus: north-eastern Uganda, northern Kenya, southern Somalia, and southern Ethiopia south to northern Tanzania. Includes *muhamedbenabdullah*. Not accepted by Urban *et al.* (1986).

Measurements

Wing, males 184–216 mm, females 170–216 mm; tail, males 85–110 mm, females 80–95 mm (Urban *et al.* 1986). Weight, males 615–896 g, average (of 173) 753 g; females 400–615 g, average (of 223) 545 g. Egg, average 45 × 34 mm (Mackworth-Praed and Grant 1952–73), estimated weight 28.7 g.

Identification

In the field (15 in.)
Associated with open bush country, woodlands, and forest margins of eastern Africa from eastern Tanzania north, where it is the only yellow-throated francolin. In flight it exhibits distinctive pale buffy 'windows' on the flight feathers, and the advertising call is a loud, grating, and repeated *graak* call, uttered especially at dawn and dusk.

In the hand
Most similar to the preceding and comparably yellow-throated species, but separable by its more buffy breast and abdomen, more reddish chin and facial skin, and the distinctive buffy margins of the remiges.

Adult female. Like the male, but smaller and lacking the tarsal spurs.

Immature. Similar to the adults, but with broad barring on the inner secondaries, and the persistent juvenal primaries blotched and mottled with buff to their tips. The throat and bare facial skin is paler yellow, and the legs are brown.

General biology and ecology

This rather widespread species is found in many habitats, but prefers light bushland plains with mixed annual and perennial grasses and with rainfall of 200–400 mm annually, and does occur in foothill areas up to about 2400 m and with rainfall up to at least 1500 mm, there occurring around forest edges and areas of cultivation. Evidently the species can survive well without any free water, presumably obtaining necessary fluids from its foods. Locally it is fairly common in woodlands having some cultivation and interspersed bush, but it disappears under increased human populations. The birds are highly sedentary, with single pairs normally occupying the same area for most of the year, and probably moving only a few hundred metres a day between roosting and foraging areas. They feed mainly on sedge (*Cyperus*) tubers, supplemented by seeds and other plant parts of various herbs and grasses, and some insects, especially termites. Following the start of the rains, the incidence of insects in the foods increase, and that of sedge tubers diminishes (Stronach 1966; Urban *et al.* 1986).

Social behaviour

These monogamous birds are usually in pairs or family groups, the pairs remaining in the territory through the breeding season, which is marked by loud calling from the male during early mornings and evenings. Males call from promontories or even from roosting trees, producing a repeated bisyllabic *kowarrk* much like that of the Swainson's francolin, which declines in pitch and volume terminally. Male displays include courtship-feeding, and assuming an upright posture with the neck arched slightly downward and the wings partially spread, exposing the buffy patches, in which posture the male runs around the female and turns repeatedly to face her (Urban *et al.* 1986).

Reproductive biology

In Kenya and northern Tanzania the breeding period extends throughout the year excepting February, with a peak from May to July, while farther north in Ethiopia, Eritrea, and Somalia the records are generally from January to June, plus November in Somalia. Typically the birds breed late in the rainy season, with the young hatching during the dry season, and particularly during the cooler dry season where rainfall is bimodal. The nest is a simple ground scrape, and the clutch-size is from three to eight eggs, typically five. The incubation period is 18–20 days, and the young are able to fly short distances when only about 2 weeks old. Spurs begin to appear on males at about 18 weeks, become sharp by a year, and continue to grow into the second year of life. The young remain with their parents until they are nearly fully grown, typically almost a year in areas of monomodal annual rainfall, but may break up after about 6 months in areas of bimodal rainfall patterns. The young are probably able to breed within a year, and perhaps earlier in areas of bimodal rainfall (Urban *et al.* 1986).

Evolutionary relationships

This is the northernmost of the members of the bare-throated francolin group, but has diverged surprisingly little from the southernmost forms except for leg and throat colour. The entire group is obviously a closely knit superspecies.

Status and conservation outlook

Still widespread and locally common, this species has disappeared over large areas of Kenya as human populations have increased and snaring is prevalent (Urban *et al.* 1986).

ERCKEL'S FRANCOLIN (Plate 55)

Francolinus erckelii (Rüppell) 1835
Other vernacular names: francolin d'Erckel (French); Erckelfrankolin (German).

Distribution of species (see Map 20)

Resident from Eritrean Ethiopia to northern Sudan. Introduced on Hawaiian Islands (main islands from Kauai eastward).

Distribution of subspecies

F. e. erckelii: northern Ethiopia (Eritrea).
 F. e. pentoni: northern Sudan (Red Sea Province).

Measurements

Wing, males 200–227 mm, females 167–194 mm; tail, males 120–142 mm, females 98–131 mm (Urban *et al.* 1986). Weight, males 1050–1590 g, females 1136 g (Urban *et al.* 1986). Egg, average 46 × 36.5 mm, estimated weight 33.8 g.

Identification

In the field (17 in.)
Associated with grasslands of the Ethiopian mountains and the coastal hills of Sudan, between 6000 and 10 000 ft, often also found in woodlands or scrub vegetation. The large size, plus the chestnut striping on the crown, neck, and underparts, should all provide for easy identification. The advertising call is a repeated *kri* or *erk*, which is initially harsh and creaking but changes to lower and softer *wa* or *kuk*

Map 20. Distribution of the montane group of *francolinus*, including Cameroon Mountain (C), chestnut-naped (Cn), Erckel's (E), handsome (H), Jackson's (J), pale-bellied (P), and Swierstra's (S) francolins.

notes toward the end; males usually call from a prominent point.

In the hand
The very large size (weight over 1 kg), the black forehead and eye-stripe, the entirely black bill, and the chestnut crown and flank streaks provide a distinctive combination of features for identification.

Adult female. Like the male, but with the upper tail-coverts, rectrices, and outer scapulars barred with buff and black, and lacking tarsal spurs (males usually have two, the upper pair longer).

Immature. Similar to the adult female, but with wider buff shaft-stripes on the outer scapulars and back, and dark barring on the outer webs of the flight feathers, the wing-coverts, rump, and tail similarly barred and streaked (Urban *et al.* 1986).

General biology and ecology

This species is generally found between 2000 and 3500 m in patches of scrub cover consisting mostly of forest remnants, but also occurs in thickets near the edge of forests, in woods or other dense cover along water courses, and extends in the higher mountains into the giant heath (*Erica arborea*) zone. Because of its affinity for heavy cover and steep, almost inaccessible slopes, it is rarely seen, but does enter fields early in the morning to glean fallen grain, and may sometimes be observed calling from a prominence. Roosting is done in trees, and the birds often go to drinking sites late in the afternoon prior to roosting. The birds are apparently highly sedentary. Foods include the usual assortment of plant materials, especially grass, herb and shrub seeds, grain, shoots and berries, and insects (Urban *et al.* 1986).

Social behaviour

Based on observations of captive birds, this is a monogamous species, with paired birds remaining constantly together. Males are known to courtship-feed their mates, but other displays are undescribed. Advertisement calls are similar to those of the other montane francolins, and a softer call is used during courtship-feeding (Urban *et al.* 1986).

Reproductive biology

Egg records in the Ethiopian highlands are for May and September–November, during the rainy months, and for northern Sudan for April and May, also during the rains. The clutch-size is of four to ten eggs, laid in a ground scrape. The incubation period is unreported, but both parents remain with the brood, forming family parties that probably remain intact until the following breeding season (Urban *et al.* 1986).

Evolutionary relationships

Part of a superspecies that includes the three following forms, this species is closely similar in appearance to the chestnut-naped francolin, with which it is closely allopatric.

Status and conservation outlook

Still locally common in Ethiopia, this species has apparently decreased as a result of widespread forest destruction over much of northern Ethiopia (Urban *et al.* 1986).

PALE-BELLIED (DJIBOUTI) FRANCOLIN
(Plate 63)

Francolinus ochropectus Dorst and Jouanin 1952
Other vernacular names: Dorst's francolin; ochre-breasted francolin; Tadjoura francolin; francolin des Somalis (French); Wacholderfrankolin (German).

Distribution of species (see Map 20)
Endemic to juniper forests of Djibouti. Rare species; no subspecies recognized. Listed as 'endangered' by Collar and Stuart (1985).

Measurements

Wing, males 210–212 mm, females 180 mm; tail, males 102–115 mm, females 71.5 mm (Dorst and Jouanin 1954). Estimated weight, males 940 g, females 940 g. Egg, no information.

Identification

In the field (12 in.)
Limited to the juniper woodlands of the Forêt du Day in the Goda Mountains of central Djibouti, where it is the only large francolin species. A prominent black stripe extends through the eye elmost to the nape, but the rest of the bird, and especially the underparts, appears quite pale. The male's advertising call is a loud *Erk-ka-ka...*, the notes becoming faster and softer toward the end, and ending in a chuckle.

In the hand
Similar to but smaller than the Erckel's francolin, with a similarly prominent black eye-stripe, but with yellow present on the lower bill and the chestnut tones replaced by ochre.

Adult female. Similar to the male, but lacking spurs (two present in males), with more rufous in the tail, and the upperparts somewhat vermiculated.

Immature. Like the female, but more barred than streaked with buff and grey (Urban *et al.* 1986).

General biology and ecology

This highly localized endemic form breeds in areas of dense, lush vegetation such as palms and ferns, in wadis of the Forêt du Day, an area of relict juniper (*Juniperus procera*) forest, the trees mostly 5–8 m tall, and the undercover mostly of *Buxus* and *Clutea*, as well as around acacia in more open areas. Where the parasitic fig (*Ficus*) is present the birds often feed on the figs along the forest edge, but retire to dense forest cover for roosting. Besides figs, the birds also feed on berries, grass, other seeds, and termites (Welch and Welch 1984b).

Social behaviour

Little is known so far of this species; it is elusive during the daylight hours, roosting above ground, and is most active for a short period around dawn. The call of the male is similar to that of the other montane francolins, and birds foraging in groups also utter conversational noises, with family parties of up to nine being recorded (Welch and Welch 1984b).

Reproductive biology

Laying evidently occurs between December and February, but no information is available on other aspects of reproduction (Urban *et al.* 1986).

Evolutionary relationships

Intermediate both morphologically and geographically between the Erckel's and the chestnut-naped francolins, this is apparently a valid if weakly differentiated species (Urban *et al.* 1986).

Status and conservation outlook

One of the rarest of the African francolins, the population may number (mid-1980s) no more than about 5000 birds, which are mostly confined to about 1400 ha. of forest habitat in the Forêt du Day, which was once about 30 times its present size. This forest was reduced in size by half between 1977 and 1983, and beyond these threats there are other disturbance problems caused by overgrazing. Some birds may also occur in the forest at Mabla, to the north-east, which is even smaller than the Forêt du Day (Welch and Welch 1984a,b).

CHESTNUT-NAPED FRANCOLIN (Plate 56)

Francolinus castaneicollis Salvadori 1888
Other vernacular names: francolin à cou roux (French; Braunnackenfrankolin (German).

Distribution of species (see Map 20)
Resident in woodlands from Ethiopia to Somalia, and south to the Kenya border.

Distribution of subspecies
F. c. castaneicollis: eastern Ethiopia. Includes *bottegi* and *gofanus*. Urban *et al.* (1986) also include the two following races.
    F. c. ogoensis: Somalia.
    F. c. kaffanus: western Ethiopia.
    F. c. atrifrons: southern Ethiopia.

Measurements

Wing, males 191–226 mm, females 169–203 mm; tail, males 123–143 mm, females 99–128 mm. Weight, males 915–1200 g, females 550–650 g (Urban *et al.* 1986). Egg, average 47 × 38 mm (Mackworth-Praed and Grant 1953–72), estimated weight 37.5 g.

Identification

In the field (12 in.)
Associated with heavy cover such as juniper, other evergreen forest, and bamboo, usually at moderate elevations. The legs and bill are red, and both sexes have a chestnut patch on the nape. The call is a loud and strident *kawar-kawar*, and a harsh *kek, kek, kek, kerak* has been heard in duet or chorus.

In the hand
Differs from the two preceding species in having a red bill and legs, with the upper spur no longer than the lower one. The large size (wing at least 170 mm), red feet and bill, rich chestnut on the nape and crown (bounded by a black forehead), mantle and wing-coverts, and underparts that are white or creamy streaked with chestnut, provide a distinctive combination of identifying traits. Similar to Jackson's francolin, but that species lacks a black forehead.

Adult female. Like the male, but smaller and unspurred (two spurs present in males).

Immature. Like the adult female, but with black and buff barring on the rump, tail-coverts, and rectrices. The bill is sepia, becoming dull red at the base.

General biology and ecology

This species occurs in forest, forest glade, and undergrowth vegetation at elevations of 2500–4000 m and above, as well as in tall heath (*Erica*), locally in burned *Erica* and grassland, and in arid juniper forests of Somalia at lower elevations (to only 1200 m). Optimum habitats are forests between 3100 and 3500 m with dense wet undergrowth of lilies (*Kniphofia*) and giant lobelias, where the undergrowth provides ready escape cover. Roosting is usually done in trees, but in *Erica* scrub the roosting may be on or close to the ground. Foraging may occur as late as midday, perhaps because it is not until then that the wet undergrowth is sufficiently dry. The birds are probably quite sedentary, and seldom move out into cultivated fields, although some feeding on fallen barley may occur. Foods are typically taken from the ground, and include various seeds, as well as insects such as termites (Urban *et al.* 1986).

Social behaviour

Little is known of the behaviour of this species, which is probably monogamous and territorial. The birds are usually found in small groups, presumably family parties, and singing often occurs in duet or chorus form. They generally call from dense cover, usually in the morning and evening hours, but at times throughout the day.

Reproductive biology

The breeding season in Ethiopia extends from October to March, with the drier months possibly preferred, while in Somalia breeding has been noted both during the wet (May) and dry (December) periods. The clutch-size is from five to six eggs, and the nest is a scrape hidden under vegetation. No details of breeding are known, but the young remain with their parents until grown, forming family parties of up to about eight birds. Most coveys average five to eight birds, suggesting good survival of the young (Urban et al. 1986).

Evolutionary relationships

Part of a superspecies with Erckel's, Jackson's, and handsome francolins, this form is especially similar to Jackson's, whose range is closely allopatric to the south, and the two must have been derived from a common ancestral type.

Status and conservation outlook

Generally common to abundant, this species is not threatened and is locally very abundant in Ethiopia (Urban et al. 1986).

JACKSON'S FRANCOLIN (Plate 57)

Francolinus jacksoni Ogilvie-Grant 1891
Other vernacular names: francolin de Jackson (French); Bambusfrankolin (German).

Distribution of species (see Map 20)
Endemic resident in montane forests of Kenya.

Distribution of subspecies
F. j. jacksoni: Aberdare mountains of Kenya. Includes gurae.
   F. j. pollenorum: Mt Kenya. Not recognized by Urban et al. (1986).

Measurements

Wing, males 203–234 mm, females 195–217 mm; tail, males 116–152 mm, females 111–121 mm. Weight, males 1164–1130 g, females no data (Urban et al. 1986). Egg, average 46 × 36 mm (Mackworth-Praed and Grant 1953–72), estimated weight 32.9 g.

Identification

In the field (13–15.5 in.)
This very large francolin is associated with bamboo, mixed, or montane forests of Kenya, and is recognized by its red bill and legs, and its strong chestnut cast, both dorsally and on the sides and underparts, which are grey, streaked with chestnut. Its call is a harsh, repeated *grrr*, commonly uttered at dusk. It is locally sympatric with the smaller montane red-winged (moorland) francolin, which lacks red on the bill and feet.

In the hand
Similar to western forms of *castaneicollis*, but more rufous on the upperparts, with little or no black on the forehead, and with chestnut and grey streaking replacing black and white streaking on the breast and underparts. The lower eyelid is also distinctively scarlet (bare eye-rings are lacking in most species of the montane group).

Adult female. Like the male, but smaller, duller, and unspurred (one or two present in males).

Immature. Similar to adults, but with a chestnut chest, black and white barring below, and with dark grey and buff barring on the inner secondaries, tail-coverts, rectrices, and remiges. The underparts are also generally paler.

General biology and ecology

This species occurs in a wide variety of forest habitats, as well as in bamboo and moorland thickets. It is especially characteristic of dense, shrubby growth rather than heavy forest, and the giant heath (*Erica*) moors of the Aberdare mountains are prime habitat. Although associated with such dense vegetation, the birds may emerge to feed on short grasses and associated insect food. Foraging is often done in the middle of the day, probably to avoid wet vegetation, but in fine weather activity may begin before sunrise. The birds are probably fairly sedentary, and consume a variety of grassy shoots and roots, as well as berries, bamboo seeds, and some invertebrates (Urban et al. 1986).

Social behaviour

These birds are monogamous, with pair-bonds possibly persisting throughout the year, and with pairs occupying their same home ranges or territories year after year. Their calls include a series of high-pitched, extremely loud cackles, similar to those of the scaly and Hildebrandt's francolins. The birds are particularly noisy during the morning and evening hours, as they are going to roost or descending from their roosts (Urban et al. 1986).

Reproductive biology

In the Aberdares and Mt Kenya breeding occurs in December and January, with a single record for August, during the driest time of the year; on the Mau Highlands the records extend from December to February, plus August and October, and in the highest mountain ranges breeding probably occurs only during the dry season. The nest and eggs are poorly described, but the clutch is of three or more eggs. As many as seven young have been seen in a brood, and the young remain with their parents for about 8 months, or until the next breeding season (Urban *et al.* 1986).

Evolutionary relationships

This is certainly a close relative of the chestnut-naped francolin, with which it is closely allopatric.

Status and conservation outlook

Reportedly this species is locally common to abundant, especially in national parks, but has suffered some loss of habitat in recent years (Urban *et al.* 1986).

HANDSOME FRANCOLIN (Plate 63)

Francolinus nobilis Reichenow 1908
Other vernacular names: Ruanda francolin; francolin nobil (French); Kiwufrankolin (German).

Distribution of species (see Map 20)
Resident in montane forests of eastern Zaire, Ruanda, Burundi, and western Uganda.

Distribution of subspecies
F. n. nobilis: eastern Zaire, Ruanda, Burundi, and western Uganda.
   F. n. chapini: Ruwenzori mountains of Zaire–Uganda border. Not recognized by Urban *et al.* (1986).

Measurements

Wing, males 206–211 mm, females 184–201 mm; tail, males 87–93 mm, females 78–85 mm (Chapin 1932). Weight, males 862–895 g, females 600–670 g (Urban *et al.* 1986). Egg, no information.

Identification

In the field (13–14 in.)
Associated with dense montane forests of western Uganda and Zaire north to Lake Albert, and with deep chestnut-maroon on the back and wing-coverts, a contrasting grey to bluish grey head and rump, and generally rich chestnut underparts. Rarely seen, but its call, a *chuck-a-rik*, *cock-rack*, or *kre* repeated some six to eight times, is often heard.

In the hand
Distinguished by its large size (wing over 180 mm), red bill, legs, and eye-ring, rich chestnut back and breast, grey head and rump, and heavily chestnut-streaked underparts.

Adult female. Like the male, but unspurred (males are usually doubly spurred, but the upper spurs are small and blunt), and also usually less red on the bill, face, and tarsus.

Immature. Much greyer in the centre of the chest and abdomen than adults, and the upperparts barred with dark grey and rufous buff.

General biology and ecology

Like the following species, this form is found in dense montane undergrowth from the lower parts of the montane forest up through the bamboo zone up to the alpine heath zone at about 3700 m. Over most of its range it is the only francolin species present. Roosting in trees or bushes, it comes out of cover to feed along road edges only in very early morning and late afternoon hours, and escapes by running into heavy cover (Chapin 1932; Urban *et al.* 1986).

Social behaviour

Found in pairs and small, presumably family, groups, this species is most apparent during morning and evening hours, when its loud crowing or squealing calls are often conspicuous. These perhaps are territorial calls, but nothing is known of the social behaviour patterns (Chapin 1932).

Reproductive biology

Breeding records for eastern Zaire are for late April–May to August–September. The nest and eggs are still undescribed (Urban *et al.* 1986).

Evolutionary relationships

This is presumably a fairly old offshoot from earlier stock that produced the other montane isolates that comprise this group, and seems to represent a very distinct taxon.

Status and conservation outlook

This elusive species is still common to locally abundant in parts of its range, although it is locally trapped by natives (Urban *et al.* 1986).

CAMEROON MOUNTAIN FRANCOLIN (Plate 58)

Francolinus camerunensis Alexander 1909
Other vernacular names: Cameroon francolin; francolin du Mont Cameroun (French); Kamerunbergwald-Frankolin; Kamerunfrankolin (German).

Distribution of species

Endemic resident on south-eastern slopes of Cameroon Mountain in montane forests and alpine heaths. No subspecies recognized. Listed as 'rare' by Collar and Stuart (1985).

Measurements

Wing, males 167–175 mm, females 157–169 mm; tail, males 79–85 mm, females 74–87 mm (Urban *et al.* 1986). Estimated weight, males 593 g, females 509 g. Egg, no information.

Identification

In the field (13 in.)
Limited to Cameroon Mountain, where it inhabits heavy forest up to the upper limit of forest growth and into the zone of alpine heaths, and is still almost totally unstudied. It and the scaly francolin (which lacks a large red facial skin patch) are the only francolins of the area that are rich dark brown to blackish above and almost plain-coloured below. It lacks strong underpart patterning except in females, which have some buffy to off-white markings. One of its calls is said to be a rather soft and high-pitched triple whistle.

In the hand
Similar to *nobilis*, but more dark brown above and with no chestnut or pale greyish patterning. Distinguished from the other montane francolins by its red bill, the rather large area of red to dull orange skin around the eye and extending back toward the ear, its uniformly dark brown wing-coverts, and its lack of chestnut tones throughout.

Adult female. Mottled and barred dorsally and ventrally with black, buff and whitish; black or dark brown with whitish U-shaped markings below. Also much less greyish and more brownish than the male, and lacking tarsal spurs (males are doubly spurred, with the upper spur short and blunt).

Immature. Similar to the adult female dorsally, but with barring of black and white from the chest to the abdomen (see plate). Flank feathers with subterminal black spots and whitish tips; bill, tarsus, and toes dusky red.

General biology and ecology

Little studied, this species occurs in the dense undergrowth of primary and secondary forests between 850 and 2100 m, apparently avoiding open areas such as montane grasslands. It eats berries, grass seeds, and insects, and is rarely seen, preferring to escape danger by running into cover (Urban *et al.* 1986).

Social behaviour

Other than occurring in pairs and small parties that presumably are family groupings, almost nothing is known of this species' behaviour. Its call is a high-pitched, musical triple whistle.

Reproductive biology

Nesting occurs from October to December, during the dry season, but nothing is known of the eggs or young of this rare species.

Evolutionary relationships

This dull-coloured francolin is not very similar in plumage to any of the other montane francolins, and has probably been isolated from all of them for a substantial time period. It is most similar in plumage to the scaly francolin, but these similarities are believed to be the result of superficial convergence.

Status and conservation outlook

Although locally common, the total range of this species is probably less than 200 km^2, and its status in some areas may have been adversely affected by forest destruction and excessive hunting (Urban *et al.* 1986).

SWIERSTRA'S FRANCOLIN (Plate 63)

Francolinus swierstrai Roberts 1929
Other vernacular names: francolin de Swierstra (French); Swierstrafrankolin (German).

Distribution of species (see Map 20)

Endemic resident in woodlands and forests of central Angola (Cuanza Sul to Huilla districts). No subspecies recognized.

Measurements

Wing, males 160–181 mm, females 167 mm; tail, males 90–105 mm, females no data (Urban *et al.* 1986). Estimated weight, males 600 g, females 560 g. Egg, no information.

Identification

In the field (13 in.)

The most southerly and easily recognized of the montane group, inhabiting thick clumps of forests in mountain gullies and cliffs. It has a distinctive white superciliary stripe above a dark eye-stripe, and a blackish breast-band around an otherwise white chest and underparts. The shrill, harsh calls are apparently much like those of the Jackson's francolin.

In the hand

The only francolin with a white breast that is interrupted by a black breast-band, and additionally has a distinctive white to buffy superciliary stripe directly above a black eye-stripe.

Adult female. Similar to the male, but unspurred (males have one or two spurs) and mainly white below, with irregular brown or black blotches or bars, which are concentrated on the upper breast to form a mottled band, and are sparse in the centre of the abdomen.

Immature. Similar to adult females, but with the mantle, wing-coverts, and innermost secondaries streaked, vermiculated, or barred with rusty red, the flanks and abdomen barred with black and white, and the throat and supercilium pale buff (Urban *et al.* 1986).

General biology and ecology

This montane-adapted species occurs in the undergrowth of evergreen forests and forest edges, as well as on rocky and grass-covered slopes of mountains, and in tall grass savannahs associated with gullies and mountain tops. Foraging is done in the undergrowth among fallen leaves; grasses, seeds of legumes, and insects are known to be among the foods consumed (Urban *et al.* 1986).

Social behaviour

Little information is available. Display in a male was observed during September, and the species is said to have a shrill, harsh cry, presumably a male advertisement call.

Reproductive biology

Little information, but breeding is believed to occur between May and July, based on reproductive condition of collected specimens. The eggs and downy young are still undescribed.

Evolutionary relationships

This is part of a superspecies that also includes *camerunensis*, which is isolated from it by the Congo Basin, and presumably the two are old isolates. They also may provide a phyletic link between the montane and scaly groups of francolins, based on plumage similarities of the females (Urban *et al.* 1986).

Status and conservation outlook

Currently the range of this form is limited to some relict patches of montane forests, such as on Mts Moco and Soque, and its habitat is being destroyed by deforestation. Probably it should be regarded as an endangered species; it was considered of 'indeterminate' status by Collar and Stuart (1985), and was listed in Appendix 2 of the 1972 Convention on International Trade in Endangered Species of Wild Fauna and Flora (CITES), of which Angola is a signatory nation (Urban *et al.* 1986).

AHANTA FRANCOLIN (Plate 59)

Francolinus ahantensis Temminck 1854
Other vernacular names: francolin d'Âhanta (French); Ahantafrankolin (German).

Distribution of species (see Map 21)

Resident in forests discontinuously from Gambia to Nigeria.

Distribution of subspecies

F. a. hopkinsoni: Gambia, Senegal, and Guinea-Bissau. Not accepted by Urban *et al.* (1986).

F. a. ahantensis: Guinea east to south-western Nigeria, discontinuously in the Ivory Coast.

Measurements

Wing, males 173–192 mm, females 167–172 mm (Bannerman 1963); tail, males 90–109 mm, females

Map 21. Distribution of the scaly group of *Francolinus*, including Ahanta (A), grey-striped (G), and scaly (S) francolins.

80–109 mm (Urban *et al.* 1986). Estimated weight, males 608 g, females 487 g. Egg, average 42 × 33 mm (Mackworth-Praed and Grant 1952–73), estimated weight 25.2 g.

Identification

In the field (14 in.)

Associated with forest habitats from Nigeria west to the Gambia, isolated from *squamatus* by the Niger River. The birds are generally dark and relatively unpatterned, with the reddish bill and legs the only colourful aspects. Calls are loud, repeated *kee-kee-keree* notes, similar to but more squealing than those of the scaly francolin. Antiphonal or simultaneous calling between males and females may also occur, the female's call with vibrant undertones.

In the hand

Upperparts closely similar to *squamatus*, but the feathers of the back of the neck are blacker and more distinctly edged with white on the sides only, and the underparts are dull brownish with distinctive whitish U-markings rather than being streaked with white.

Adult female. Like the male, but unspurred (males are doubly spurred, the lower spur longer), with more black on the innermost secondaries, and with buff spots on the wing-coverts.

Immature. Underparts ashy, with dull white streaks, and with some arrow-shaped black markings on the mantle, scapulars, and innermost secondaries.

General biology and ecology

This is a forest ecotone species, occurring along forest edges and clearings, in secondary growth, in dense cover near water and overgrown cultivation, and in tangled scrub occurring between gallery forests and farmed savannahs. The birds feed on a variety of seeds, beans, cassava, and large fruits, as well as insects including termites. They occur in small coveys or pairs, and presumably are fairly sedentary, although little is known of their ecology (Urban *et al.* 1986).

Social behaviour

Presumably territorial and monogamous, the apparent advertisement call is a loud and high-pitched series of about three squealing notes. Additionally the male may call in concert with a female, presumably its mate, the presumed female producing a simultaneous *ker-week* with the male's call, or replying antiphonally to his call (Urban *et al.* 1986).

Breeding biology

Breeding in the Senegambia region has been reported for January and September, and a breeding record for Ghana is for late December. Little is known of the breeding, but the nest is a scrape in heavy cover, and the clutch-size is said to be of four to six eggs, with up to 12 (probably the work of two females) recorded (Urban *et al.* 1986).

Evolutionary history

This species, *squamatus*, and *griseostriatus* form a superspecies of similarly appearing birds, isolated by the Niger and Congo rivers. The first two of these seem to be particularly closely related.

Status and conservation outlook

This form varies from rare and local (north-western populations) to common (eastern populations), according to Urban *et al.* (1986).

SCALY FRANCOLIN (Plate 60)

Francolinus squamatus Cassin 1857
Other vernacular names: Schuett's francolin (*schuetti*); Francolin écaille (French); Schuppenfrankolin (German).

Distribution of species (see Map 21)

Widespread resident of forest and brush from eastern Nigeria and northern Angola eastward across central Africa to central Ethiopia, southern Tanzania, and western Malawi.

Distribution of subspecies

F. s. squamatus: southern Nigeria to northern Zaire. The following races are all included in *squamatus* by Urban *et al.* (1986).

F. s. schuetti: north-eastern Angola to Uganda, western Kenya, and central Ethiopia. Includes *tetraoninus*, *zappeyi*, and *dowashanus*.

F. s. maranensis: southern Kenya to the Tanzania border. Includes *kapitensis*, *keniensis*, and *chyuluensis*.

F. s. usambarae: Usambara mountains of Tanzania.

F. s. uzungwensis: Uzungwe mountains of Tanzania.

F. d. doni: Vipya plateau of western Malawi.

Measurements

Wing, males 159–184 mm, females 147–182 mm; tail, males 81–115 mm, females 73–110 mm. Weight, males 372–565 g, females 377–515 g (Urban *et al.* 1986). Egg, average 41 × 31 mm (Mackworth-Praed and Grant 1952–73), estimated weight 21.7 g.

Identification

In the field (13.5 in.)

Associated with forest and thick bush habitats from Sudan to Zambia, and its loud, guttural, and repeated *ke-rak*, *hu-hu-hu-hurrrr*, or *kew-koo-wah* calls often indicate its presence. It is the most uniformly dark-coloured of all francolins, but has contrasting bright red leg and bill coloration.

In the hand

This plain but distinctive species is almost entirely uniformly dull brown, with faint greyish scalloped markings on the upperparts, and with a dingy grey throat. The underparts may be rather uniformly dull brown or irregularly streaked with whitish.

Adult female. Like the male, but unspurred or with a single rudimentary spur (males are doubly spurred, the lower longer), smaller, and tending to have more vermiculations.

Immature. With black markings and buff spots on the mantle, wing-coverts, and secondaries, and with black and white barring on the underparts.

General biology and ecology

Tropical evergreen forests, with a dense undergrowth about 3 m high, and moderate elevations of from 800 to 3000 m represent this species' preferred habitat. It also occurs in forest clearings, in the dense undergrowth of abandoned plantations or other secondary growth, and in scattered remnants of forests. It is a secretive species, rarely seen and usually seeking refuge in trees or running into cover when threatened. It also roosts in heavy tree cover, and is evidently highly sedentary. Its foods are the usual assortment of fruits, seeds, cultivated crop plants (cassava, sweet potatoes, peanuts, rice), and a variety of insects and other invertebrates (Urban *et al.* 1986).

Social behaviour

Probably monogamous and presumably territorial, inasmuch as the birds have a loud chorus of advertisement calls, typically uttered at dusk as the birds settle into roosting areas, and again at dawn, often from a promontory. The typical call is a high-pitched and nasal *ke-rak* (see variants above), which is repeated up to 12 times with increasing volume.

Reproductive biology

Over much of eastern Africa breeding occurs throughout the year, with little or no seasonality evident, but breeding has only been recorded during August in Malawi. In Zaire breeding has been noted between January and June, in southern Cameroon during the dry season from October to December, and in Gabon from June to August. The nest is a scrape usually hidden under vegetation, and the clutch is of three to eight eggs, typically six. The incubation is probably by the female alone, but is of unknown length. Nothing is known of the development of the young, but mongooses and genets are probably important predators (Urban *et al.* 1986).

Evolutionary relationships

This species occurs geographically and perhaps phyletically between the Ahanta and grey-striped francolins, with both of which it forms a superspecies (Urban *et al.* 1986).

Status and conservation outlook

Generally locally common to abundant, this forest-adapted species is likely to suffer from the effects of deforestation, although it can survive in forest remnants and in secondary growth. Its reclusive habits probably remove it from most direct hunting pressures.

GREY-STRIPED FRANCOLIN (Plate 61)

Francolinus griseostriatus Ogilvie-Grant 1890
Other vernacular names: francolin à banades grises (French); Graustreifenfrankolin (German).

Distribution of species (see Map 21)

Endemic resident in forests of northern and western Angola, including southern Cuanz and western Milanje, and a second population in southern Benguela and extreme north-eastern Huila districts. Not recorded since 1954. No subspecies recognized.

Measurements

Wing, males 139–161 mm, females 144–153 mm; tail, males 87–102 mm, females 85–96 mm (Urban *et al.* 1986). Estimated weight, males 430 g, females 390 g. Egg, no information.

Identification

In the field (13 in.)

In its very limited Angolan range this brush-adapted species occurs in remnant patches of forests in the western escarpment. It differs from its near relatives in having grey-edged chestnut coloration throughout the anterior upperparts, and shows dark rufous to tawny wing coloration in flight. Both the bill and feet are bright reddish. Its calls are still poorly known, but include a high-pitched, rasping *kerak*.

In the hand

Differs from all the preceding species of this group in having the breast, mantle, and wing-coverts chestnut, broadly edged with grey. The underparts are buffy, with wide dull chestnut stripes, and the tail and remiges are also chestnut to dark rufous brown, narrowly marked with black.

Adult female. Like the male, but unspurred or with only a rudimentary spur (males have single spurs of more than 5 mm in length).

Immature. Similar to adults, but with narrower markings on the more whitish underparts, the upperparts richer cinnamon, and breast less chestnut, and the feathers up the upperparts with blackish rather than chestnut centres.

General biology and ecology

This species is limited to the few areas of dense undergrowth associated with gallery and secondary forests between 800 and 1200 m. From these it moves out into grassy areas and abandoned cotton fields to forage in morning and afternoon, flying back to the forest when disturbed. Foods include green shoots, seeds, insects, and other arthropods. Roosting is done in tree cover. Nothing is known of its mobility or other aspects of ecology (Urban *et al.* 1986).

Social behaviour

Almost totally unknown, but presumably monogamous and territorial, as other forest-adapted francolins.

Reproductive biology

No information.

Evolutionary relationships

Part of a superspecies with the Ahanta and scaly francolins, this is the southernmost. It is isolated geographically by the Congo River from *squamatus*, its probable nearest relative.

Status and conservation outlook

This perhaps is an endangered species, as it has not been reported for more than 30 years, and may be threatened by forest destruction over its very small known range (Urban *et al.* 1986).

DOUBLE-SPURRED FRANCOLIN (Plate 62)

Francolinus bicalcaratus
Other vernacular names: common bushfowl; francolin à double éperon (French); Dopplespornfrankolin (German); francolin biespolada (Spanish).

Distribution of species (see Map 22)

Resident of varied open-country habitats from western Morocco and Senegal south to Cameroon, over most of the Niger Basin.

Distribution of subspecies

F. b. ayesha: western Morocco. This is the only non-nominate race accepted by Urban *et al.* (1986).
 F. b. bicalcaratus: Senegal to Niger and northern Nigeria.
 F. b. thornei: Sierra Leone to Benin.
 F. b. adamauae: northern Nigeria and Cameroon.
 F. b. ogilviegranti: Cameroon.

Measurements

Wing (*bicalcaratus*), males 170–185 mm, females 161–178 mm; tail, males 65–75 mm, females 56–65 mm (Cramp and Simmons 1980). Weight, five

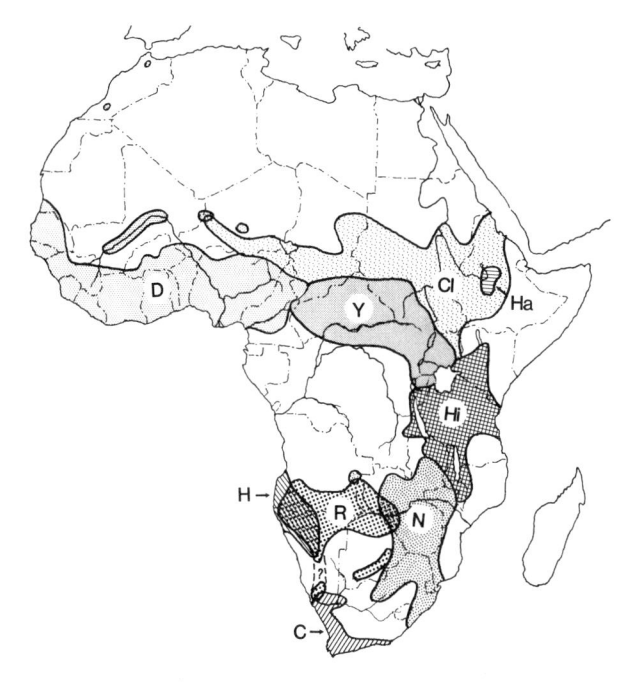

Map 22. Distribution of the vermiculated group of *Francolinus*, including Cape (C), Clapperton's (Cl), doubled-spurred (D), Hartlaub's (H), Harwood's (Ha), Hildebrandt's (Hi), Natal (N), red-billed (R), and yellow-billed (Y) francolins.

males from Senegambia, 507 g, three females 381 g (Urban *et al.* 1986). Egg, average 42 × 33 mm, estimated weight 25.2 g.

Identification

In the field (14 in.)

One of the most widespread and abundant francolins in western Africa north of the Equator, in habitats ranging from fallow fields to open woodlands, and often in areas of cultivated or recently abandoned lands. Variable in appearance, but with a conspicuous black-edged white superciliary stripe, a variably bright rufous-tinged neck, and cream-coloured underparts that are streaked with black and bright rufous. The male advertising call is a loud, harsh, and repeated *ke-rak* or *kor-ker*, usually uttered at the rate of about two per second; it is especially noisy at dawn and dusk.

In the hand

The combination of a black-edged white superciliary stripe, a chestnut-tinged hindneck and rear crown, becoming black on the forehead, and dull, greenish legs and bill colour serve to identify this species. The black-edged 'windows' in the middle of the large side and flank feathers are also distinctive.

Adult female. Like the male, but unspurred with a rudimentary spur (males are doubly spurred, the lower spur longer), and usually showing more chestnut on the breast feathers.

Immature. Similar to adults, but duller, more marked with black on the flanks and underparts, and with more pointed primaries (Cramp and Simmons 1980). The flight feathers are most distinctly barred with buff, while the upperparts have indistinct buff barring and streaking. The legs of young males are yellow, with only a single spur (Urban *et al.* 1986).

General biology and ecology

The wide variety of habitats used by this species include nearly all habitats except very thick bush and forest, ranging from natural ones such as fairly moist savannahs, woodlands, arid savannahs, and *Cistus* heaths to derived habitats such as dryland farmlands, shrubby pastures, former cultivations, and similar ones influenced by humans. Often the birds forage on grains or other cultivated crops, generally feeding in early mornings and late afternoons, seeking shady areas around midday, and proceeding at dusk to roosting trees after drinking. Their foraging is opportunistic and predominantly (about 80 per cent) from plant materials, with the remainder comprising invertebrates (especially insects such as termites), grasshoppers, and the like. Little is known of their mobility, but they are probably quite sedentary, usually flying less than 200 m when flushed, and preferring to escape by running into cover (Cramp and Simmons 1980; Urban *et al.* 1986).

Social behaviour

This species may be more gregarious than most francolins, and has been observed in flocks of up to about 40 birds, although group sizes are usually of 12 or less, probably representing a few family groups. The dominant males may lead such groups as the birds move about, and these groups are likely to break up well before the breeding season, as territorial limits are established. Male advertising calls are uttered from small promontories such as termite hills, and are most common around dawn and dusk. There is apparently a good deal of individual variation in such advertising calls, which probably facilitates easy recognition. The birds occur in apparently monogamous pairs during the breeding season, which is quite restricted in some areas but may occur almost year-around in others (Cramp and Simmons 1980; Urban *et al.* 1986).

Reproductive biology

Breeding periods are highly variable, ranging from any time during the year in the southern Sahel (probably opportunistically) to during the rainy season in waterless habitats of the Senegambia, and peaking during the dry season in rather wet climatic areas such as Nigeria and the southern Cameroon. As usual, the nest is a simple scrape under vegetation, and the clutch ranges from five to seven eggs, averaging 5.5. The incubation period is unknown, but an average of about four chicks was raised in one study (Urban et al. 1986).

Evolutionary relationships

Part of a superspecies that includes the next two species and probably also *harwoodi* (Urban et al. 1986), this one is probably most similar to *clappertoni*, with which it is essentially parapatric.

Status and conservation outlook

Except for the small disjunctive population of *ayesha* in north-western Morocco, which is rare if not threatened, the general status of this species is fairly good, with the birds generally common to very abundant, although they have locally been reduced by hunting effects (Urban et al. 1986).

YELLOW-BILLED (HEUGLIN'S) FRANCOLIN (Plate 64)

Francolinus icterorhynchus Ogilvie-Grant 1890
Other vernacular names: francolin à bec jaune (French); Heuglinfrankolin (German).

Distribution of species (see Map 22)
Resident of grassland habitats from Central African Republic to Zaire, and east to Sudan and Uganda.

Distribution of subspecies
F. i. icterorhynchus: Central African Republic to south-western Sudan. Includes *grisescens*.
  F. i. dybowskii: north-eastern Zaire, western and central Uganda. Includes *ugandensis*. Not recognized by Urban et al. 1986.

Measurements

Wing, males 158–181 mm, females 145–171 mm; tail, males 75–93 mm, females 73–85 mm. Weight, males 504–588 g, average (of seven) 571 g, females 420–462 g (Urban et al. 1986). Egg, average 40 × 30 mm (Mackworth-Praed and Grant 1953–72), estimated weight 19.9 g.

Identification

In the field (11–12 in.)
Associated with savannah woodlands and open grassy country with scattered trees and anthills. Similar in appearance to the double-spurred francolin, but parapatric with it, it has more yellow to orange-coloured legs and bill, it is more dusky brownish (less greyish) on the back, lacks chestnut tinges on the neck and nape, and has little or no chestnut on the flanks. Both species have similar conspicuous white superciliary stripes. The male's advertising call is a harsh, slowly repeated *kerak-kerak-kek* or *k-rack-krack-k*, usually uttered in morning and evening, or after rain showers.

In the hand
Similar to *bicalcaratus*, but more vermiculated and less V-patterned above, and blotched or spotted with dark brown or blackish below, and with dusky yellowish to orange rather than greenish bill and leg colour. There is also an area of bare yellowish skin behind the eye that is lacking in *bicalcaratus*, and the superciliary stripe is not distinctly outlined above and below with black.

Adult female. Like the male, but usually unspurred (males are doubly spurred); one or two rudimentary spurs may be present.

Immature. Similar to the adult, but more distinctly barred both above and below, and probably with shorter spurs in males.

General biology and ecology

This species ranges from about 500 to 1500 m in open grasslands, scrubby grasslands, lightly wooded savannahs, and extending out into cultivated lands. Roosting is probably done in trees where these are available, and foods include the usual array of plant materials such as seeds and berries, plus green matter and arthropods such as termites. The birds fly for only short distances, and presumably are fairly sedentary, preferring to escape by running into heavier cover (Urban et al. 1986).

Social behaviour

Covey sizes up to about five birds, probably family groups, are typical, but the birds are in pairs throughout much of the year. Calling is done from small promontories such as termite mounds, and nearby males often call in response to one another, probably as territorial challenges. Most calling occurs in early morning or late afternoon, or following rains (Urban et al. 1986).

Reproductive biology

Nesting over eastern Africa shows no correlation with rainfall, with dates ranging from February to October, but in Zaire breeding occurs from September to November, during the latter part of the rains. The nest is placed under dense cover, and the clutch is of six to eight eggs. Little else is known of the breeding (Urban *et al.* 1986).

Evolutionary relationships

This species occurs geographically between *bicalcaratus* and *clappertoni*, and is part of the same superspecies of geographic replacement forms, which according to Urban *et al.* (1986) also includes the more distinctively different *harwoodi*.

Status and conservation outlook

This species is fairly widely distributed, and still common to abundant over much of its range.

CLAPPERTON'S FRANCOLIN (Plate 65)

Francolinus clappertoni Children 1826
Other vernacular names: Gedge's francolin (*gedgi*); francolin de Clapperton (French); Clapperton-frankolin (German).

Distribution of species (see Map 22)
Resident of grasslands from Mali east to Sudan, Uganda, Kenya, and Ethiopia.

Distribution of subspecies
F. c. clappertoni: Mali to western Sudan. Urban *et al.* (1986) do not recognize any of the following subspecies.
   F. c. heuglini: south-western Sudan.
   F. c. gedgii: south-eastern Sudan to northern Uganda. Includes *cavei*.
   F. c. sharpii: eastern Ethiopia. Includes *testis*.
   F. c. nigrosquamatus: western Ethiopia.

Measurements

Wing, males 170–193 mm, females 150–178 mm; tail, males 77–96 mm, females 70–91 mm. Weight, average of 12 males 604 g, 10 females 463 g (Urban *et al.* 1986). Egg, average 43 × 33 mm (Mackworth-Praed and Grant 1952–73), estimated weight 25.8 g.

Identification

In the field (12–13 in.)
Similar to and parapatric or perhaps even locally sympatric with the yellow-billed and double-spurred francolins, and occupying similar semi-arid grassland habitats. The underparts of this pale-coloured species are distinctively creamy white, with large black or brownish oval markings; a large red eye-ring and a white superciliary stripe may also be evident. The males reportedly crow at dawn and dusk, uttering loud, grating *kerak* calls in series of four to six.

In the hand
Similar (especially in western forms) to *icterorhynchus* above, but more 'scaly' above and orange-brown in general tone, and resembling *bicalcaratus* below, but the feathers are marked with maroon rather than chestnut, and have no pale 'windows' in the feather centres. Differs from both in its large red eye-ring (sometimes enlarged to form a conspicuous facial patch) and legs (shown as yellow on the plate). The base of the lower mandible is also usually red (shown as yellow on the plate); the bill is otherwise black.

Adult female. Like the male, but smaller unspurred or with rudimentary spurs (males are doubly spurred).

Immature. Similar to the adult, but with the innermost secondaries barred, the markings of the upperparts less clear, and the underpart spotting less dark.

General biology and ecology

Associated with semi-arid, often sandy grasslands with some bushes and trees, and ranging from sea-level to 2300 m, this species also occurs on rocky hillsides and in cultivated areas. It apparently eats various seeds, berries, and probably other plant materials, as well as some invertebrates such as insects and molluscs. It roosts in tall trees, and forages during daylight hours. Little else is known of its biology (Urban *et al.* 1986).

Social behaviour

Apparently monogamous, these birds occur in pairs or small coveys, probably family units. Courtship-feeding has been observed in captive birds, and calling, presumably by territorial males, occurs from small promontories (Urban *et al.* 1986).

Reproductive biology

Breeding dates for the species' entire range extend from February to December, with little indication of any specific peak period. The clutch-size is unknown, but no more than four young have been observed with their parents (Urban *et al.* 1986).

Evolutionary relationships

Part of a superspecies including the two preceding species and *harwoodi* (Urban *et al.* 1986), this form has the palest plumage of the group, and is probably the most arid-adapted. It evidently is in local contact with *icterorhynchus* at the south-western limit of its range (southern Sudan), but is not known to hybridize with it.

Status and conservation outlook

Although patchily distributed, this form is locally common to abundant over much of its range.

HILDEBRANDT'S FRANCOLIN (Plate 66)

Francolinus hildebrandti Cabanus 1878
Other vernacular names: Fischer's francolin (*fischeri*); Johnston's francolin (*johnstoni*); francolin de Hildebrandt (French); Hildebrandtfrankolin (German).

Distribution of species (see Map 22)
Resident of brushy habitats from south-eastern Zaire to Kenya, Malawi, and northern Mozambique.

Distribution of subspecies
F. h. altumi: western Kenya.
 F. h. hildebrandti: eastern Kenya to north-eastern Zambia and western Malawi. Includes *helleri*; Urban *et al.* (1986) regard all variation as clinal and accept no races.
 F. h. fischeri: central Tanzania.
 F. h. johnstoni: southern Tanzania, eastern Zambia, southern Malawi, and northern Mozambique. Includes *grotei* and *lindi*.

Measurements

Wing, males 158–189 mm, females 151–179 mm; tail, males 92–118 mm, females 86–110 mm. Weight, males 600–645 g, females 430–480 g (Urban *et al.* 1986). Egg, average 42 × 32 mm (Mackworth-Praed and Grant 1952–73), estimated weight 23.7 g.

Identification

In the field (13.5 in.)
Associated with rocky ground and river valley or hillside thickets, such as acacia or *Brachystegia* woodlands. Males resemble the preceding (closely allopatric) species in having pale underparts, but exhibit rounded and more isolated black spotting, while the distinctive female is ochre- to orange-coloured below. Both sexes have red bills and legs and have very indistinct pale superciliary stripes. The male's advertisement call is a loud, repeated *kok* note.

In the hand
Upperparts similar to *icterorhynchus*, but with more solid blotchy blackish markings below in males, the females distinctively orange-ochre below, and both sexes with a red rather than yellow bill. Also very similar to *natalensis*, but the underparts of males are more heavily spotted with black, and females are ochre-coloured below. Also similar to *hartlaubi*, but both sexes have red rather than yellow legs, males are spotted rather than streaked with black below, and females have a reddish rather than ochre-coloured bill.

Adult female. Similar dorsally to the male but almost uniformly cinnamon-brown below, including the chest, and with one or two vestigial tarsal spurs (males usually are doubly spurred, but may have one or rarely three).

Immature. Buff below like the female, but with dark spotting and shaft-streaks (probably in males only), and with black and buff barring or black arrow-shaped markings in the upperparts.

General biology and ecology

This species occurs between 2000 and 2500 m in dense scrub, thickets, or bushy grassland on rocky hillsides, and as high as the lower heath zone in some areas such as Arusha National Park, Tanzania. Its foods include plant materials, insects and their larvae, and perhaps other arthropods, but are little studied. The birds are extremely wary and found on steep, sometimes inaccessible slopes. They are known to perch and roost in trees, and usually are found in pairs throughout the year (Urban *et al.* 1986).

Social behaviour

Almost unstudied, but a high-pitched cackle, *kek-kekek-kek-kerak*, is uttered mainly at dawn and dusk, sometimes in chorus, and presumably serves in territorial advertisement. Presumably monogamous, the males are said to be highly pugnacious during the early part of the breeding season (Urban *et al.* 1986).

Reproductive biology

Breeding shows no clear correlation with rainfall patterns, and in East Africa is scattered from January to December. In Malawi the spread is from April to November, with a peak in June and July. The nest is a well-hidden scrape, and the clutch is of four to eight eggs (Urban *et al.* 1986).

Evolutionary relationships

Apparently forming a closely related superspecies with *natalensis*, this species is also part of the rather large vermiculated group and shows some similarities to *hartlaubi*, including the pattern of sexual dimorphism in plumage.

Status and conservation outlook

This species is uncommon and local through its range.

NATAL FRANCOLIN (Plate 67)

Francolinus natalensis Smith 1834
Other vernacular names: francolin du Natal (French); Natalfrankolin (German).

Distribution of species (see Map 22)
Resident of scrub and bush habitats from Zambia to Natal.

Distribution of subspecies
F. n. neavei: northern Zambia to western Mozambique.
  F. n. natalensis: southern Zambia to Natal. Includes *thamnobium*. Not recognized by Urban *et al.* (1986).

Measurements

Wing, (*natalensis*), males 177–189 mm, females 160–164 mm; tail, males 88–102 mm, females 82–91 mm (Clancey 1967). Weight, males 485–723 g, average (of 10) 606 g, females 370–482 g, average (of five) 426 g (Newman 1983). Egg, average 40 × 35 mm (Mackworth-Praed and Grant 1953–72), estimated weight 27.0 g.

Identification

In the field (13.5 in.)
Associated with rocky scrub and bush, especially riverine bush, immediately south of the range of *hildebrandti*, and differing from it in having an entirely red bill and much more intricately patterned with dark markings on the sides and underparts. The male's advertisement call is a repeated, harsh crowing, sometimes in duet form, and sounding like *graa-che-che-chee* or *ker-kik-kik-kik*, with the last three notes emphasized, or loudest in the middle and then slowing and fading away. The crested francolin also occurs in the same general area, but is more buffy below and has a white throat; the red-necked, Swainson's, and red-billed francolins are all distinctly darker in overall colour.

In the hand
Males are similar above to *hildebrandti*, but with black mottling or striping on the back and scapulars, the bill brighter orange-red, with no black on the culmen, and usually with the sides and underparts much more intricately patterned with dark barring and U-shaped markings. Most similar to *hartlaubi* in male plumage, but that species lacks red on the bill and feet. Separable from *clappertoni*, the only other similar red-billed species, by its lack of a clear white superciliary stripe. The tarsus is bright red (shown as orange on the plate), and the bill is also orange-red, with no black on the culmen, and becoming dull greenish basally and on the cere.

Adult female. Like the male, but unspurred (males are doubly or more commonly singly spurred).

Immature. Similar to adult females, but with broader black markings on the mantle and scapulars, and with more black bars and spots below, as well as being paler and more greyish brown above, with indistinct rufous buff barring. The bill is greenish, and the legs flesh-coloured, with poorly developed spurs in males.

General biology and ecology

This species is associated with a broad array of habitat types occurring between sea-level and 1800 m, including dense thickets on hillsides, dry riverine forest or bush, woodlands with good ground cover, and acacia scrub on rocky ground. Where cover is adequate the birds also venture out into cultivated areas, feeding at times on fallen grain. They also consume a wide variety of plant materials such as seeds, roots, bulbs, berries, and the like, as well as arthropods and molluscs. The species often associates with the crested francolin, and probably competes to some extent with it. Hybrids with both the Swainson's and red-necked francolins have also been reported. They are probably fairly sedentary, roosting in trees or tall brush at night and retiring to heavy cover during the hottest part of the day (Clancey 1967; Urban *et al.* 1986).

Social behaviour

Presumably monogamous, these birds utter apparent territorial advertisement calls at dawn and dusk, with the female often joining in a duet. Calling by one male may also stimulate responses by other nearby birds. The coveys generally number less than 10 birds, apparently mostly of family groups. Nothing is known of their pair-forming displays (Clancey 1967; Urban *et al.* 1986).

Reproductive biology

Breeding records from South Africa are irregular, but peak in January–February and from April to July, while in Zimbabwe they peak from March to May, the same period for which breeding records from Zambia are available. The nest is a well-concealed scrape, and the clutch-size is normally of five to eight eggs, with occasional larger clutches probably the result of two females. Incubation is by the female alone, and probably lasts 20 days, based on data from captivity (Newman 1983; Urban *et al.* 1986).

Evolutionary relationships

This species forms a closely related and parapatric superspecies with *hildebrandti* (Urban *et al.* 1986). It is probably slightly less closely related to *hartlaubi*, from which it is more strongly separated geographically.

Status and conservation outlook

Although generally still common to abundant locally, in South Africa the numbers have been reduced through land development (Urban *et al.* 1986).

HARTLAUB'S FRANCOLIN (Plate 63)

Francolinus hartlaubi Bocage 1869
Other vernacular names: francolin de Hartlaub (French); Bergfrankolin (German).

Distribution of species (see Map 22)
Resident of rocky and hilly habitats from Angola to central Namibia.

Distribution of subspecies
F. h. hartlaubi: Angola. The two following races are not accepted by Urban *et al.* (1986).
  F. h. bradfieldi: northern Namibia. Includes *ovambensis*.
  F. h. crypticus: central Namibia (Onguato area).

Measurements

Wing (*bradfieldi*), males 144–148 mm, females 134–138 mm; tail, males 75–89 mm, females 73–80 mm. Weight, males 245–290 g, females 210–240 g (Clancey 1967). Egg, average 43 × 29 mm (Mackworth-Praed and Grant 1952–73), estimated weight 20.0 g.

Identification

In the field (10–11 in.)
Associated with rocky habitats of Angola and Namibia, where its fairly small size, underparts that in males are pale buffy, streaked with sepia, and in females are rich tawny-cinnamon, and a disproportionately large bill in both sexes, are distinctive. Its calls include squeaky alarm notes and antiphonal duetting by paired birds. The male's black and white under tail-coverts are conspicuous in flight and during sexual display.

In the hand
Differs from the other members of its group in its small size (wing under 150 mm), yellowish to horn-coloured and relatively large bill (culmen 19.5–25 mm from cere), and in having distinctive rufous mottling on the mantle and wing-coverts. Females have cinnamon-coloured rather than sepia-streaked underparts, as in females of *hildebrandti* and both sexes of *finschi*, but they differ from the former in their yellowish legs and from the latter in their lack of rufous chestnut on the flight feathers.

Adult female. Similar to male on the upperparts, wing, and tail, but the sides of the face fawn to greyish brown; chin and throat vinaceous fawn; lower throat and chest drab, the feathers with indistinct fawn shaft-streaks; rest of underparts all unstreaked tawny-chestnut, the under tail-coverts fringed with white. Spurs are lacking (males may lack spurs or have one or two blunt tarsal knobs, the lower better developed).

Immature. Immature males have pale shaft-streaks above and narrower streaks below than do adult males; young females are like adult females, but have the webs of the primaries peppered and barred with buff markings.

General biology and ecology

Associated with dry ridges, hills, and kopjies having a combination of dense, mixed grasses and shrubs and abundant undergrowth, together with a sandy soil. This combination favours the growth of food plants such as *Cyperus edulis*, the bulbs of which are major foods, and the birds' large bills are apparently adapted for digging out such subsurface plant materials. They also consume plant seeds, fruits, berries, and insects such as termites. Population densities tend to be low, with each pair occupying about 0.5 km^2, and the birds

Plate 64. Yellow-billed francolin, pair. Painting by H. Jones.

Plate 65. Clapperton's francolin (*sharpii*), pair. Painting by H. Jones.

Plate 66. Hildebrandt's francolin (*johnsoni*), pair. Painting by H. Jones.

Plate 67. Natal francolin, pair. Painting by H. Jones.

Plate 68. Harwood's francolin, male. Painting by H. Jones.

Plate 69. Red-billed francolin, pair. Painting by H. Jones.

Plate 70. Cape francolin, pair. Painting by H. Jones.

Plate 71. Crested francolin, pair. Painting by H. Jones.

Plate 72. Ring-necked francolin, pair. Painting by H. Jones.

Plate 73. Montane red-winged francolin, pair. Painting by H. Jones.

Plate 74. Shelley's francolin, pair. Painting by H. Jones.

Plate 75. Grey-winged francolin, pair. Painting by H. Jones.

Plate 76. Acacia francolin (*jugularis*), pair. Painting by H. Jones.

Plate 77. Levaillant's red-winged francolin (*kikuyuensis*), pair. Painting by H. Jones.

Plate 78. Finsch's francolin, male. Painting by H. Jones.

Plate 79. Coqui francolin, pair. Painting by H. Jones.

Plate 80. Schlegel's banded francolin, male. Painting by H. Jones.

Plate 81. White-throated francolin, pair. Painting by H. Jones.

Plate 82. Latham's forest francolin, pair. Painting by H. Jones.

Plate 83. Grey francolin, pair. Painting by H. Jones.

Plate 84. Swamp francolin, pair. Painting by H. Jones.

Plate 85. Grey partridge, male and females. Painting by H. Jones.

Plate 86. Daurian partridge, pair. Painting by H. Jones.

Plate 87. Tibetan partridge, pair. Painting by H. Jones.

Plate 88. Long-billed wood-partridge, pair. Painting by H. Jones.

Plate 89. Madagascar partridge, pair. Painting by H. Jones.

Plate 90. Black wood-partridge, pair. Painting by H. Jones.

Plate 91. European migratory quail, pair. Painting by H. Jones.

Plate 92. Asian migratory quail, pair. Painting by H. Jones.

Plate 93. African harlequin quail, pair. Painting by H. Jones.

Plate 94. Black-breasted quail, pair. Painting by H. Jones.

Plate 95. Stubble quail (*pectoralis*), pair. Painting by H. Jones.

are strongly territorial, maintaining and defending their territories all year round. Thus they are usually found in pairs or, at most, in family groups of up to about four birds. These groups roost among the rock precipices, call from various promontories, and find shelter there under ledges from the midday heat. They sometimes forage in open terrain at the base of kopjies (Komen and Myer 1984).

Social behaviour

Both sexes contribute to the defence of the permanent territory, which is done by a complex series of postures and vocalizations, including at least 12 calls. One important vocalization is antiphonal calling, consisting of four 'call units' that are precisely uttered by the sexes alternately. This serves to maintain and advertise the pair-bond (and territory), and perhaps helps to confuse potential predators by making the call more difficult to localize. Males lack the sharp spurs of most francolins and, instead of overt fighting, most territorial disputes are settled by duetting contests, with the duration and loudness of the calling apparently influencing the outcome. When soliciting a female, the male approaches her while slowly raising and lowering his tail, simultaneously fanning it, swaying his body from side to side, and apparently exposing his flank and underpart plumage; the female may respond with calling. During courtship both birds may circle one another, stooping low, drooping their wings, and ruffling their feathers. Pair-bonds are evidently thus established without the usual male domination typical of most phasianids (Komen and Myer 1984).

Reproductive biology

Breeding in Namibia has been reported in May, June, and November, suggesting breeding during the early cool season, and nests are apparently well hidden in hollows on cliff ledges, inasmuch as only one has so far been described. The clutch-size is probably three to eight eggs (one known clutch has three eggs), judging from brood sizes, and the young and parents remain together for an extended period (Clancey 1967; Urban *et al.* 1986).

Evolutionary relationships

Although generally similar to the last two species in appearance, this form is considered somewhat aberrant by Komen (1986) on the basis of its behaviour and vocalizations.

Status and conservation outlook

This species is distinctly local in distribution, and ranges in abundance from common to uncommon (Urban *et al.* 1986).

HARWOOD'S FRANCOLIN (Plate 68)

Francolinus harwoodi Blundell and Lovat 1899
Other vernacular names: francolin de Harwood (French); Harwoodfrankolin (German).

Distribution of species (see Map 22)
Endemic to Blue Nile gorge and its tributaries in Ethiopia. No subspecies recognized.

Measurements

Wing, males 177–187 mm, females 161–165 mm; tail, males 75–86 mm, females 69–73 mm. Weight, males 545 g, females 413–446 g (Ash 1978; Urban *et al.* 1986). Egg, no information.

Identification

In the field (14 in.)
Limited to the streamside habitats among cliffs of the Blue Nile gorge south-west of Lake Tana, and unique in having dark scalloping on the breast and underparts, and red on the base and underside of the bill, legs, and around the eyes. Calls include a loud, rasping *koree* similar to that of *clappertoni*, which occurs in the same general area but in different habitats.

In the hand
Most similar to *natalensis*, but with more defined U-shaped patterning on the hindneck and upper mantle, and darker scalloped patterning on the breast and underparts, except for the centre of the abdomen, which is cream-coloured.

Adult female. Very similar to the adult male, but apparently slightly browner, with less contrasting patterning on the breast feathers, the centres of these feathers having small V-shaped rather than large U-shaped marks; these tend to form bars on the flanks. No tarsal spurs, compared with two on males (Ash 1978).

Immature. In young birds the outer two primaries are relatively pointed and are retained through the first prebasic moult. First-year females resemble adults, but have less distinct barring (Ash 1978).

General biology and ecology

This species occurs in dense and extensive beds of cattails (*Typha*) that occur along shallow streams with scattered trees. These beds are used both for escape cover and for escaping the midday heat, while roosting is done in trees as well as in beds of *Typha*. The birds may also occur in woodlands associated with high grasslands and in areas of mixed shrubs and cultivated areas around Gibe Gorge and Dembidollo (based on sight records). Known foods include tubers, grass, and other seeds, berries and berry-like fruits, sorghum grain, and termites (Urban *et al*. 1986).

Social behaviour

No specific information.

Reproductive biology

Nests and eggs remain undescribed, but a brood of three young seen in late February suggests mid-December laying (Urban *et al*. 1986).

Evolutionary relationships

Considered by Urban *et al*. (1986) to be part of a superspecies that also includes *bicalcaratus*, *clappertoni*, and *icterorhynchus*, but quite distinct in plumage from all of these, and seemingly not closely related to any single one of them.

RED-BILLED FRANCOLIN (Plate 69)

Francolinus adspersus Waterhouse 1838
Other vernacular names: close-barred francolin; francolin à bec rouge (French); Rotschnabelfrankolin (German).

Distribution of species (see Map 22)
Resident of semi-arid habitats from southern Angola to Zambia, Botswana, Zimbabwe, and southern Namibia. No subspecies recognized. Introduction recently attempted in the Hawaiian Islands; success uncertain.

Measurements

Wing, males 181–201 mm, females 164–176 mm; tail, males 95–105 mm, females 85–93 mm (Clancey 1967). Weight, males 340–635 g, average (of 12) 465 g, females 340–549 g, average (of 24) 394 g (Newman 1983). Egg, average 43 × 33 mm (Mackworth-Praed and Grant 1952–73), estimated weight 25.8 g.

Identification

In the field (12.5–13 in.)
A common resident in such semi-arid and often thorny habitats as the Kalahari thornveld, occurring in riverine thickets and scrub, especially near water, and identified by its red bill, bare yellow eye-ring, and narrowly barred underpart patterning. Its advertising call is a harsh crowing, consisting of a long series of *cho* or *chorr* notes that tend to lengthen and diminish in volume toward the end.

In the hand
Rather easily identified by the densely barred black and white patterning of the underparts and the finely vermiculated upperparts, together with the yellow eye-ring (shown as red on the plate) and red bill and legs.

Adult female. Like the male, but unspurred or the spur rudimentary (males are singly or sometimes doubly spurred).

Immature. Browner than adults, with faint quail-like pattern of buffy barring and streaking, the scapulars blotched with black at their tips, the underparts brownish white with fine black vermiculations, and some of the abdomen and flank feathers widely barred with black.

General biology and ecology

Associated with a variety of habitats, but often near water and with scrub, thickets, or woodlands such as acacia–veld interspersed with more open ground. The birds use the trees for nocturnal roosting and forage in clearings and old cultivated lands, where they consume a wide variety of plant materials as well as insects such as termites and some small molluscs. The birds are sedentary, but move to higher ground during floods, and are usually found in groups of up to about 10 individuals, but they sometimes form coveys of up to about 20 during the non-breeding seasons. They are active mainly in early mornings and late afternoons, usually drinking in late afternoon, and call primarily around dawn and dusk (Clancey 1967; Urban *et al*. 1986).

Social behaviour

Apparently monogamous and distinctly territorial, the males call to announce territories and probably to attract females in the case of unmated mates. The male's courtship display is the usual later posturing with wings fanned, similar to that of *Gallus*. The territorial call is highly variable, but is relatively low-pitched and hoarse, resembling hysterical laughter, the individual notes repeated up to 10 times, and often trailing off (Clancey 1967; Urban *et al*. 1986).

Reproductive biology

In northern Namibia and Botswana the peak breeding period is from April to July, but extends from January to August, while it is more extended but peaks between December and April. In Zimbabwe the period is from January to August; breeding is evidently set to coincide with the end of the rains and the early dry season. The clutch-size ranges from four to ten, with six clutches averaging 6.7 eggs. In captivity the incubation period has been determined as 22 days. Young raised in captivity acquired adult plumage by about 3 months, and males began to develop spurs at about 5 months (Urban *et al.* 1986).

Evolutionary relationships

This species is not obviously a close relative of any of the others of the vermiculated group, but is perhaps most closely related to the Cape francolin, which it resembles in several plumage features. Hall (1963) believed that it might have been derived from an earlier isolate than any of the other members of this group.

Status and conservation outlook

Generally common to abundant through its range, this is not a species needing conservation attention.

CAPE FRANCOLIN (Plate 70)

Francolinus capensis (Gmelin) 1789
Other vernacular names: francolin criard (French); Kapfrankolin (German).

Distribution of species (see Map 22)
Endemic in woodlands of south-western Cape Province, from the Orange River Valley to the eastern Cape. No subspecies recognized.

Measurements

Wing, males 218–230 mm, females 195–220 mm; tail, males 94.5–116 mm, females 97–108 mm (Clancey 1967). Weight, males 600–915 g, females 435–659 g (Newman 1983). Egg, average 46.8 × 36.4 mm (Schönwetter 1967), estimated weight 31.6 g.

Identification

In the field (16–18 in.)
This is the largest of the South African francolins, and a resident of riverine scrub and rocky outcrops that support woods. It is generally dark-coloured, with prominent whitish streaking on the flanks and underparts. Its call is a loud, crowing cackle of double *ka-keek* notes, the second accented, and the series initially rising and then diminishing in volume.

In the hand
Similar to *natalensis* and *harwoodi* in having V-shaped and U-shaped brown and white patterning on the back and underparts, but also with conspicuous whitish shaft-streaks evident on the underparts. Also similar to the red-billed francolin, but with the upper mandible brownish and reddish only basally, no yellow eye-ring, and the underpart patterning more U-shaped than barred. The very large size (wing 195–230 mm) and prominent white edging and shaft-streaking on the feathers of the underparts provide a distinctive combination.

Adult female. Like the larger male, but usually with a single short spur of less than 4 mm (males have one or usually two longer and sharper spurs), and with the tarsus and lower mandible a duller orange.

Immature. Juveniles have conspicuous white shaft-streaks, especially below, and older birds in full first-basic plumage also have more sharply defined shaft-streaks than do adults. The wings are more greyish brown, with lighter transverse barring evident. The tarsus is dark, and the spurs of young males are much shorter than in adults (Clancey 1967; Urban *et al.* 1986).

General biology and ecology

The preferred habitat of this species is scrubby heath, particularly coastal or montane fynbos vegetation, and also sheltered riverine scrub. However, the birds also move into cultivated areas to forage, sometimes in company with domestic fowl, and consume a wide array of vegetational materials, including waste grain, fallen fruit, and natural vegetation, plus animal materials such as termites and other insects. They occur in pairs and small coveys, the latter probably composed of family groups, and evidently are quite sedentary, moving into heavy tree cover to roost at night; they are generally reluctant to fly (Clancey 1967; Urban *et al.* 1986).

Social behaviour

Presumably monogamous and territorial, males advertise during the breeding season with dawn and dusk calling, but nothing specific has been written on pair-bonding or display behaviour.

Reproductive biology

Breeding occurs between July and February, peaking in September–October, during the late winter rains or the early dry season. Nests are well hidden under vegetation, and clutch-sizes are normally from six to eight eggs, with the average of 25 clutches being 7.4 eggs. The incubation period is unreported, but the average size of 45 families was 4.2 chicks (Urban *et al.* 1986).

Evolutionary relationships

This is a fairly isolated form, both geographically and morphologically, in the vermiculated group, and its ancestral population was probably cut off from the rest of the members of this group earlier than any of the others except *adspersus* and *harwoodi* (Hall 1963). Komen (1986) regards *capensis* as more closely related to the montane group of francolins than to the vermiculated group, based on vocal and behavioural characteristics.

Status and conservation outlook

This form is locally uncommon and not threatened at present (Urban *et al.* 1986).

CRESTED FRANCOLIN (Plate 71)

Francolinus sephaena Smith 1836

Other vernacular names: crested partridge; Grant's francolin (*grantii*); Kirk's francolin (*rovuma*); Smith's francolin; spotted francolin (*spilogaster*); Francolin huppé (French); Schopffrankolin (German).

Distribution of species (see Map 23)

Widespread resident in eastern and southern Africa from southern Sudan, Ethiopia, and Somalia south to Angola, Namibia, and Natal.

Distribution of subspecies

F. s. somaliensis: Somalia (coastal plains, possibly extirpated). Not recognized by Urban *et al.* (1986).

F. s. granti: Ethiopia to central Tanzania. Includes *schoanus*, *ochrogaster*, *delutescens*, and *jubaeensis*.

F. s. spilogaster: Somalia to southern Kenya (probably a hybrid population of *granti* and *rovuma*).

F. s. rovuma: eastern Tanzania to eastern Mozambique.

F. s. sephaena: eastern Zimbabwe to Mozambique and northern Natal. Includes *zluensis*.

F. s. zambesiae: Namibia to southern Malawi. Includes *thompsoni*, *chobiensis*, and *mababiensis*.

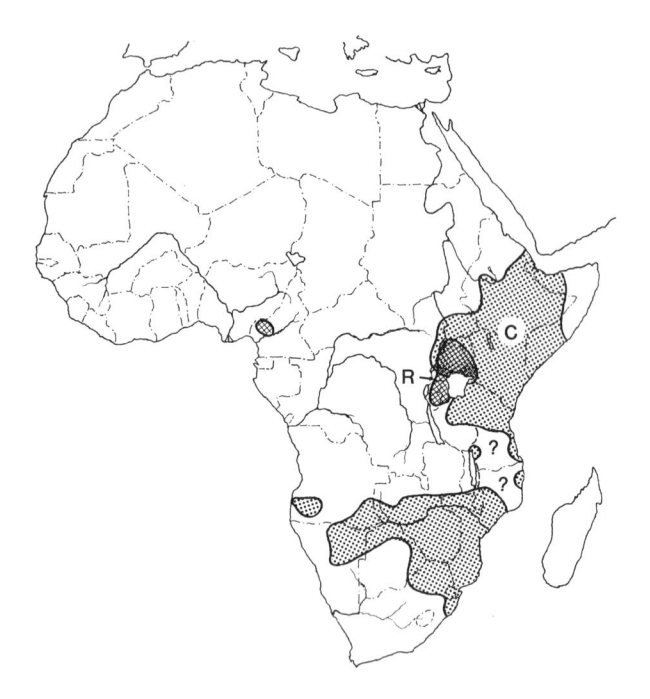

Map 23. Distribution of the striated group of *Francolinus*, including crested (C) and ring-necked (R) francolins.

Measurements

Wing (all races), males 124–170 mm, females 128–166 mm; tail, males 85–112 mm, females 83–89 mm (Newman 1983; Urban *et al.* 1986). Weight (all races), males 265–482 g, average of *sephaena* 387 g, of *granti* 288 g, females 225–352 g, average of *sephaena* 315 g, of *granti* 240 g (Clancey 1967; Urban *et al.* 1986). Egg, average 40 × 28 mm (Mackworth-Praed and Grant 1952–73), estimated weight 17.3 g.

Identification

In the field (13–14 in.)

Associated with riverine thickets, the vicinity of rock outcrops, and moister areas of veld, and often found in company with the Natal francolin. The fairly heavy streaking on the neck and chest, which rather abruptly terminates on the buffy lower breast, and the similar contrasting white shaft-streaking on the dark mantle and wing feathers, provide for easy distinction, as does the contrasting black crown and white superciliary stripe. The ring-necked francolin of the same general area has much more brown on the sides of the head, and the Natal francolin has mostly white rather than yellowish buffy underparts. Calls include an excited, high-pitched squealing *kerra-kreek*, as well as possible antiphonal duetting *che-che-chirra* between the pair members.

In the hand
Distinguished from other members of its group by the distinctive chestnut and white blotching on the neck and breast, the broad black-edged white shaft-streaks on the wing-coverts and mantle, and the black crown, eye-stripe, and sometimes also a black malar stripe, contrasting with the otherwise pale head. The legs are red, and the bill dull red in both sexes.

Adult female. Similar to the male, but with the middle of the back, scapulars, and wing-coverts more densely covered with wavy narrow bars of buff and lines of black; the lower back, rump, and upper tail-coverts are mottled with black and whitish buff. Unspurred (males are singly spurred).

Immature. Young males resemble females, but have small tarsal spurs. Some juvenal tertials and primaries are retained in the first-winter (first-basic) plumage, and young males may exhibit vermiculations on the scapulars and long upper tail-coverts. Young of both sexes have broader and more clearly defined buffy shaft-streaks dorsally (Clancey 1967; Urban *et al.* 1986).

General biology and ecology

This species occupies a rather wide range of thickety and brushy habitats, as well as forest edges, woodlands with sparse grassy cover, and dry river beds, but is usually found near surface water. It shares its habitats with coqui, Natal, red-necked, and Swainson's francolins, and probably competes to some extent with all of them, as its foods are the usual francolin array of plant materials, especially sedge bulbs, and insects, especially termites. Roosting is done in trees, and during daylight hours the birds take cover during the hottest part of the day. They are found in pairs and presumed family groups of up to about seven birds, and are probably fairly sedentary, quickly running for cover when alarmed (Clancey 1967; Urban *et al.* 1986).

Social behaviour

Presumably monogamous and territorial, the birds are especially noisy at roosting time, when two or more birds may sing simultaneously, the male producing a resonant whistling sound that is quite distinctive. The pair may also call antiphonally, with the female adding the second part of the 'song', but when she is incubating eggs this antiphonal singing is lacking. Territorial males also respond strongly to recorded playbacks of the advertisement call (Clancey 1967; Urban *et al.* 1986).

Reproductive biology

The nesting season is quite variable and often prolonged over this species' broad range, and in South Africa and Zimbabwe extends from October, with the start of the main summer rains, to May, with no marked peak. Records from Zambia and Malawi are from November to February, and across much of East Africa are quite varied, perhaps concentrating between January and July, but also including autumn and winter in Somalia. The nest is usually hidden under vegetation, and the clutch is of five to six eggs in Somalia, and in southern Africa four to nine, usually six. The incubation is probably by the female alone, and its length is about 19 days (Clancey 1967; Urban *et al.* 1986).

Evolutionary relationships

This species and the following one comprise the 'striated group' of Hall (1963), and she considered them quite distinct, both morphologically and ecologically, their ancestral populations perhaps splitting in the area between Uganda and Ethiopia during a humid period of forest spreading. Crowe and Crowe (1985) consider this species to be a link between the Latham's forest francolin and the red-tailed and red-winged groups of francolins, with *streptophorus* a part of the latter group.

Status and conservation outlook

This widespread species is generally locally common to abundant, and is not a conservation problem (Urban *et al.* 1986).

RING-NECKED FRANCOLIN (Plate 72)

Francolinus streptophorus Ogilvie-Grant 1891
Other vernacular names: francolin à collier (French); Kragenfrankolin (German).

Distribution of species (see Map 23)

Resident in open grasslands of Kenya, Uganda, and Tanzania; a disjunct population occurs in the Cameroon highlands (Foumban area). No subspecies recognized.

Measurements

Wing, males 141–167 mm, females 139–160 mm; tail, males 67–80 mm, females 74–83 mm. Weight, two males 364–406 g (Urban *et al.* 1986). Egg, no information.

Identification

In the field (11 in.)

Associated with sparsely vegetated stony hillsides of equatorial Africa, and prone to run rather than fly. The birds are sometimes found in the same areas as *sephaena*, but occupy savannah grasslands rather than scrubby thornbush. Their call is a distinctive melodious trill or whistle, typically preceded by two low notes. If seen, the birds are notable for their conspicuous white superciliary stripes and brownish face colouring, terminated by a band of black and white barring extending entirely around the neck. The flanks are tawny, with heavy dark spotting, and the legs are yellowish. The advertisement call consists of two soft, dove-like cooing notes, the first at a lower pitch, followed by a piping trill.

In the hand

Similar to the preceding species, but generally darker, with fewer mantle striations, a definite barred breast-band and 'necklace', and a bright chestnut face. The superciliary stripes are extended back to the nape; between this stripe and the white throat the face is mostly rufous-toned, and the crown is also brownish, rather than black as in *sephaena*. Spurs are lacking in both sexes or rudimentary in males (shown as spurred in Urban *et al.* 1986).

Adult female. Similar to the male, but with the back, rump, and upper tail-coverts barred or streaked with buff, and the wing-coverts spotted with buff. The outer webs of the inner primaries and secondaries variably clouded along the basal margin with buff, and the crown feathers are more contrastingly edged with lighter edges.

Immature. Undescribed.

General biology and ecology

Found in thin, rocky grasslands at 1050–1200 m elevation in Cameroon, and in wooded grasslands and sparsely covered stony hillsides at 600–1800 m in East Africa, with occasional visits to cultivated areas, presumably to feed on weeds, crops, and insects, although little is known of its preferred foods. It may visit roadside areas in the early morning, but spends the hottest part of the day in heavy cover, in which it probably also roots. It typically also flees into heavy cover when alarmed (Urban *et al.* 1986).

Social behaviour

Presumably monogamous and territorial, these birds usually occur in pairs or small coveys, and males utter advertisement calls from prominent points during the early morning hours.

Reproductive biology

In Uganda the breeding records are for April, early in the rainy period, and in western Kenya they are from December to March, during the dry season. The nest is often located close to a rock, and the usual clutch is of four to five eggs (Urban *et al.* 1986).

Evolutionary relationships

Considered by Hall (1964) to comprise with *sephaena* a separate 'vermiculated' group, it is regarded by Crowe and Crowe (1985) as part of a superspecies in the red-winged group, which also includes *africanus*, *levaillantii*, and *finschi*.

Status and conservation outlook

The disjunct Cameroon population is small and possibly in some danger, and the East African population varies from uncommon to locally abundant, although patchily distributed (Urban *et al.* 1986).

MONTANE RED-WINGED (MOORLAND) FRANCOLIN (Plate 73)

Francolinus psilolaemus G. R. Gray 1867
Other vernacular names: Harris' francolin; montane francolin; francolin montagnard (French); Berghalde-frankolin (German).

Distribution of species (see Map 24)

Endemic resident in mountain grasslands and heaths of Uganda, Kenya, and Ethiopia. Sometimes considered part of *africanus*.

Distribution of subspecies

F. p. psilolaemus: central Ethiopia.

F. p. ellenbecki: southern Ethiopia. Considered part of *psilolaemus* by Urban *et al.* (1986).

F. p. elgonensis: Mt Elgon, on Kenya–Uganda border. (Often considered part of *shelleyi*.)

F. p. theresae: Mt Kenya, Aberdare Mountains, and Mt Mau, Kenya. Considered part of *elgonensis* by Urban *et al.* (1986).

Measurements

Wing, males 150–180 mm, females 151–173 mm; tail, males 76–101 mm, females 81–96 mm. Weight, two males, 510–530 g, two females 370–510 g (Urban *et al.* 1986). Egg, no information.

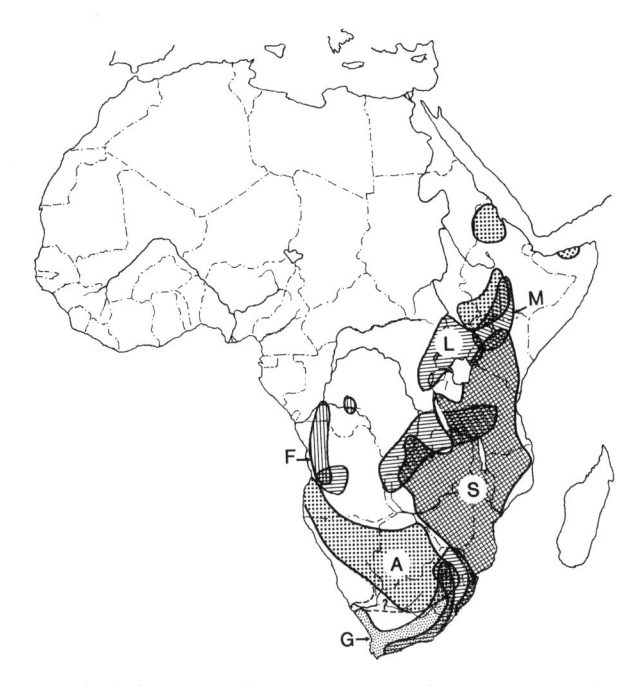

Map 24. Distribution of the red-winged group of *Francolinus*, including acacia (A), Finsch's (F), grey-winged (G), Lavaillant's (L), montane (M), and Shelley's (S) francolins.

Identification (see Fig. 18)

In the field (12–13 in.)

Associated with montane grasslands and heaths usually above 2000 m in eastern Africa. Exhibits a buff-striped *Coturnix*-like pattern above, and rich chestnut on the wings; ventrally it has a rufous breast (more tawny in *psilolaemus*) grading below into a buff to greyish buff belly, and bordered above by a white throat that may be heavily black-spotted (*psilolaemus*) or clear, bounded below with black spotting (*elgonensis*). The sides are also spotted with chestnut. The call consists of three or four strident, grating notes, nearly identical to that of *shelleyi* (which is sympatric but ecologically isolated at lower elevations below montane forests in central Kenya), and the birds typically squeal when taking flight, as do others in the red-winged group. It occurs sympatrically with the Levaillant's red-winged francolin, which lacks black spots on its chestnut breast feathers and has distinct black and white neck barring.

In the hand

Similar to *shelleyi*, but distinctive in its fuscous-tipped chestnut feathers on the breast. The throat feathers are usually also extensively tipped with brownish black (lacking in *elgonensis*, which is more rufous-toned throughout) rather than being surrounded by a distinct black 'necklace', and it is unique among the red-winged group in having some black barring at the tips of the primaries, which are otherwise chestnut-coloured, becoming more greyish toward the tips. Like the next two species, the flanks are blotched with chestnut, and finely barred with black. The legs are yellowish, and the relatively small bill is variably yellow, at least on the lower mandible.

Adult female. Like the male but lacking spurs (males have single spurs).

Immature. Undescribed.

General biology and ecology

This species occurs in montane heath (*Erica*) moors and grasslands between 1800 and 4000 m, occurring in pairs or small groups that probably are comprised of family units. Its foods are little known, but include vegetable materials such as bulbs and roots or grasses and sedges, seeds, and insects (Urban *et al.* 1986).

Social behaviour

Probably monogamous and occurring in pairs during the breeding season, these birds utter advertisement calls that often result in contagious calling on the part of others.

Reproductive biology

Breeding evidently occurs mainly during the first half of the year, but includes all months except September–October. One nest found in rough grass cover contained three eggs plus two chicks, and there are usually four to five chicks present in young broods, suggesting an average clutch of about five eggs. However, little else is known of the reproductive biology (Urban *et al.* 1986).

Evolutionary relationships

This is obviously a very close relative of *shelleyi*, and the southern populations of it have generally been included within that species in the past, differing mainly in the breast spotting and having a more ill-defined facial and neck pattern. Hall (1963) suggested that it evolved from ancestral red-winged stock isolated on Mt Kenya or Mt Elgon during an early warm climatic period.

Status and conservation outlook

This species is locally common in Ethiopia, and uncommon to rare in Kenya (Urban *et al.* 1986).

164 *The Quails, Partridges, and Francolins of the World*

Fig. 18. Adult plumage variation in: the montane red-winged francolin, subspecies (A) *psilolaemus* and (B) *elgonensis*; (C) the grey-winged francolin; and the Shelley's francolin, subspecies (D) *uluensis*, (E) *shelleyi*, and (F) *whytei*.

SHELLEY'S FRANCOLIN (Plate 74)

Francolinus shelleyi Ogilvie-Grant 1890
Other vernacular names: Ulu francolin (*uluensis*); francolin de Shelley (French); Shelleyfrankolin (German).

Distribution of species (see Map 24)

Widespread resident in grasslands of eastern Africa, discontinuously extending from Kenya south to Natal and Transvaal.

Distribution of subspecies

F. s. uluensis: central Kenya. (Often considered part of *africanus*.) This and the following form are included in *shelleyi* by Urban *et al.* (1986).

F. s. macarthuri: Chyulu Hills, southern Kenya. (Often considered part of *africanus*.)

F. s. shelleyi: Uganda (one record) to Tanzania and Natal. Includes *canidorsalis*, *trothae*, and *sequestris*.

F. s. whytei: south-eastern Zaire, Zambia, and northern Malawi.

Measurements

Wing, males 152–177 mm, females 145–172 mm; tail, males 76–97 mm, females 74–93 mm (Urban *et al.* 1986). Weight, males 397–600 g, average (of seven) 497 g, females 411–600 g, average (of 10) 482.4 g (Newman 1983). Egg, average 40 × 28 mm (Mackworth-Praed and Grant 1952–73), estimated weight 17.3 g.

Identification

In the field (11–12 in.)

Associated with open grasslands and savannahs with scattered trees, and similar to but ecologically separated from the last species, occurring below montane forests but on uplands up to 3000 m. The birds have a breast that is streaked with chestnut but little if at all spotted with black, which instead forms a narrow, indefinite 'necklace' around the white throat, and on the lower flanks and abdomen there is black barring interspersed with chestnut streaking. The call consists of two low notes followed by two higher ones, sounding like 'I'll drink yer beer', which may be repeated up to seven times. In flight the primaries appear mostly grey, with rufous evidently only at the base, and a shrill alarm squeal is usually uttered. The other species in the group most likely to be encountered within its range is *levaillantii*, which lacks barred underparts that are buffy yellow rather than white, and has more red visible on the wings in flight.

In the hand

Identified by the rather heavily marked (reduced in *whytei*) underparts of black and white, which tend toward a narrowly barred pattern, and the white (buffy in *whytei*) throat with only a narrow black necklace, the latter separated from the black ventral barring by uniformly chestnut breast markings that become streaks extending from the breast to the flanks. The rufous buffy superciliary stripe is extended back as a diffuse band of sometimes black-spotted feathers that pass down the sides of the neck, and the primaries are reddish brown for only about half their length, becoming greyish terminally. The legs are dull yellow, with a single spur, and the bill is yellowish only at the base of the lower mandible.

Adult female. Like the male, but normally unspurred (sometimes with a rudimentary spur that is much shorter than in males).

Immature. Very similar to the adult, but with the outermost primaries more sharply pointed, and the outer vane barred with vinaceous buff. The birds are generally paler than adults, with the rufous buff on the upperparts reduced or lacking (Clancey 1967; Urban *et al.* 1986).

General biology and ecology

Generally found where there is a combination of fairly rich grass cover and a sprinkling of trees such as acacia or *Brachystegia*, and also on the fringes of cultivated areas, in stony or rocky areas, and on open montane grasslands up to at least 3000 m, but also to coastal areas. It feeds on a variety of plant materials, but apparently prefers roots and bulbs, which it digs out, forming cone-shaped holes. It is found in pairs and small coveys of up to eight birds, probably family groups, and is relatively independent of surface water (Clancey 1967; Urban *et al.* 1986).

Social behaviour

Apparently monogamous and territorial, uttering advertising calls at dawn, dusk, and sometimes during moonlit nights, frequently in chorus, but silent during most of the day. Male calling is highly contagious, with other birds quickly responding to calling by one (Clancey 1967; Urban *et al.* 1986).

Reproductive biology

The breeding season is quite variable across this species' range, extending from August to January in South Africa, peaking during March–April and September–October in Zimbabwe, and breeding during the dry season in Tanzania and central Kenya, between March and July. The clutch-size varies from four to seven eggs, the average of 15 clutches in southern Africa averaging 4.8 and possibly somewhat less in equatorial Africa. The incubation period is about 22 days, and is probably by the female alone (Newman 1983; Urban *et al.* 1986).

Evolutionary relationships

Regarded as part of a superspecies with *psilolaemus* and *levaillantoides* by Crowe and Crowe (1985), this species was reorganized by Hall (1963) to include some populations previously considered part of *africanus*. She believed that these two forms to be just sufficiently different as to warrant specific separation, whereas Crowe and Crowe associated *africanus* with another superspecies group that includes *finschi* and *levaillantii*. Relationships in these francolins are thus obviously complex and perhaps still unsettled.

GREY-WINGED FRANCOLIN (Plate 75)

Francolinus africanus (Latham) 1790
Other vernacular names: grey-wing; francolin à ailes grises (French); Grauflügelfrankolin (German).

Distribution of species (see Map 24)

Resident in coastal and montane grasslands of Transvaal, Orange Free State and Cape Province; local in Natal. Considered monotypic by Hall (1963); the species has often included several north-eastern forms (*gutturalis*, *archeri*, and *lorti*) that are here considered part of *F. levaillantoides*.

Measurements

Wing, males 155–175 mm, females 150.5–167 mm; tail, males 69.5–79 mm, females 63.5–74 mm (Clancey 1967). Weight, males 345–539 g, average (of 13) 422.8 g, females 354–369 g, average (of three) 359 g (Newman 1983). Egg, average 37 × 29 mm (Mackworth-Praed and Grant 1952–73), estimated weight 17.2 g.

Identification (see Fig. 18)

In the field (13 in.)

Associated with mountainous and coastal areas of grassland in South Africa. Distinctive in having the lower breast and abdomen narrowly barred, and the whitish throat extensively flecked but not margined with black. The upperparts are also less contrasting than in the allopatric *shelleyi*, and less black striping is apparent on the sides of the head. The call is high-pitched and squeaky, and consists of two to seven introductory *ki* notes, followed by a strident four-noted phrase, *wip'-ki-wee'-oo*. A loud squealing note is uttered on flushing, and reddish brown colour is evident only at the base of the otherwise greyish buff primaries.

In the hand

Most similar to *shelleyi*, but with much finer underpart patterning, less rufous in the wings, and with a black-flecked white throat that is not distinctly formed into a spotted 'necklace' pattern. The feathers of the white post-ocular stripe are also only narrowly edged rather than spotted with black. The upperparts are similar to *psilolaemus*, but only the outer webs of the inner primaries are tinted basally with chestnut, and the secondaries are duller brown, barred with rufous buff.

Adult female. Like the male, but unspurred (males are singly spurred).

Immature. Similar to adults, but generally duller, with the middle underparts less uniformly vermiculated and more spotted. The outer juvenal remiges are also retained during the post-juvenal moult (Clancey 1967).

General biology and ecology

This is a low scrub- and grassland-adapted species, avoiding recently burned areas, those with heavy litter cover, and those with tall grasses. The birds feed primarily on the underground storage structures of plants such as corms and tubers, especially during winter. They occur in pairs and small coveys that consist of pairs and their most recent offspring, in rather low densities (up to 0.03 birds per hectare in one Natal Drakensberg study area). Although they share their habitats with the Levaillant's red-winged francolin the two are commonest in areas of sympatry and apparently reduce competition by habitat differences (the grey-winged using sparser and shorter grass at higher altitudes) and to some degree by food differences (Mentis 1973, 1979; Mentis and Bigalke 1973, 1979, 1980, 1981, 1985).

Social behaviour

These birds are monogamous and apparently territorial, with pairs having non-overlapping areas of activity. They may remain territorial to some degree all year, as they appear to be quite attached to particular areas. Males advertise by uttering calls from prominent locations during early morning and evening hours, in the usual francolin manner (Mentis and Bigalke 1980; Urban *et al.* 1986).

Reproductive biology

In the Cape Province breeding occurs from July to December, and in the Natal area it is from August to March, peaking in November–December. It is possible that some birds breed when they are less than 5 months old, and that some pairs may raise more than one brood in a single breeding season. Clutch-sizes are usually of five to ten eggs, the average of five clutches being 7.2. The incubation period is unreported, but mean covey sizes range seasonally from 2.7 to 7.4, suggesting that the average size of small broods is about five chicks. The estimated annual adult mortality rate is 50 per cent (Mentis and Bigalke 1980; Urban *et al.* 1986).

Evolutionary relationships

Considered by Crowe and Crowe (1985) to be part of a superspecies with *streptophorus*, *levaillantii*, and *finschi*, Hall (1963) believed it a monotypic species derived from a common ancestor with *shelleyi*.

Status and conservation outlook

This South African endemic is declining in numbers, and may need conservation attention (Urban *et al.* 1986).

ACACIA (ORANGE RIVER) FRANCOLIN
(Plate 76)

Francolinus levaillantoides Smith 1836
Other vernacular names: Archer's grey-winged francolin; Büttikofer's francolin (*jugularis*); Gariep francolin; Rüppel's francolin (*gutturalis*); francolin d'Archer (French); Rebhuhnfrankolin (German).

Distribution of species (see Map 24)

Widespread resident of eastern and southern African grasslands, occurring disjunctively from Ethiopia and Somalia south-west to northern Uganda, and from Angola south-east to Orange Free State.

Distribution of subspecies

F. l. gutturalis: northern Ethiopia (Eritrea). Possibly deserves specific distinction.

F. l. archeri: Uganda, Sudan, and western Somalia. Includes *stantoni* and *friedmanni*. Included in *lorti* by Urban *et al.* (1986).

F. l. lorti: eastern Somalia. (These first three northern races have sometimes been included in *africanus*.)

F. l. jugularis: southern Angola to the Cunene. Includes *cunenensis* and *stresemanni*.

F. l. pallidior: Namibia. Includes *watti*; included in *levaillantoides* by Urban *et al.* (1986).

F. l. kalaharica: Botswana (Kalahari Desert). Includes *langi*; included in *levaillantoides* by Urban *et al.* (1986).

F. l. levaillantoides: eastern Botswana, south-western Transvaal, and Orange Free State. Includes *ludwigi* and *garipensis*.

Measurements

Wing (*levaillantoides*), males 161–172 mm, females 157–167 mm; tail, males 74–80 mm, females 71–76 mm (Clancey 1967). Wing (*gutturalis*), males 151–169 mm, females 148–158 mm (Urban *et al.* 1986). Weight, southern races, males 370–528 g, females 379–450 g (Newman 1983); northern races, males 340–397 g, females 400 g (Urban *et al.* 1986). Egg, averages range from 37.2 × 29.4 mm in southern races to 41.2 × 31.2 mm in northern (Urban *et al.* 1986); estimated weight 16.7–20.8 g.

Identification (see Fig. 19)

In the field (13–14 in.)

Associated with fairly heavy cover in grassland habitats, from open plains to montane areas. In flight the primaries appear rufous for about half of their length, and are greyish brown toward their tips. The lack of black and white patterning on the underside helps distinguish it from Shelley's and grey-winged francolins, and it has less red in the wings than does Levaillant's red-winged francolin. The throat and the post-ocular stripes are white, the former is bordered below with a narrow black necklace. The flanks are rather irregularly barred and spotted with dark chestnut. The call is a repeated four-noted *ki-bi-til-ee* phrase that is rapidly uttered, and is accented on the third note. It is similar to that of the Shelley's francolin, but is faster and has shorter intervals between notes.

In the hand

Similar to *psilolaemus*, but with an unspotted throat and generally less rufous on the wings, and less strongly patterned on the abdomen than *shelleyi*. Usually with a clearly marked black 'necklace' around the throat (absent in *gutturalis*). Differs from *levaillantii* in that the white or buffy superciliary stripe merges with black and white speckling on the sides of the neck rather than fusing at the back of the neck in a speckled area (not evident in *gutturalis*), and has a shorter bill (exposed culmen to 30 mm).

Adult female. Like the male, but unspurred (males usually have a single spur).

Immature. Similar to adults, but retaining some juvenal remiges. The facial striping and necklace are less well defined, and the underparts are irregularly barred with black and buff (Clancey 1967; Urban *et al.* 1986).

General biology and ecology

Occurring in open fields, plains with fairly rich grassy cover, hillsides or mountain slopes with low thorn or other scrub, and dry woodlands, this species forages on roots, corms, and bulbs of plants such as *Moraea*, plant shoots, grass seeds, waste grain, and some larvae and adult insects. The birds occur in pairs and coveys of up to about a dozen birds, and are probably fairly sedentary. They are sympatric in southern Africa with the grey-winged and Levaillant's red-winged francolins, and probably compete with both, but the red-winged francolin usually occurs in somewhat moister habitats (Clancey 1967; Urban *et al.* 1986). Some contact with *shelleyi* may occur in

Fig. 19. Adult plumage variation in: the Levaillant's red-winged francolin, subspecies (A) *kikuyuensis*, (B) *crawshayi*, and (C) *levaillantii*; acacia francolin, subspecies (D) *gutturalis*, (E) *jugularis*, and (F) *levaillantoides*.

the Transvaal, but that species also prefers somewhat wetter climates.

Social behaviour

Evidently monogamous, the birds are territorial during the breeding season, and react strongly to playbacks of their advertisement call, which is normally uttered in early morning and evening hours, often by several birds in concert (Clancey 1967; Urban *et al.* 1986).

Reproductive biology

Breeding probably varies with seasonal rainfall in much of this species' range, and in South Africa occurs from September to May, while in Namibia and Angola it has been reported for June and August, and in Ethiopia during February, April, and August. The clutch-size is normally of four to eight eggs, but little else is known of the breeding (Clancey 1967; Urban *et al.* 1986).

Evolutionary relationships

This form is clearly part of a superspecies that includes *shelleyi* and *psilolaemus*, the three forming a generally allopatric series of populations (with some contact between *shelleyi* and *levaillantoides* in the Transvaal), although *levaillantoides* is unusual in that the northern part of its range is separated from the southern by the intervening population of *shelleyi*.

LEVAILLANT'S (RED-WING) FRANCOLIN
(Plate 77)

Francolinus levaillantii Valciennes 1825
Other vernacular names: Levaillant's red-winged francolin; francolin de Levaillant (French); Rotflügelfrankolin (German).

Distribution of species (see Map 25)
Local resident of montane or upland grasslands from north-western Kenya discontinuously to western Angola and the Cape.

Distribution of subspecies
F. l. kikuyuensis: Uganda and Kenya south to Angola and Zambia. Includes *muleme*, *adolffriederici*, *benguellensis*, *clayi*, and *momboloensis*.
   F. l. crawshayi: northern Malawi. Included in *levaillantii* by Urban *et al.* (1986).
   F. l. levaillantii: Transvaal, Natal, and Cape Province.

Measurements

Wing (*levaillantii*), males 168–179.5 mm, females 160–172 mm; tail, males 67.5–73 mm, females 60–73 mm (Clancey 1967). Weight, males 359–567 g, average (of three) 463 g, females 354–454 g, average (of four) 401 g (Newman 1983). Egg, average 40 × 31 mm (Mackworth-Praed and Grant 1952–73), estimated weight 21.2 g.

Identification (see Fig. 19)

In the field (13 in.)
A grassland species found on rocky hillsides and mountains from about 600 to 2100 m. The very heavy and long bill is a good field mark, supplemented by underparts that are mostly reddish brown to dark brown, rather than mostly white as in the previous species. The ochre stripe down the side of the neck, and the rather broken black and white 'bib' provide additional field marks. In flight the primaries appear almost completely reddish, with slightly greyer tips. The call consists of about five rapidly uttered high-pitched and musical notes, followed after a slight pause by a sixth slurred note, the last two notes being the loudest. It is very similar to that of the grey-winged francolin but somewhat less musical, and may be repeated several times, perhaps antiphonally.

In the hand
Differs from the other red-winged francolins in its heavy and long bill (exposed culmen at least 30 mm), and in having an ochre collar and with ochre on the edges of the throat, inside a black and white bib-like border (the black poorly developed in *crawshayi* and *kikuyuensis*). Two paired ochre post-ocular stripes also pass back to spread out and join or nearly join at the nape, or reach downward along the sides of the neck.

Adult female. Like the male, but unspurred (males are usually bluntly spurred).

Immature. Young birds have the black and white markings on the lower foreneck replaced by brownish spots, and retain some juvenal remiges into the first-basic plumage (Clancey 1967; Urban *et al.* 1986).

General biology and ecology

This species is adapted to moist montane grasslands, usually occurring at lower elevations and taller, denser cover than the grey-winged francolin, although they do overlap ecologically. It also occurs in grasses along streams, woodland, and scrub patches, in grassy clearing of second-growth, and in cultivated areas. Roosting is done communally on the ground, by social groups ranging in size from a pair to coveys of up to about 10 birds. The species apparently has rather low population densities in such grassland habitats (up to about 0.07 birds per hectare in Natal Drakensberg), and like the grey-winged francolin consumes a large proportion of subsurface bulbs and corms among its foods, especially during winter. Its unusually long and robust bill is certainly a digging adaptation associated with this type of foraging. Most of the rest of the species' foods are of animal materials such as insects, particularly during summer months (Mentis 1973; Mentis and Bigalke 1981, 1985).

Social behaviour

Monogamous and apparently territorial, pairs occupy non-overlapping home ranges, which for part of the year they share with their most recently raised brood. Seasonal variation in average group sizes thus range from two to about four birds, suggesting that reproductive rates are not very high. Advertisement calling is usually done in early morning and evening, and may be at least in part antiphonal. The relatively poor development of male spurs in this species, together with the close similarity of the sexes in size and plumage, suggests that the female may play an important role in territorial defence and advertisement, as is the case in the Hartlaub's francolin (Clancey 1967; Urban *et al.* 1986).

Reproductive biology

In the eastern Cape area breeding extends from March to July, but in Natal and areas to the north this species is mainly a spring and early summer nester, with nesting usually starting with the onset of October rains, and lasting until about February, perhaps to avoid grass fires. In Zaire breeding occurs from July to December during the dry season, and in East Africa during the rains or the early part of the dry season. The clutch-size ranges from three to twelve eggs, but averages about five. During the autumn season in Natal the immature component makes up about half of the population, suggesting an approximate 50 per cent annual adult mortality rate (Clancey 1967; Urban *et al.* 1986).

Evolutionary relationships

This species is part of a superspecies that also includes *psilolaemus* and *finschi* (Urban *et al.* 1986), which form an allopatric series of populations in southern and eastern Africa.

Status and conservation outlook

This form varies from locally common to rather uncommon, and it utilizes habitats that often are subjected to overgrazing or burning effects (Urban *et al.* 1986).

FINSCH'S FRANCOLIN (Plate 78)

Francolinus finschi Bocage 1881
Other vernacular names: francolin de Finsch (French); Finschfrankolin (German).

Distribution of species (see Map 24)
Endemic resident in woodlands and grasslands of western Angola (Cuanza Norte and Milanje south to northern Huila) and south-western Zaire (lower Congo River and Kwango Province). No subspecies recognized.

Measurements

Wing, males 162–170 mm, females 158–174 mm; tail, males 77–97 mm, females 80–99 mm (Urban *et al.* 1986). Estimated weight, both sexes 560 g. Egg, no information.

Identification

In the field (14–15 in.)
Limited to montane *Brachystegia* woodlands and scrub, and probably also adjacent grasslands of western Angola and southern Zaire. It is relatively long-billed, and has a grey breast. In flight the primaries appear to be mostly rufous except at their tips. Unlike the sympatric (at the southern part of its range) Levaillant's francolin it lacks any black and white patterning on the head and neck, although both species have long bills and ochre-tinted faces. Calls are still poorly known, but include a loud *whit-u-wit* usually uttered at dusk.

In the hand
Distinguished from the other red-winged francolins by its greyish brown breast, which shades to buffy and spots of pale chestnut on the abdomen, and by an ochre face that is not separated from the white throat by a black band, or with any other black markings on the face.

Adult female. Very similar to the male, but with small and rounded spurs averaging under 5 mm (the male is singly spurred, the spur at least 8 mm).

Immature. Undescribed.

General biology and ecology

In Zaire this species occurs in grasslands close to gallery forests at no more than 600 m, while in Angola it is associated with wooded savannahs, open *Brachystegia* woodlands, and bare slopes above timber-line at about 2100 m. It has been observed foraging among burnt brasses and leaves, and is known to consume seeds, beetles, and insect larvae, but little else is known of its ecology (Urban *et al.* 1986). It has a relatively long and robust bill, suggesting that it probably digs for subsurface plant storage organs such as bulbs and tubers.

Social behaviour

Nothing is known of this, other than that the birds are usually found in pairs.

Reproductive biology

Breeding records for Zaire extend from January to July, and for Angola for June and July. The nest is hidden in ground vegetation, and the clutch-size is of about five eggs (Urban *et al.* 1986).

Evolutionary relationships

The plumage similarities between this species and Levaillant's francolin suggest that they are closely related, and Urban *et al.* (1986) consider these two species and *africanus* to comprise a superspecies.

Status and conservation outlook

In Angola this species is uncommon to rare, but it is at least locally common in Zaire (Urban *et al.* 1986).

COQUI FRANCOLIN (Plate 79)

Francolinus coqui Smith 1836
Other vernacular names: Hubbard's francolin (*hubbardi*); francolin coqui (French); Coquifrankolin (German).

Distribution of species (see Map 25)

Widespread resident of woodlands, savannahs, and grasslands of sub-Saharan Africa from Mali east to Ethiopia, and south to Botswana and Natal.

Distribution of subspecies

F. c. spinetorum: Mali to Nigeria. Urban *et al.* (1986) also include the two following forms.
  F. c. buckleyi: Ghana to southern Nigeria.
  F. c. angolensis: Gabon to Angola and Zambia.
  F. c. maharao: southern Ethiopia. Urban *et al.* (1986) also include the three following forms.
  F. c. ruahdae: southern Uganda.
  F. c. hubbardi: western Kenya.
  F. c. thikae: central Kenya.
  F. c. coqui: Kenya to Botswana and Natal. Includes *stuhlmanni*, *lynesi*, and *campbelli*. Urban *et al.* (1986) also include the three following forms.
  F. c. kasaicus: central Zaire.
  F. c. vernayi: Botswana.
  F. c. hoeschianus: northern Namibia.

Measurements

Wing, males 127–152 mm, females 123–147 mm; tail, males 65–95 mm, females 63–91 mm (Urban *et al.* 1986). Weight, males of *coqui* 227–289 g, average (of four) 262 g, females 218–259 g (Newman 1984; Urban *et al.* 1986). Egg, average c. 35.5 × 28.5 mm (Mackworth-Praed and Grant 1952–73), estimated weight 15.9 g.

Identification

In the field (8–10 in.)

Associated with grasslands, savannahs, cultivated areas, and open brushy habitats such as *Protea* veld or *Brachystegia* woodlands, and one of the smallest of the francolins. In flight the primaries appear uniform greyish brown. It is a small and relatively pale-coloured francolin, with a rather bright buff head and whitish to buffy underparts, contrasting with black barring on the lower neck and chest and sometimes the entire underparts. The advertising call is a rasping, grating *kee-tirra, tirra*..., the earlier notes in the series of 7–10 notes the loudest, and the latter ones weakening, or a similarly repeated but squeaky *co'-qui, co'-qui, co'qui*, and the birds also utter plaintive *kwee-kit* calls.

In the hand

Distinctive in its small size (wing to 152 mm) and the uniformly ochre-brown head colour of males, which is sharply set off from the variably barred (least in xeric areas) lower neck and underparts pattern. Females are less ochre-tinted on the head, with a more distinctly white throat and black eye-stripe, and have a barred breast (most forms) or flank and underpart (*coqui* group) pattern.

Adult female. Differs from the male in having a black streak extending from the eye down the side of the neck, a second passing along the sides of the throat (these most evident in southern populations), circumscribing the buffy throat with a black line;

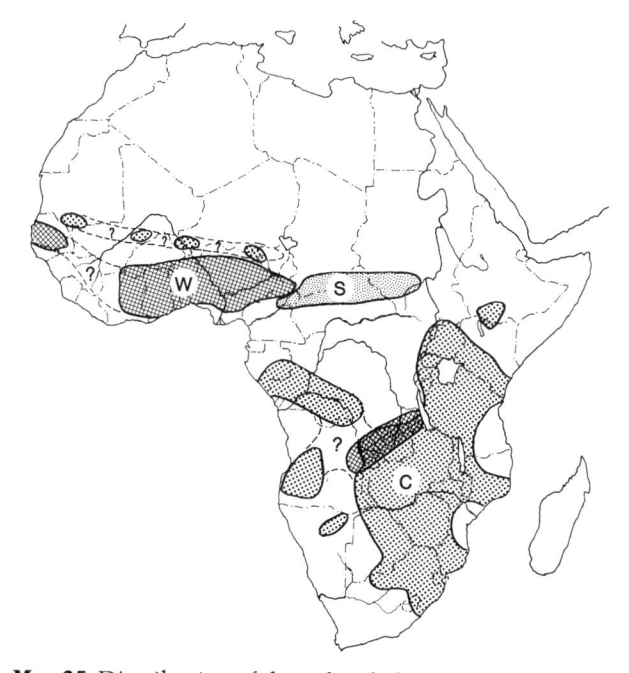

Map 25. Distribution of the red-tailed group of *Francolinus*, including coqui (C), Schlegel's (S), and white-throated (W) francolins.

neck and chest dull chestnut, mixed with grey and with white shafts; the rest of the upperparts somewhat greyer, with the bars and markings narrower and less regular. Tarsal spurs rudimentary (the male is singly spurred).

Immature. Similar to adults, but more buffy and less blackish above, and the males with the hindneck and upper mantle having brown rather than black barring. The underparts are also more buffy, with only faint barring evident (Clancey 1967; Urban *et al.* 1986).

General biology and ecology

This species occupies a wide variety of usually grassy habitats, including well-grassed woodlands such as miombo woodlands, as well as sandy areas with brush cover, thinned plantations, crop lands, and even rather thick bushveld, where there is a good grassy undergrowth. It occurs in coveys of up to about eight birds, presumably family groups, and roosts on the ground rather than in trees, using them only for perching when flushed by dogs. It consumes grass and other sedges, leaves, and arthropods such as insects, but apparently does not dig for bulbs as do many larger francolins. Surface water is possibly not required (Clancey 1967; Urban *et al.* 1986).

Social behaviour

Monogamous and probably territorial during the breeding season, based on observations of male aggressiveness during this time. During the early part of the breeding period the male's advertisement calls may be heard throughout the day, but is most common in morning and evening (Clancey 1967; Urban *et al.* 1986).

Reproductive biology

The breeding period in southern Africa is quite extended, peaking in Natal in November–February, in Zimbabwe in December–February during the rainy season, in Zambia mainly from November to February, in eastern Africa during the rains and the early part of the dry season, in Zaire between late August and March (during the rainy period and early dry season), and in Ethiopia possibly during May to June. The clutch-size is seemingly quite variable, from two to eight eggs, but usually five in southern Africa and two to four in equatorial Africa. The incubation period is unreported, but the young remain with their parents for an extended period, probably until the next breeding cycle (Clancey 1967; Urban *et al.* 1986).

Evolutionary relationships

This is clearly part of a superspecies with the next two forms, which form a distinctive 'red-tailed' group that are mostly allopatric except in a few areas where *coqui* and *albogularis* overlap locally. Hall (1963) postulated that this group originated in savannah habitat, perhaps in north-eastern Africa, with an early eastern isolate separating *coqui* stock from that which produced the other two species.

Status and conservation outlook

This form varies from local and uncommon to fairly common, and it may be sensitive to overgrazing and burning effects (Urban *et al.* 1986).

SCHLEGEL'S BANDED FRANCOLIN
(Plate 80)

Francolinus schlegelii Heuglin 1863
Other vernacular names: Schlegel's francolin; francolin de Schlegel (French); Schlegelfrankolin (German).

Distribution of species (see Map 25)

Resident of semi-arid habitats from Cameroon and Central African Republic east to south-western Sudan. No subspecies recognized, but the Cameroon population may be distinct (Traylor 1960); sometimes considered a subspecies of *coqui*.

Measurements

Wing, males 121–133 mm, females 118–126 mm; tail, males 59–71 mm, females 63–71 mm (Urban *et al.* 1986). Estimated weight, males 251 g, females 223 g. Egg, average 33 × 24 mm (Mackworth-Praed and Grant 1952–73), estimated weight 10.5 g.

Identification

In the field (8–10 in.)

About as small as the previous and following species, but allopatric with both, and found through many equatorial regions in grassy plains, savannahs, low scrub, and woodlands. Both sexes have ochre on the sides of the face and throat, rufous chestnut on the mantle and scapulars, and narrowly barred breast and flanks. Their call is a trumpet-like and rapidly repeated *ter, ink, terrra*, uttered mostly at dusk. The wings and tail are distinctly reddish when visible in flight.

In the hand

Very similar to *coqui*, but usually more rufous above, with reduced *Coturnix*-like patterning. Males have chestnut streaks on the flanks that are usually lacking in *coqui*, and the black and white breast barring does not extend around the nape. In females the breast and abdomen are both barred, unlike the more banded condition of female *coqui*. Similar dorsally to *albogularis*, but always barred on the flanks and breast.

Adult female. Differs from the male in having the chest feathers edged and barred with black, the back more blotched with black, and in having no spurs (males are singly spurred). Females are more irregularly barred below, almost lack white shaft-streaks dorsally, and have some dorsal transverse barring.

Immature. Differs from adults in having the mantle and scapulars barred with rufous buff and blackish.

General biology and ecology

Associated with grassy wooded savannahs, especially where the tree *Isoberinia doka* is present, and usually found in pairs or small groups that are probably family units. Roosting is done on the ground, with the birds often sleeping quail-like, the birds huddled close together and facing in opposite directions. Foods are mostly leaves, leaves of the *Isoberinia* tree, grain, and some insects (Urban *et al.* 1986).

Social behaviour

Little known, other than that the species is apparently monogamous.

Reproductive biology

Breeding in the Sudan apparently occurs between September and November. The clutch-size is probably of two to five eggs, with one probably aberrant record of 10 eggs that presumably was from two females (Urban *et al.* 1986).

Evolutionary relationships

An obvious close relative of *coqui*, and sometimes considered as a subspecies of that form, Hall (1963) considered it specifically distinct and perhaps showing closer affinities with *albogularis*.

Status and conservation outlook

This endemic form is generally uncommon to rare, and local in distribution (Urban *et al.* 1986).

WHITE-THROATED FRANCOLIN (Plate 81)

Francolinus albogularis Hartlaub 1854
Other vernacular names: francolin à gorge blanche (French); Weisskehlfrankolin (German).

Distribution of species (see Map 25)
Resident of brushy habitats from Senegal to Angola and Zambia.

Distribution of subspecies
F. a. albogularis: Senegal and Gambia.
　F. a. buckleyi: Ghana to Cameroon. Includes *gambagae*.
　F. a. dewittei: south-eastern Zaire.
　F. a. meinertzhageni: eastern Angola, north-western Zambia. Included in *dewittei* by Urban *et al.* (1986).

Measurements

Wing, males 133–143 mm, females 128–133 mm (Bannerman 1963); tail, males 68–69 mm, females 60–70 mm (Urban *et al.* 1986). Estimated weight, males 263 g, females 222 g; eight unsexed birds averaged 276 g (Urban *et al.* 1986). Egg, average 32 × 26 mm (Mackworth-Praed and Grant 1952–73), estimated weight 11.9 g.

Identification

In the field (9 in.)
This small francolin is associated with brushy growth and edges of cultivation, often in areas with a light scrub cover. Variable in appearance, but with a pale superciliary stripe and throat, and pale buff (males) to barred (females) underparts. In flight the chestnut wing-coverts and primaries are visible, and the tail is also chestnut. The call is a high-pitched, trumpet-like *ter-ink-inkity-ink* very much like that of the coqui francolin, but uttered more rapidly, and a double *ter-ink, ter-ink*, also faster than the coqui's corresponding call.

In the hand
Similar to *coqui*, but never with black flank barring and more rufous above. Males have white throats and are mostly chestnut (heavier in *dewittei*) and ochre on the breast and abdomen, the latter always lacking dark barring. The wings are generally more rufous-toned than in *coqui*, and the dorsal *Coturnix*-like patterning is less apparent, especially in females. The ground colour of the breast is buff or creamy, rather than white as in *schlegelii*, and the birds are unbarred on the sides and abdomen.

Adult female. Differs from the male in having a diffuse 'necklace' of black barring extending from the lower neck to the chest, some black barring on the underparts and flanks, and having narrower shaft-stripes and transverse barring dorsally. Unspurred (males are singly spurred).

Immature. Similar to adults, but usually with some barring on the breast as in adult females, and the ventral barring sometimes more extensive.

General biology and ecology

This is a savannah-adapted species, occurring especially in recently burned areas where green grass and insects are probably abundant, and also feeding on grass seeds. It has also been seen on open, sandy, and sparsely treed habitats. As with the previous species, it has also been reported from thick woodland where the tree *Isoberinia doka* is abundant. It occurs in pairs and small coveys, remaining hidden in the grass so far as possible, and probably roosting there as well. Its ecology is only poorly known, but probably is similar to that of the coqui francolin (Hall 1963; Urban *et al.* 1986).

Social behaviour

Little known, but probably monogamous.

Reproductive biology

Breeding in the Senegambia region occurs in September and October, in Nigeria during June, and in southern Zaire during October to December, early in the rainy season. The clutch is usually of six eggs, but ranges from four to seven. Little else is known of breeding (Urban *et al.* 1986).

Evolutionary relationships

This is certainly closely related to the two previous species, and Hall (1963) regarded it as especially similar to *coqui*.

LATHAM'S FOREST FRANCOLIN (Plate 82)

Francolinus lathami Hartlaub 1854
Other vernacular names: forest francolin; Latham's francolin; francolin de Latham (French); Waldfrankolin (German).

Distribution of species (see Map 26)
Resident in forests from Sierra Leone south to Gabon and east to south-western Sudan and Uganda.

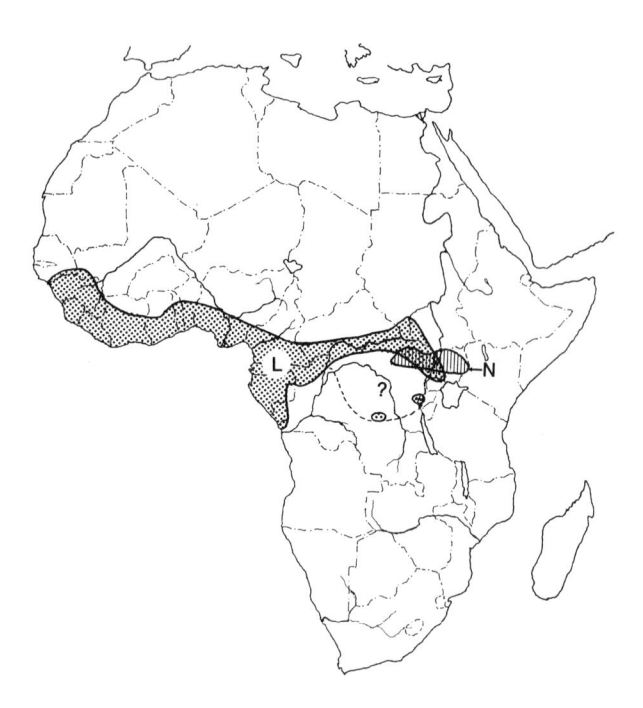

Map 26. Distribution of the Latham's (L) and Nahan's (N) francolins.

Distribution of subspecies
F. l. lathami: Sierra Leone to Gabon and north-western Zaire.
F. l. schubotzi: north-eastern Zaire to south-western Sudan and Uganda.

Measurements

Wing, males 128–143 mm, females 127–143 mm; tail, males 67–87 mm, females 66–78 mm. Estimated weight, males 254 g, females 284 g (Urban *et al.* 1986). Egg, average 40 × 26 mm (Mackworth-Praed and Grant 1953–72), estimated weight 14.9 g.

Identification

In the field (12–13 in.)
Associated with dense tropical forest, rarely if ever found in clearings. One of the darkest of the francolins, but with a contrasting superciliary stripe, greyish cheeks, and white spotting below; the chin and throat are black. The usual call is a prolonged series of rather uniform, high-pitched whistles, also described as a dove-like *kwee-coo-coo* that is repeated in extended series of eight units; clucking and 'flute-like' sounds have also been heard. Rarely flies unless nearly stepped on.

In the hand

The very dark colour, including a blackish chin and throat, but with contrasting white spots below, comprise a unique combination of francolin traits.

Adult female. Differs from the male in having the ground colour of the upperparts and crown olive-brown, the feathers lacking white markings but faintly barred with rufous buff and black; many of the scapulars and wing-coverts blotched with black, and the chest and breast feathers margined with brown. Small spurs are present (males are singly spurred).

Immature. Juveniles of both sexes have a white throat and more white on the underparts. Young males resemble females but have reddish brown upperparts that are mottled with black, the crown mottled with black and brown, the breast and underparts brown with cross-shaped markings outlined in black, and the flanks brown with white shaft-streaks. Young females also resemble adult females but have more rufous upperparts, the crown feathers tipped with black, the breast brown with white streaks, and the flanks rufous with black barring (Mackworth-Praed and Grant 1952–73; Urban *et al.* 1986).

General biology and ecology

Associated mainly with lowland primary forests, less often in dense secondary forests, and locally in gallery forests. The birds occur in pairs or small groups that are extremely difficult to observe, as they remain in heavy cover. They forage by scratching in the forest litter, primarily on arthropods, especially termites and ants, and only to a minor degree (about 10 per cent of food) on plant materials, mostly fruits of the oil palm (*Elaeis*) and other materials (Urban *et al.* 1986).

Social behaviour

Very little known, but probably monogamous.

Reproductive biology

Breeding records for Cameroon are from December and February, and for Zaire extend from December to April. There is a record of August breeding in Uganda, and birds in breeding condition from May to June. The eggs are deposited on a leafy base between the buttresses of forest trees, and the usual clutch is of two eggs, rarely three. One study suggests a fairly high (80 per cent) hatching success rate, perhaps because of the placement of the nest (Brosset 1974).

Evolutionary relationships

Regarded by Hall (1963) as possibly related to *schlegelii* but not assigned to any specific group; Crowe and Crowe (1985) confirmed these speculations that this species is related to the red-tailed group of francolins.

Status and conservation outlook

This form is uncommon to locally common and, of course, is dependent upon preservation of adequate forest habitat for its survival.

NAHAN'S FRANCOLIN (Plate 63)

Francolinus nahani Dubois 1905
Other vernacular names: Nahan's forest francolin; francolin de Nahan (French); Iturifrankolin (German).

Distribution of species (see Map 26)

Resident in forests of north-eastern Zaire and western Uganda; also local in south-central Uganda. No subspecies recognized.

Measurements

Wing, males 135–148 mm, females 135–151 mm; tail, males 60–77 mm, females 65–79 mm (Chapin 1932). Weight, two males 308–312 g, three females 234–260 g (Urban *et al.* 1986). Egg, average 36 × 26 mm (Mackworth-Praed and Grant 1952–73), estimated weight 13.4 g.

Identification

In the field (13 in.)

Associated with heavy forests, and appropriately dark-bodied, the blackish underparts having white spots or streaks, and with some rather large white spots on the flanks. Very similar to the previous species, and possibly overlapping somewhat in range and habitats with it, but with red rather than yellow legs. Calls are as yet undescribed.

In the hand

Very similar to *lathami*, namely blackish brown above, and blackish streaked and spotted with white below and on the flanks, but the back is mottled brown and black, without white shaft-streaks. Also separable from *lathami*, the only similar species of dark-bodied and white-spotted francolin, by the large red orbital skin area and red legs of *nahani*.

Adult female. Smaller, with the rufous blotching and

streaking extending to the forehead and sides of the head, the throat rufous buff, the upper breast grey with rufous mottling, and the rest of the underparts also rufous buff, becoming blotched with black on the under tail-coverts. Both sexes are unspurred (Urban *et al.* 1986).

Immature. Generally darker above than the adults, and the spotting of the neck does not extend to the upper side (Mackworth-Praed and Grant 1952–73). Also has a more diffuse general pattern, and the feathers of the flanks and breast with brownish markings and edgings (Chapin 1932).

General biology and ecology

Found only in dense primary forests up to about 1400 m, and very rarely seen. The birds feed on the forest floor in the leaf litter, eating insects, other invertebrates, and various plant materials (Urban *et al.* 1986).

Social behaviour

Unstudied, but probably monogamous.

Reproductive biology

Only a single nest has been found, which was located in a tree hollow about a metre above ground, and contained four eggs. Collected birds suggest an extended breeding season (Urban *et al.* 1986).

Evolutionary relationships

Hall (1963) did not assign this species to a particular group, but suggested that it might be an offshoot of the scaly group. Crowe and Crowe (1985) confirmed that speculation and placed it sequentially at the end of that group.

Status and conservation outlook

This is apparently a distinctly rare species, although no numerical information on its status is available. It was listed as 'rare' by Collar and Stuart (1985).

GREY FRANCOLIN (Plate 83)

Francolinus pondicerianus Gmelin 1789
Other vernacular names: grey partridge; francolin gris (French); Graufrankolin (German).

Distribution of species (see Map 27)

Resident in grasslands and semi-arid thorny scrub habitats from eastern Iran east throughout India and Sri Lanka. Local (apparently introduced) in northern Oman. Also established locally on the Hawaiian Islands (Molokai, Kanai, Maui, Hawaii) and several other islands (Amirante, Mauritius, Réunion,

Map 27. Distribution of the grey (G) and swamp (S) francolins.

Seychelles, and Rodrigues). Introduced but possibly now extirpated from Andaman Islands. Introduced but of doubtful survival in various parts (Oklahoma, Nevada, Texas, etc.) of the south-western USA (Bump and Bump 1964; Long 1981).

Distribution of subspecies (excluding introductions)

F. p. mecranensis: south-eastern Iran and southern Pakistan, grading into *interpositus* in Sind.

F. p. interpositus: north-western India and Pakistan from Sind to West Bengal, south to the Deccan plateau and the Dodavery River. Includes *paganus, titar,* and *prepositus.*

F. p. pondicerianus: southern India south of *interpositus,* and Sri Lanka (north-western coast).

Measurements

Wing (*interpositus*), males 144–160 mm, females 134–146 mm; tail, males 80–94 mm, females 79–91 mm (Ali and Ripley 1978). Wing (*mecranensis*), males 138–151 mm (Meinertzhagen 1954). Weight (*interpositus*), average of 114 males, 274 g, 91 females 228 g (Bump and Bump 1964). Egg (*mecranensis*), average 34 × 26 mm (Baker 1935), estimated weight 12.7 g.

Identification

In the field (13 in.)

Associated with dry, hilly plains and thorny scrub, and sometimes occurring near crop lands or villages. This is the most uniformly and finely patterned below of the Asian francolins, and the one with the most distinctively demarcated yellowish throat, which is bounded by a narrow black 'necklace'. In flight the chestnut tail is conspicuous. The male's calls consist of repeated *titur* or *ke-titur* notes, which when advertising typically begin with about six preliminary well-spaced hiccough-like notes that are followed by a very loud series of repeated *titur-titur-titur* notes. When separated from its mate or group, a similar series of *ke-titur, ke-titur, ke-titur* notes may be uttered, and the female may respond with a series of single high-pitched notes interspersed among the male's calls (Bump and Bump 1964).

In the hand

The combination of chestnut-coloured outer rectrices and a whitish to yellow-rufous throat patch that is bounded by a narrow black band provides for easy recognition of this francolin species.

Adult female. Like the male, but usually unspurred (males are singly spurred).

Immature. Like the adult, but less rufous on the forehead and a paler throat patch, the latter with little or no black border (Ali and Ripley 1978). Males usually have small, blunt spurs by three months of age (Bump and Bump 1964).

General biology and ecology

Small coveys or flocks consisting of family groups are the usual social pattern, with the birds typically roosting in small trees or among thick, spiny shrubs. At dawn they move from their roosts to foraging areas, obtaining their food both from the surface and often digging with their bills for root stalks or corms. A wide variety of plant materials are eaten, with weed seeds and cultivated grains being important components. Insects are eaten in quantity during summer months. The birds can survive rather long periods without drinking water, and evidently are more drought-adapted than either black francolins or chukar partridges (Bump and Bump 1964).

Social behaviour

The coveys of this species are not migratory, but do undertake local movements of at least 20 miles. Likewise, liberated birds are known to have moved as far as about 50 miles in the course of a few years. In some areas semi-isolated habitats of less than 100 acres in extent have been found to support apparently resident populations, and at times a pair will raise a brood in city gardens of only a few acres in area (Bump and Bump 1964).

Reproductive biology

The breeding season of this species in India is extremely prolonged, with eggs having been recorded every month of the year, and it has been suggested that females raise two or perhaps even three broods in a single year there. Nests are usually hidden under brush, but sometimes are placed on quite bare ground. Shrubby ravines, hedges, isolated clumps of cacti, and crop lands have all been mentioned as favoured locations for nest sites. The usual clutch is of six to nine eggs, and in captivity single females have been found to average about 37 eggs per year. The incubation period is 18–19 days, and is performed by the female alone. However, it is not unusual to see both parents attending a brood, so males probably remain close by during the incubation period and join the family after hatching. Parents and their brood remain together at least into the autumn period (Bump and Bump 1964).

Evolutionary relationships

Hall (1963) suggested that this species may have affinities with both the red-tailed and striated groups of African francolins, probably having separated from the African stock before these two groups diverged.

Status and conservation outlook

According to Bump and Bump (1964), this species is by far the most common resident game bird over most of its range, which is probably a reflection of its remarkable adaptability to arid habitats as well as its behavioural adaptability and its high reproductive potential.

SWAMP FRANCOLIN (Plate 84)

Francolinus gularis Temminck 1815

Other vernacular names: kyah; swamp partridge; francolin multirale (French); Sumpffrankolin (German).

Distribution of species (see Map 27)

Resident in swamps of the terai of northern India and Nepal from the Kumaon terai and Uttar Pradesh east to Bangladesh and Assam. No subspecies recognized.

Measurements

Wing, both sexes 162–186 mm; tail, both sexes 101–127 mm. Weight, males *c.* 510 g (Ali and Ripley 1978). Egg, average 39.4 × 30 mm (Baker 1935), estimated weight 19.6 g.

Identification

In the field (15 in.)

This large francolin is invariably associated with swampy habitats, and often even wades about in shallow water. It is difficult to flush, but in flight exhibits chestnut outer tail feathers. It also has strongly contrasting white-streaked flanks and a chestnut-coloured chin and throat patch that is not bounded by a black line. Its vocalizations include repeated harsh *chukeroo* calls and occasional loud *qua* notes that ascend in tone.

In the hand

Readily separated from the other Asian francolins by its linear white flank markings, and its rusty red coloration on the chin, throat, and foreneck.

Adult female. Like the male, but with no spur or only a rudimentary one, and duller tarsus colour (Ali and Ripley 1978).

Immature. No specific information.

General biology and ecology

This is by far the most water-dependent species of francolin, often wading about in shallow marshy areas, or even climbing on reed stems where water may be too deep in which to wade. Roosting occurs in thorny trees growing in swampy jungles, or on flattened reeds or elephant grass (*Typha elephantina*) in flooded areas. However, during mornings and evenings the birds may move out into open grasslands, into rice fields, or even onto freshly ploughed fields. They consume virtually every kind of seeds and grains, as well as insects and sometimes even small crustaceans (Baker 1930; Ali and Ripley 1978).

Social behaviour

Almost nothing is known of the social behaviour of this species, owing to its inaccessible and overgrown habitats. It is known to be highly pugnacious, and as a result is often maintained in captivity for fighting purposes. The male's advertising call is usually preceded by several single chuckling or croaking sounds, and consists of a rather harsh series of *chuckeroo-chuckeroo-chuckeroo* notes that apparently sound very much like those of the grey francolin (Baker 1930).

Reproductive biology

Nesting occurs from about the end of February or early March until May, with most laying occurring during late March and April. The nest is always placed in reeds, which are usually actually standing in water, ranging in depth from a few inches to as deep as 18 in. or more. The nest under such conditions is a large, thick pad of rushes and similar vegetation, but when it is located on a swamp margin or dry bank area it is usually not so well constructed. The clutch is small, typically consisting of only three or four eggs, with five uncommon and six apparently distinctly rare. The incubation period is still uncertain.

Evolutionary relationships

Hall (1963) suggested that this form may have become ecologically isolated before the link between Asian and African francolins was broken, since in both its morphology and ecology it is sufficiently distinct as to suggest a divergence that occurred relatively long ago. However, on her evolutionary dendrogram Hall shows it diverging from a common origin with the spotted group, and I believe that this group does indeed represent the nearest phyletic relatives of *gularis*. Crowe and Crowe (1985) have also included from a phenetic analysis that the nearest relatives of *gularis* are the species comprising Hall's spotted group. They have also suggested that the ancestral francolins were very quail-like, and that the coqui francolin is the living species that most closely approximates this ancestral type in both its external morphology and its skeletal structure, as well as in having the least complex advertisement calls.

Status and conservation outlook

Because of drainage of its swampland habitats, this species is now becoming scarce in many areas where it was formerly abundant.

18 · Genus *Perdix* Brisson 1760

The typical partridges are medium-sized grassland-adapted Palaearctic species with rounded and chestnut-coloured tails of 16–18 rectrices that are slightly more than half as long as the wing. The wing is rounded, with the seventh primary the longest. The post-juvenal primary moult is incomplete, and the tail moult is perdicine (centrifugal). The sexes are somewhat dimorphic, with males (and sometimes females) having brown to blackish breast patches, and the tarsus is unspurred in both sexes. Three allopatric species are recognized here:

    P. perdix (L.): grey partridge
        P. p. perdix (L.) 1758
        P. p. sphagnetorum (Altum) 1894
        P. p. armoricana Hartert 1917
        P. p. hispaniensis Reichenow 1892
        P. p. italica Hartert 1917
        P. p. lucida (Altum) 1894
        P. p. robusta Homeyer and Tancre 1883
        P. p. canescens Buturlin 1906
    P. dauuricae Pallas (= *P. barbata* Verreaux and De Murs 1863): Daurian partridge
        P. d. dauuricae Pallas 1811
        P. d. suschkini Poliakov 1915
    P. hodgsoniae Hodgson: Tibetan partridge
        P. h. caraganae R. and A. Meinertzhagen 1926
        P. h. hodgsoniae Hodgson 1857
        P. h. sifanica Przhevalski 1876

KEY TO THE SPECIES OF TYPICAL PARTRIDGES (*PERDIX*)

A. Rectrices 16; with white throat and superciliary stripe: Tibetan partridge (*hodgsoniae*)
AA. Rectrices 18; no white on throat or superciliary stripe
  B. Chest uniformly grey; 'horseshoe' patch on anterior abdomen chestnut-brown, or if absent then lightly barred below: grey partridge (*perdix*)
  BB. Chest with medial yellow band; patch on anterior abdomen black or blackish brown, or rather heavily barred below: Daurian partridge (*dauuricae*)

GREY PARTRIDGE (Plate 85)

Perdix perdix (L.) 1758
Other vernacular names: common partridge; European partridge; Hungarian partridge; Hun partridge; perdrix grise (French); Rebhuhn (German); seraya kuropatka (Russian); perdix pardilla (Spanish).

Distribution of species (see Map 28)
Resident of Eurasian grasslands from Scandinavia south to northern Spain and Italy, east to the western Sayans in Siberia, Tuva ASSR, Xinjiang, and south-east to north-western Iran. Introduced widely in North America and locally introduced or reintroduced in Europe.

Distribution of subspecies (excluding introductions)

P. p. perdix: British Isles and southern Scandinavia south to France and east across south-eastern Europe to Greece, except in areas of four following subspecies.
    P. p. sphagnetorum: local on moors of northern Holland and north-western Germany.
    P. p. armoricana: local in France.
    P. p. hispaniensis: local in central Pyrenees, west to the Cantabrian Mountains.
    P. p. italica: Italy.
    P. p. lucida: Finland east to the Urals, and south to the Black Sea, Crimea, and northern Caucasus, intergrading to the east with *robusta* and to the south with *canescens*.
    P. p. robusta: USSR, from about the Urals east across Siberia to the Russian Altai, western Sayans, and Tuva ASSR, and south to the coasts of the Caspian and Aral seas and to northern and western Xinjiang.
    P. p. canescens: Turkey east to the Caucasus, Transcaucasia, and western and north-western Iran. Includes *furvescens*.

Measurements

Wing (*perdix*), males 154–166 mm, females 151–160 mm; tail, males 76–83 mm, females 68–76 mm. Weight (*perdix* and *sphagnetorum*), males 325–455 g, females 310–450 g (Cramp and Simmons 1980). Egg, average 36 × 27 mm (Cramp and Simmons 1980), estimated weight 14.6 g.

Identification

In the field (12 in.)

Inhabits grasslands, small-grain crop lands, and similar open-country habitats. Both sexes have tawny heads that contrast with more greyish breasts, and in flight exhibit bright chestnut outer tail

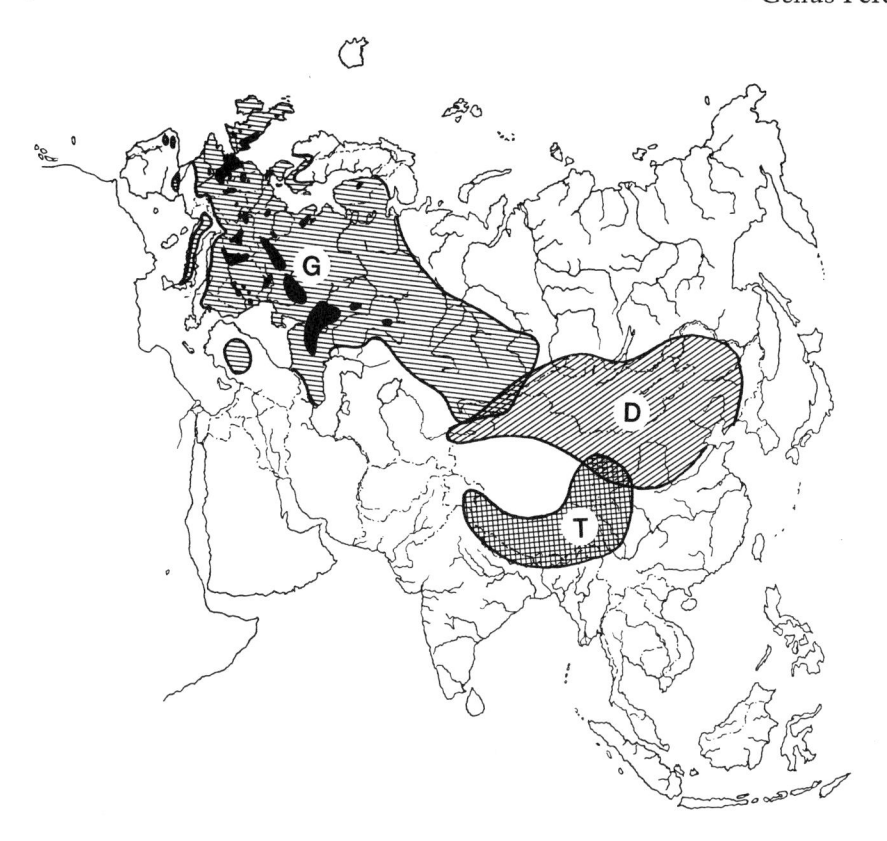

Map 28. Distribution of the Daurian (D), grey (G), and Tibetan (T) partridges. See also Map 10 for introduced North American distribution of grey partridges, both of which show denser populations by inking.

feathers similar to those of most *Alectoris* species, but the latter are always patterned with black on the throat and flanks. The advertising call is a sharp *kree-arit*, mainly heard in spring.

In the hand
Distinguished from the other species of *Perdix* by its brownish abdominal spots, its lack of a definite 'beard', and lack of ochre in the mid-breast area.

Adult female. Differs from the male in that the sides of the neck are brownish, the lesser and median wing-coverts and the scapulars have the ground-colour mostly black, with wide-set transverse buff bars, and the abdominal horseshoe mark is usually reduced or absent.

Immature. Immatures have the two outermost (juvenal) primaries more pointed and frayed, the legs more brownish, and the barring on the outer primaries not so reduced as in adults.

General biology and ecology
Within its native Eurasian range, the grey partridge favours temperate and steppe climatic zones. Vegetation favoured is that of a continuous grassy or herbaceous nature, not much taller than the bird itself, but interspersed with or within a few hundred metres of taller and denser cover of brushy or woody nature, such as hedgerows, woodland verges, or scrub. Over most of its range it has adapted to land that has been largely given over to agriculture, especially the cultivation of cereal grains, but also meadows and pastures are regularly exploited. Access to drinking water and dusting areas are also part of the species' needs, but lakes, rivers, and coastlines are generally avoided, as are uninterrupted woodlands (Cramp and Simmons 1980). In North America the introduced populations have survived best in areas of continental climate and grasslands or crop lands having very fertile soils, fairly short growing seasons, and limited precipitation during the incubation and brooding periods. Severe winters are usually not a serious limiting factor unless the amount of snowfall is so great as to make grain or other food sources unavailable (Westerskov 1965; Johnsgard 1973). Native grasslands or hay fields are favoured nesting habitats, but brushy areas may be used for winter cover, and nests are sometimes placed near brushy edges. Brood-

ing cover is much like nesting cover, and includes hay fields, grain fields, and natural grasslands. Yocom (1943) found that most of the nests in his study area were situated in hay fields, especially alfalfa (lucerne) fields, and McCabe and Hawkins (1946) similarly found that alfalfa provided immediate cover for about half of the nests that they found, which were often near the edges of such fields. Foods have been extensively analysed in England and Europe (Dwenger 1973; Glutz von Blotzheim 1973; Cramp and Simmons 1980). These studies also indicate that three major food sources are used, including grain and weed seeds, cereals and clovers, and the green leaves of grasses. These same categories of foods form the major basis of the diet in North America, with the grains varying by locality, but often including oats, barley, wheat, and corn (maize). These foods are especially significant during winter months, while weed seeds are often consumed in spring, and green leafy materials such as grasses are also consumed as soon as they emerge in spring (Yocom 1943; Westerskov 1966). The birds are fairly sedentary during winter, with only small home ranges typical of coveys and of individual birds (Yocom 1943; Johnson 1969). In one study, the winter ranges of eight coveys varied from 4.9 to 34.4 ha., and winter mobility from 0.3 to 1.3 km (Schulz 1980), while the late winter and spring prenesting range of nine radio-tagged females was found to average 212 ha., suggesting substantially greater mobility during the period of covey break-up and nest-site selection (Church *et al.* 1980).

Social behaviour

Covey sizes are fairly consistently uniform in grey partridges, and usually average 10–15 birds during the autumn and early winter months, declining to about five to seven birds by late winter and early spring, through both the separation of already paired birds from the winter groups and normal attrition by various mortality factors. At least in some areas winter mortality rates are quite high, and at times may exceed 50 per cent (Ratti *et al.* 1983), although other studies (e.g. Weigand 1980) have suggested that winter losses are only slightly higher than at other seasons. The studies by Jenkins (1961) are the classic observations on the social behaviour of this species, and the general mechanisms of covey organization and pair formation were discussed in Chapters 2 and 5. By early spring males begin uttering their distinctive 'rusty-gate' advertisement calls, which consist of a metallic *keee-uck*', with the second syllable accented. This call is uttered in an upright posture (see Fig. 8), and it serves both as a self-advertisement and threat. A lateral 'courtship upright' display, without obvious wing-lowering but with exhibition of the flank feathering, is used by males toward females, and in contrast to other genera so far studied there is a 'sinuous neck' display of the female during which she circles the male and alternately raises and lowers her outstretched neck (see Fig. 8). Pairing may occur as long as four months prior to breeding, but at times short-lived pairings occur, especially among young birds. Most pairings occur between birds of different social groups, facilitating outbreeding, and when both members of a pair survive over the following winter they commonly reunite. Pairing brings about territorial-like dispersion of pairs, but these pairs do not defend definite territorial boundaries, and instead like the winter flocks they exhibit a proximity-tolerance to near-by birds. By the end of March their breeding territories are topographically established, and by late April and May nest-site selection is well under way (Blank and Ash 1956; Cramp and Simmons 1980).

Reproductive biology

In central Europe the breeding season extends from early May to June, in Scandinavia from the end of May to June, and in Britain from latter April onward. Renestings extend the season considerably, sometimes to August or even September (Cramp and Simmons 1980). The eggs are laid at intervals of about two every three days, and as noted in Chapter 2 there are marked geographic and seasonal variations in clutch-size, but these average about 15–17 eggs for first clutches in Europe. In North America the average clutch-sizes are very similar to these (McCabe and Hawkins 1946; Hupp *et al.* 1980). Incubation is normally by the female alone, with the male possibly helping on rare occasions (more probably simply sitting beside the female near the time of pipping), and lasts 24–25 days. Studies in North America (summarized by Hupp *et al.* 1980) suggest that the nesting success (percentage of initiated nests hatching) averages only about 30 per cent, but renesting efforts undoubtedly bring overall female breeding success to a much higher level. Brood losses are often substantial during the first week or two after hatching, and are greatly affected by unfavourable weather, particularly its effects on chilling losses and on insect food availability for chicks during this critical period (Yocom 1943; Blank *et al.* 1967). When the chicks are about 13 days old they are able to fly for short distances, and by the time they are about 120 days old their outermost (eighth) first-winter primaries have attained maximum length (Willard 1973; Demers and Garton 1980). By about the same age their weight stabilizes at about that typical of adults (Dementiev and Gladkov 1952).

Evolutionary relationships

Beyond the obvious affinities with the other species of *Perdix* (see below), the relationships of the genus *Perdix* are not at all clear. There are few records of intergeneric hybridization, in spite of the frequency with which these birds have been raised and bred in captivity. Presumed but not proven hybrids with *Alectoris* have been reported, as well as some reputed but poorly documented captive-bred hybrids with various pheasants and grouse (Gray 1958). The seemingly distinctive courtship behaviour of *Perdix* also suggests that it is not very close to *Alectoris* or other seemingly 'main-line' perdicines.

Status and conservation outlook

In Europe this important sporting species has shown a steady decline in population throughout its entire range since about 1950, varying from 50 to 90 per cent, apparently in connection with the advent and spread of such modern agricultural practices as intensive cultivation, widespread use of pesticides, and the like. In areas where hunting is regulated and formalized it accounts for about 20–30 per cent of the annual European mortality, and in England winter losses average about 45 per cent of the total (Cramp and Simmons 1980). In the United States and Canada the autumn to spring mortality factors remove about 50 per cent of the population, representing most of the total average overall mortality rate of about 70 per cent. Many of these losses apparently occur during winter and the following period of prenesting dispersal, with hunting apparently usually responsible for a relatively small percentage of the overall annual mortality (Schulz 1977; Weigand 1980). In the late 1960s the estimated harvest of this species in North America was about 650 000 birds (Johnsgard 1973). The annual harvest during the 1970s approximated 970 000 birds, with substantial increases over the past decade occurring in four states and one Canadian province (Peterson and Nelson 1980). In part these harvest trends reflect concurrent declines in the ring-necked pheasant population in the same areas, and the resulting greater attention being paid by hunters to this small and previously secondary game species, rather than to measurable increases in grey partridge populations.

DAURIAN PARTRIDGE (Plate 86)

Perdix dauuricae Pallas 1811
Other vernacular names: bearded partridge; pan-chi'ih shan-chun (Chinese); perdrix barbue; perdrix de Daourie (French); Bartrebhuhn (German); borodataya kuropatka (Russian).

Distribution of species

Resident of Asian grasslands from Russian Altai east to Transbaicalia, Amurland, and Ussuriland, and south to Manchuria, Mongolia, Nei Monggol (Inner Mongolia), Gansu, and Qinghai. Introduced locally to Manilla area of Philippines.

Distribution of subspecies

P. d. dauuricae: Kirgiza and Xinjiang east to Transbaicalia and perhaps to western Amurland, south to Tuva ASSR, and Mongolia. Includes *turcomana*.

P. d. suschkini: Manchuria, Amur Valley, and Ussuriland south through Nei Monggol and west to Gansu and Qinghai. Includes *przewalskii* and *castaneothorax*.

Measurements

Wing, males 137–141 mm, females 133–145 mm; tail, males 76–98 mm, females 73–92 mm. Weight, males 294–300 g, females 200–340 g (Cheng *et al.* 1978). Egg, average 35.2 × 26.5 mm. Estimated weight 12.8 g.

Identification

In the field (12 in.)

Similar to the previous species and occupying comparable (but somewhat drier and more rocky) habitats, but overlapping only slightly with it and locally hybridizing with it in the Russian Altai and around Zaysan Nor. Its calls are also essentially the same, but it has a blackish abdominal patch, and is more ochre-toned in the middle of the breast.

In the hand

Separated from the previous species by its distinctive 'beard' of stiffened feathers along the sides of the chin (during autumn and winter), its black abdominal patch, and an ochre band connecting this patch with the tawny throat.

Adult female. Differs from the adult male in having the black horseshoe reduced or absent, much less bright buff on the chest and breast, and with a chestnut patch on the cheek below the eye.

Immature. First-year males have the black horseshoe much reduced, and the upperparts more coarsely marked. Both sexes probably retain the two outer juvenal primaries through the first year.

General biology and ecology

In the USSR this species occupies grassy and wooded steppes, up to at least 3000 m elevation, but avoids steep mountain slopes as well as deserts. The belt of foothills and adjacent plains, with meadows, plantations, and river valleys, is favoured, but the birds also extend up open mountain slopes into subalpine meadows. On grassy slopes of foothills and mountain meadows the birds are more widely distributed than is the grey partridge, and it may be also better able to move out into gravel- and stone-covered steppes than is that species (Dementiev and Gladkov 1952). There are probably minor winter movements, and it has been reported that birds that had been released into an unoccupied area dispersed up to 200 km from the release point. This was part of an apparently unsuccessful effort to establish this species in the central USSR (Long 1981). Foods are little studied, but are apparently much like those of the grey partridges, with cereal grains, weed seeds, berries, and the like eaten for much of the year, and insects eaten in quantity during the summer (Dementiev and Gladkov 1952).

Social behaviour

Apparently, like the grey partridge, this species is found in flocks of 15–30 individuals during the autumn months. In early winter these flocks sometimes increase to groups of as many as 200 birds. During spring these flocks dissolve into monogamous pairs, with pairing beginning by mid-March in the Altai region, and laying beginning in late April and early May (Dementiev and Gladkov 1952). Nothing specific has been written on the vocalizations of this species, but they are apparently much like those of the grey partridge.

Reproductive biology

Very few nests have apparently been found, but clutches of 16 and 20 eggs have been reported for May and June, and very small chicks have been reported in mid-June and mid-July (Dementiev and Gladkov 1952). In all likelihood the nests, incubation characteristics, and development of the young are very similar to those of the grey partridge.

Evolutionary relationships

Although this form is essentially parapatric with *perdix*, their ranges come into slight contact in western Siberia (Johansen 1961). Together with the following form, these three taxa comprise a closely related species group that probably originated in northern Asia. Glaciation may well have isolated this ancestral form into a westerly component that gave rise to *perdix*, an easterly form that produced *dauuricae*, and a southern and more montane-adapted population in the Himalayas that produced *hodgsoniae*.

Status and conservation outlook

No detailed information is available. Much of this species' range is in an area of low human occupation and activity, and it has very probably been affected much less by civilization than has that form. In some areas of eastern Siberia it is an important game bird, and to a smaller degree it is exploited in western Siberia and even less so in Kazakhstan (Dementiev and Gladkov 1952).

TIBETAN PARTRIDGE (Plate 87)

Perdix hodgsoniae Hodgson 1857
Other vernacular names: Hodgson's partridge; Ladakh partridge; perdrix de Hodgson (French); Tibetrebhuhn (German).

Distribution of species (see Map 28)
Resident from montane grasslands from Ladakh, Kumaon, and Nepal north-east through Xinzang (Tibet) to eastern Qinghai and Gansu, in montane habitats up to about 5500 m.

Distribution of subspecies

P. h. caraganae: eastern Kashmir east to extreme western Xinzang, where it grades into *hodgsoniae*.

P. h. hodgsoniae: Himalayas from western Nepal to Assam and eastern Xinzang, where it intergrades with *sifanica*.

P. h. sifanica: eastern Xinzang and adjacent central and southern Sichuan, north to central Quinghai, and Gansu. Includes *koslowi*.

Measurements

Wing (*hodgsoniae*), males 155–165 mm, females 149–155 mm; tail, both sexes 86–91 mm (Ali and Ripley 1978). Weight, males 294–370 g, females no data (Cheng *et al.* 1978). Egg (*hodgsoniae*), average 37.6 × 27.2 mm (Ali and Ripley 1978), estimated weight 15.4 g.

Identification

In the field (12 in.)
Occupies dry hillsides with scattered bushes or low junipers at altitudes of 3600 m or higher, where it is the only partridge-like bird. The chestnut collar and black abdominal patch are distinctive, and the calls include harsh, grating, and repeated *chee* (when flushed) or *schrrrreck* (during morning and evening) notes. When uttered as separation calls, the buzzing

sounds reportedly are similar to those made when a finger is repeatedly drawn over the teeth of a comb, or like the creaking of the lid of a wicker basket.

In the hand
Separated from the other species of *Perdix* by the white chin and the chestnut patch on the hindneck.

Adult female. Apparently differs from the male only in being slightly smaller.

Immature. Juveniles lack a blue-grey tinge dorsally, and lack chestnut everywhere. Dull earth buff below, with pale striations and indefinite narrow blackish bars (Ali and Ripley 1969). Older birds developing their first-basic plumage lack the orange nape-patch, and the feathers of the hindneck and mantle have well-developed buffy shaft-streaks. The head plumage of young males lacks the black cheek-patch, and the bill is black.

General biology and ecology

During the summer months this species occurs primarily between about 12 000 and 15 000 ft (3600–4500 m), but occasionally wanders as high as nearly 19 000 ft (5750 m), and in winter may occur as low as about 9000 ft (2700 m), usually remaining below 14 000 ft (4200 m). The birds are sometimes found in desolate habitats near the snow line, where little food or cover is to be seen, but perhaps more often where they can find crops, grasses, or brush for cover. In the Tsangpo Valley they occur in grass and bush cover around crop lands and yak pastures, and elsewhere in Tibet they are found on rocky hillsides and plateaux where scattered furze (gorse) bushes (*Caragana spinosa*) and dwarf junipers and rhododendrons occur (Ali and Ripley 1978). In Gansu they have been found in the rhododendron subalpine zone and where low tufts of *Potentilla tenuifolia* cover the mountainsides, occasionally descending to high plains at about 10 000 ft (3000 m) elevation. Like the more widespread grey partridge the birds apparently eat a wide variety of seeds and insects (Baker 1930).

Social behaviour

Presumably this species is similar to the grey partridge in its social organization; at least it occurs in coveys of about 10–15 birds, and when such groups are scattered they call to one another with a curious buzzing call that probably serves as a rally call. This call is said to resemble that made by the creaking lid of a lunch basket. However, the ordinary male breeding call is said to be scarcely separable from that of the grey partridge (Baker 1930).

Reproductive biology

In Tibet this species breeds commonly on high plains between about 11 000 and 15 000 ft (3300–4500 m), usually in grassy cover or among rocks or bushes, but sometimes nesting directly on bare soil. Typically the nest is placed on the leeward side of the hill, and to the lee of whatever immediate cover is available. The coveys in the Tibetan region break up about March and April. There the breeding season begins about the end of May, and eggs are laid into early July, with some late nests persisting into August. In Gansu nesting evidently begins in early May, and young birds have been seen as late as August. The usual clutch is of 8–10 eggs, with 6–12 less frequent and more than 14 quite rare (Baker 1930; Ali and Ripley 1978). There is no record of the incubation period or other aspects of hatching and brood-rearing.

Evolutionary relationships

This species is probably not in significant sympatric content with *dauuricae* in central China, although their ranges must very closely approach one another if not actually overlap in central Qinghai and Gansu. Cheng *et al.* (1978) provided an evolutionary dendrogram suggesting that these two species have a common origin, and an earlier separation of proto-*perdix* stock. I have suggested a similar phylogeny in the previous species' account.

Status and conservation outlook

This species occurs broadly over such a remote area, where high pastures and meadows are abundant and probably are so far little affected by human impact, that it seems likely that it is relatively secure.

19 · Genus *Rhizothera* G. R. Gray 1841

The long-billed wood-partridge is a large tropical forest-adapted species of the Malayan Peninsula and the Greater Sundas with a rounded tail of 12 rectrices and slightly more than half as long as the wing. The wing is rounded, with the fifth primary the longest. The bill is unusually long, heavy, and somewhat decurved. The sexes are dimorphic, but both sexes are mostly rufous to chestnut-coloured, have short tarsal spurs, and relatively short claws, especially on the hallux. A single polytypic species is recognized:

R. longirostris (Temminck): long-billed wood-partridge
    R. l. longirostris (Temminck) 1815
    R. l. dulitensis Ogilvie-Grant 1895

LONG-BILLED WOOD-PARTRIDGE (Plate 88)

Rhizothera longirostris (Temminck) 1815
Other vernacular names: rouloul à long bec (French); Langschnabelwachtel (German).

Distribution of species (see Map 29)
Resident of tropical forests of northern Borneo, Sumatra, and Malaya north to Tenasserim and southern Thailand (Pakchan estuary and Rajburi).

Distribution of subspecies
R. l. longirostris: range of the species excepting Borneo.
  R. l. dulitensis: mountains of Sarawak (Mt Dulit and Buto Song) east to Barito drainage.

Measurements

Wing, males 189–211 mm, females 180–202 mm; tail, males 80–90 mm, females no data (Baker 1935). Estimated weight, males 800 g, females 697 g. Egg, 36.7 × 26.9 mm (Schönwetter 1967), estimated weight 14.6 g.

Identification

In the field (14 in.)
Associated with drier forests, especially bamboo, up to about 1200 m elevation. The long, heavy bill and generally rusty to greyish coloration serve to separate this species from all other partridges of the area except perhaps the ferruginous wood-partridge, which is strongly spotted with black. The typical call is a loud, clear whistle, often heard at night, and additionally the birds are said to perform a bell-like duet, with one bird uttering three notes on one pitch, and the other responding with two rising notes.

In the hand
The remarkably long (at least 35 mm from gape to tip) and somewhat decurved bill is unique to this partridge.

Female. Differs from the male in having the neck and breast rufous chestnut, the lower back, rump,

Map 29. Distribution of the black (broken line) and long-billed (dotted line) wood-partridges.

and upper tail-coverts buff to rusty without much grey, and the underparts also more rusty. Tarsal spurs may be present but are probably smaller than in males.

Immature. Young males are very similar to the adult female, but have a more chestnut breast, and some barring on the flanks and breast. The bill of immatures is brownish (in dried skins), and some buffy shaft-stripes are present on the mantle and lower neck feathers. Tarsal spurs are small and are lacking in immatures of both sexes.

General biology and ecology

In the Malayan Peninsula this little-known species is said to occur in heavy jungle, usually dry submontane jungle in which there is much bamboo present, up to 4000 ft (1200 m). It is ground-dwelling and rather crepuscular in its activities, occasionally calling at night. The birds are extremely wary, and tend to run rather than fly when disturbed, so that they can hardly be seen among the dead leaves of the jungle floor (Robinson and Chasen 1936).

Social behaviour

No specific information. The birds are probably territorial, inasmuch as the Malayan natives are reported to snare them by calling them up to traps. The call is a shrill whistling, said to be harsher and less sustained on one note than that of the crested wood-partridge, and also described as a far-carrying double whistle, with the second note higher in pitch. It is uttered around dusk and dawn, and sometimes throughout the night as well (Robinson and Chasen 1936). Pairs sometimes call in duet, with the two birds producing a rising, four-noted sequence, sounding something like 'I'm a tur-key' (Medway and Wells 1976).

Reproductive biology

Only a single nest has ever been found in nature, where a nest with two eggs was located in a bamboo thicket during February of 1934 (Coomans de Ruiter 1946). Apparently no additional nests have been found nor has any other information on the natural breeding of this species since materialized. However, it has recently been kept in captivity and is reported under such conditions to lay from two to five eggs, and to have an incubation period of 18–19 days (Robbins 1984).

Evolutionary relationships

No speculations on this species' relationships have been made; its downy young are still undescribed. Its adult sexually dimorphic greyish to rufous chestnut plumage pattern is distinctive, but probably is at least partly related to concealing coloration in a shaded environment. The tarsi are relatively long and stout, which together with the unusually long and somewhat decurved beak of the species suggests that it is adapted for digging food from the ground. Perhaps it evolved from early forest-adapted *Arborophila*-like stock that became specialized to this end. A recent phenetic analysis by Crowe and Crowe (1985) suggests possible affinities with *Francolinus*.

Status and conservation outlook

No information exists on the status of the elusive species.

20 · Genus *Margaroperdix* Reichenbach 1853

The Madagascar partridge is a medium-sized grassland-adapted species endemic to that island with a wedge-shaped tail of 12 rectrices that is about half as long as the wing. The wing is relatively pointed, with the eighth and ninth primaries the longest and the outermost not much shorter. The sexes are dimorphic, the males relatively contrastingly coloured, with strong head-striping, and the females similar but mostly cryptic brown. Neither sex has tarsal spurs. A single monotypic species is recognized:

M. *madagarensis* (Scopoli) 1786: Madagascar partridge

MADAGASCAR PARTRIDGE (Plate 89)

Margaroperdix madagarensis (Scopoli) 1786
Other vernacular names: caille de Madagascar (French); Perlwachtel (German); tsipoy ou traotrao (Madagascar).

Distribution of species (see Map 33)
Resident of brush and grassland habitats of Madagascar. Introduced on Réunion. Also introduced on Mauritius, but subsequently extirpated. No subspecies recognized.

Measurements

Wing, males 119–131 mm, females 121–131 mm (Frost 1975); tail, both sexes 69 mm (Ogilvie-Grant 1893). Estimated weight, both sexes 220 g. Egg, 38.1 × 29.2 mm (Schönwetter 1967), estimated weight 17.9 g.

Identification

In the field (12 in.)
Associated with heath areas in mountains, secondary brush, weedy cultivated lands, and grasslands. This is the only Madagascan partridge with a strongly striped head, and is considerably larger than the native *Coturnix* forms. A low *peet* alarm call is uttered on flushing, and the usual (advertising?) call is a repeated and loud *cou* note.

In the hand
Readily distinguished by its rudimentary hind claw (under 5 mm long) and its wedge-shaped tail of 12 rectrices that is about half the length of the wing.

Female. Easily separated from males by the lack of a strongly patterned facial pattern.

Immature. The juvenal plumage is undescribed, but a male in nearly adult plumage still had a nearly female-like dorsal pattern (cross-barred feathers with buffy shaft-streaks), a poorly defined facial pattern, with the white replaced by dark-edged buffy feathers and the black by fulvous, and scattered feathers on the underparts that approach the female pattern.

General biology and ecology

This species evidently occupies a variety of habitats, but especially secondary brush and neighbouring grassland. It also occurs in heather habitats in mountainous country, in weedy cultivated fields, and in rice fields. It occurs from sea-level up to about 1800 m, but is absent from the extensive scantily grass-covered areas of interior Madagascar (Rand 1936). It is apparently sedentary, probably feeding on a variety of seeds and insects (Long 1981).

Social behaviour

Almost nothing is known of this, although the birds are said to occur solitarily, as pairs, trios, or small coveys of up to about 12 birds in the non-breeding season. Frost (1975) suggested that the strong sexual dimorphism in plumage might indicate a polygynous mating system, but provided no evidence for this. The usual call, which is said to carry some distance, is a repeated *cou* note (Rand 1936).

Reproductive biology

The birds are reported to nest from March to June, and to construct their nests on the ground, well hidden in a tuft of grass. A surprisingly large clutch-size of 15–20 eggs has been reported (Milon *et al.*

1973). There is no information on the incubation period, but the downy young are quite similar to those of *Coturnix* (Frost 1975).

Evolutionary relationships

Frost (1975) proposed that this species is perhaps a near relative of *Coturnix* that has increased in bodily size and undergone some ecological release as a result of reduced interspecific competition from other quails. However, Crowe and Crowe (1985) have suggested that it is actually a very close relative of the francolins, and may well approximate the ancestral francolin type, which they believed may have originally been a *Coturnix*-like migratory or nomadic form that traversed the considerable oceanic gaps occurring between Africa, Madagascar, and Asia some 28–30 million years ago, at the time of the presumed diversification of the francolin group.

21 · Genus *Melanoperdix* Jerdon 1864

The black wood-partridge is a medium-sized tropical forest-adapted species endemic to Borneo with a soft and rounded tail of 12 rectrices that is less than half as long as the wing. The wing is rounded, with the sixth primary the longest. The bill is unusually short and heavy, and the hallux has only a rudimentary claw. The sexes are dimorphic, the males almost entirely glossy black and the females mostly dark chestnut; both sexes lack spurs. A single polytypic species is recognized;

   M. nigra (Vigors): black wood-partridge
      M. n. nigra (Vigors) 1829
      M. n. borneensis Rothschild 1917

BLACK WOOD-PARTRIDGE (Plate 90)

Melanoperdix nigra (Vigors) 1829
Other vernacular names: black partridge; rouloul noir (French); Schwarzwachtel (German).

Distribution of species (see Map 29)
Resident of lowland tropical forests of Malaya, Sumatra, and western Borneo.

Distribution of subspecies
M. n. nigra: Malayan Peninsula (from Perak south) and Sumatra.
   M. n. borneensis: southern and western Borneo (including western Sarawak to Baram district).

Measurements

Wing, males 131–143 mm, females 131–144 mm; tail, males 59–70 mm, females 58–67 mm (Riley 1938). Estimated weight, both sexes 260 g. Egg, average 40.5 × 32.6 mm (Schönwetter 1967), estimated weight 23 g.

Identification

In the field (9.5–10.5 in.)
Associated with forests having dense undergrowth of stemless palms. The nearly black males are unmistakable, and females are a similar dark chestnut, with a paler throat and underparts. Very shy and rarely seen; its calls include a double whistle similar to that of a crested wood-partridge.

In the hand
Separated from the other partridges by its bill, which is unusually short (culmen length 18–21 mm) and thick (upper mandible 9.5–10.5 mm deep at base).

Female. Distinguished from the male by its lack of black feathers.

Immature. The young male is female-like, but gradually acquires mixed glossy black feathers. Young females have paler upperparts than do adults, with irregular blackish vermiculations and usually with pale buffy feather tips (Robinson and Chasen 1936).

General biology and ecology

This species occurs in dense jungle from low altitudes up to about 2000 ft (600 m), usually being found in forests having many spiny stemless palms (bertam) in the undergrowth, the apparent basis for its Malay name 'Burong bertam'.

Social behaviour

Almost nothing is known, but the species is said to occur singly or in pairs rather than in large coveys. The call is reportedly a double whistle similar to that of *Rollulus* (Robinson and Chasen 1936), but has also been described as a low creaking sound, sounding like the opening of an old door (Medway and Wells 1976).

Reproductive biology

Only a few nests have been found, including three clutches of five eggs each, found during September in the Baram district of northern Borneo (Robinson and Chasen 1936). Two additional clutches of two and three eggs were reported by Coomans de Ruiter (1946) as being obtained during January and May in Borneo. The species has only rarely been maintained in captivity (Reinhard 1981; Robbins 1984), but is reported as laying five to six eggs, and having an incubation period of 18–19 days.

Evolutionary relationships

The very heavy bill of this species must be adapted for crushing large seeds or possibly handling some type of hard fruit, but the bird's diet in nature is still unknown. Although the male's black plumage is unique, the adult female is not very different from *Arborophila* and *Rhizothera* in overall patterning. I have not seen the downy young, but believe the species to be related to the general *Arborophila*-like assemblage of forest-adapted partridges. A recent phenetic analysis places it near *Arborophila* and *Rollulus* (Crowe and Crowe 1985).

Status and conservation outlook

No information is available on the status of this species.

22 · Genus *Coturnix* Bonnaterre 1791 (including *Synoicus* Gould 1843 and *Excalfactoria* Bonaparte 1856)

The typical quails are small and widespread grassland-adapted Old World species having soft, rounded tails of 8–12 feathers that are less than half as long as the wing, and are largely hidden by the upper tail-coverts. The wing is relatively pointed, with the eighth and ninth primaries the longest. Most species are migratory or nomadic. The post-juvenal primary moult is incomplete, terminating in at least three species (*coturnix*, *japonica*, and *novaezeelandiae*) prior to the eighth primaries, and the tail moult is perdicine (centrifugal). The sexes are alike or dimorphic, with distinctively patterned and cryptic dorsal plumage, and the tarsus is unspurred in both sexes. The genera *Synoicus* and *Excalfactoria* are here considered synonyms of *Coturnix*, although Dr C. Sibley (personal communication) has recently obtained data from DNA hybridization studies suggesting that *Synoicus* may be a valid genus, having phyletic affinities both with *Coturnix* and the *Alectoris–Francolinus* complex. Eight species are recognized here, including two allospecies that are often considered subspecies:

C. [*coturnix*] *japonica* Temminck and Schlegel 1849: Asian migratory quail
C. *coturnix* (L.): European migratory quail
   C. c. *inopinata* Hartert 1917
   C. c. *confisa* Hartert 1917
   C. c. *coturnix* (L.) 1758
   C. c. *erlangeri* Zedlitz 1912
   C. c. *africana* Temminck and Schlegel 1849
C. [*coromandelica*] *delegorguei* Delegorgue: African harlequin quail
   C. d. *histrionica* Hartlaub 1849
   C. d. *delegorguei* Delegorgue 1847
   C. d. *arabica* Bannerman 1929
C. *coromandelica* (Gmelin) 1789: black-breasted quail
C. *novaezeelandiae* Quoy and Gaimard: stubble quail
   C. n. *novaezeelandiae* Quoy and Gaimard 1830
   C. n. *pectoralis* Gould 1837
C. (= *Synoicus*) *ypsilophorus* Bose: brown quail
   C. y. *ypsilophorus* Bose 1792
   C. y. *australis* (Latham) 1801
   C. y. *queenslandicus* Mathews 1912
   C. y. *cervinus* (Gould) 1865
   C. y. *dogwa* Mayr and Rand 1937
   C. y. *plumbeus* (Salvadori) 1894
   C. y. *saturatior* (Hartert) 1930
   C. y. *mafulu* Mayr and Rand 1937
   C. y. *lamonti* Mayr and Gilliard 1954
   C. y. *monticola* Mayr and Rand 1937
   C. y. *pallidior* (Hartert) 1897
   C. y. *raaltenii* (Müller) 1842
C. (= *Excalfactoria*) *chinensis* (L.): Asian blue quail
   C. c. *chinensis* (L.) 1766
   C. c. *trinkutensis* (Richmond) 1902
   C. c. *palmeri* (Riley) 1919
   C. c. *lineata* (Scopoli) 1786
   C. c. *lineatula* (Rensch) 1931
   C. c. *lepida* (Hartlaub) 1879
   C. c. *papuensis* Mayr and Rand 1937
   C. c. *novaeguinae* Rand 1942
   C. c. *colletti* (Mathews) 1912
   C. c. *australis* (Gould) 1865
C. [*chinensis*] *adansonii* J. and E. Verreaux 1851: African blue quail

KEY TO THE SPECIES OF TYPICAL QUAILS (*COTURNIX*)

A. Wing under 85 mm; third primary from outside the longest ('*Excalfactoria*')
  B. Underparts of males blue and chestnut, wing-coverts of females less distinctly barred with black: Asian blue quail (*chinensis*)
  BB. Underparts of males entirely bluish, wing-coverts of females more distinctly barred with black: African blue quail (*adansonii*)
AA. Wing over 85 mm; outermost or second primary the longest
  B. Crown and superciliary stripes indistinct; head markings obscure
    C. Underparts and flanks with fine black barring; axillaries grey: brown quail (*ypsilophorus*)
    C. Underparts and flanks with larger black streaks; axillaries white: stubble quail (*novaezeelandiae*)

BB. With clear buffy crown and superciliary stripes, and more distinct dark head striping
  C. Outer webs of primaries unbarred; males with variably black underparts
    D. Black extensive over anterior underparts in males, underparts more rufous and wing-coverts with blackish grey in females: African harlequin quail (*delegorguei*)
    DD. Black limited to centre of breast in males, underparts pale buff and wing-coverts with sandy buff in females: black-breasted quail (*coromandelica*)
  CC. Outer webs of primaries barred; males lacking black on underparts
    D. Wing over 105 mm; upper throat feathers rounded in autumn and winter; spring birds have pale buffy throats and (in males) black chins: European migratory quail (*coturnix*)
    DD. Wing under 105 mm; upper throat feathers pointed in autumn and winter; more rufous throats and generally more contrasting dorsally and more rufous below in spring: Asian migratory quail (*japonica*)

EUROPEAN MIGRATORY QUAIL (Plate 91)

Coturnix coturnix (L.) 1758

Other vernacular names: African quail (*africana*); Cape quail (*capensis*); common quail; European quail; Egyptian quail; grey quail; Madeira quail (*confisa*); Messina quail; pharaoh quail; an-chun (Chinese); caille des blés (French); Wachtel (German); perepel (Russian); codorniz (Spanish).

Distribution of species (see Map 30)

Seasonal resident of open habitats of Africa, Madagascar, their associated islands (Comoros, Cape Verde, Azores, Madeira, Canaries), and Eurasia from the Atlantic coast (and the British Isles) east to India, western Xinjiang, northern Mongolia, and Lake Baikal. Migratory populations from Europe winter south to about the Equator in Africa, and eastward to Saudi Arabia and India. Birds breeding in southern Africa may winter north-westward toward the Equator.

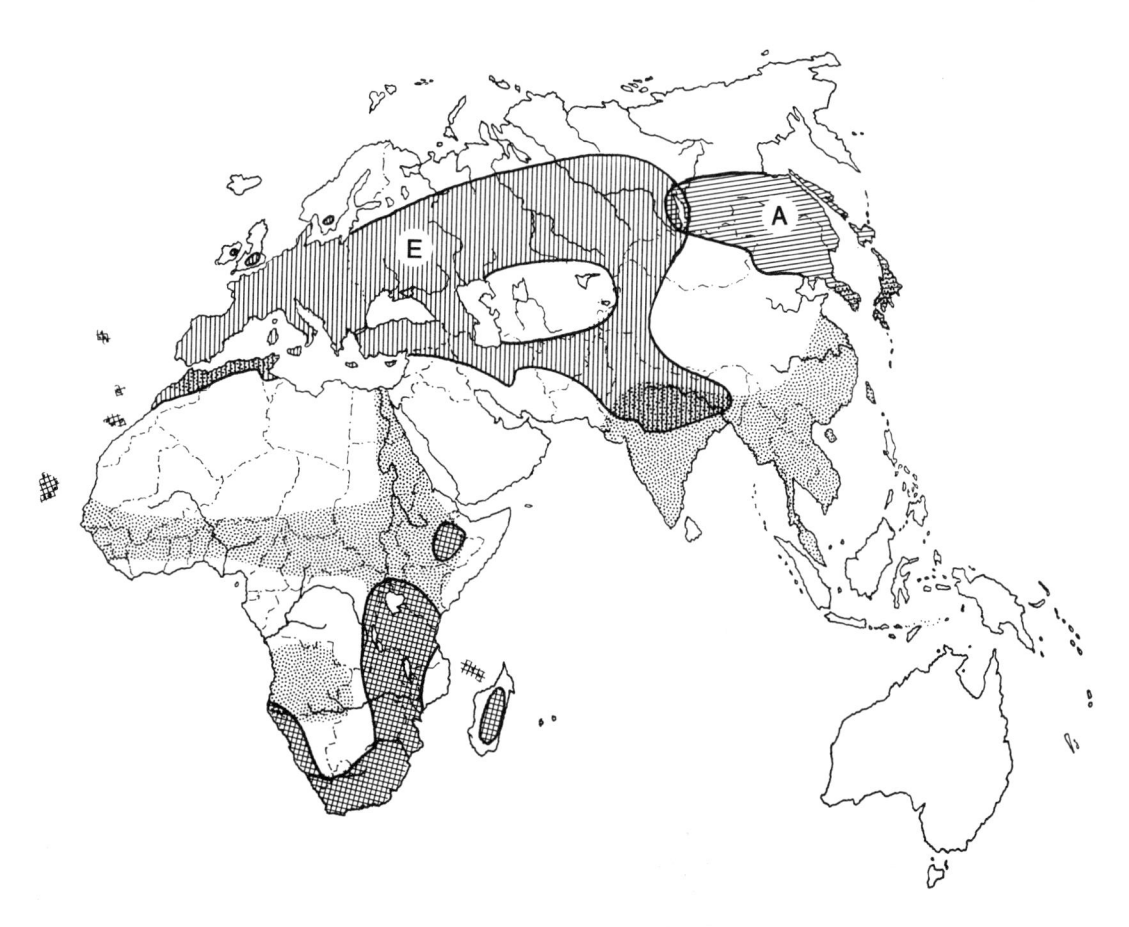

Map 30. Distribution of the European (E) and Asian (A) migratory quails. Insular and endemic African races of *C. coturnix* are shown by cross-hatching; stippling indicates wintering range of *C. c. coturnix* and *C. japonica*.

Distribution of subspecies

C. c. inopinata: Cape Verde Islands.

C. c. confisa: Canaries, Madeira, and the Azores. Includes *conturbans*.

C. c. coturnix: Eurasia, from the vicinity of Lake Baikal, Mongolia, and Xinjiang, west locally or sporadically to southern Sweden, Great Britain, the Mediterranean and its larger islands, north-western Africa, Asia Minor, and through the Near East southeast to western India (Maharastra). Includes *orientalis*.

C. c. erlangeri: highlands of Ethiopia.

C. c. africana: Africa, from Kenya, Uganda, and Zaire south to southern Angola and Cape Province; also Madagascar and Mauritius. Included in *erlangeri* by Urban et al. (1986).

Measurements

Wing (*coturnix*), males 110–115 mm, females 107–116 mm; tail, males 31–38 mm, females 36–44 mm (Ali and Ripley 1978). Weight (various European areas), males 70–140 g, females 70–155 g (Cramp and Simmons 1980); (Africa), males 76–111 g, average (of 144) 90 g, females 81–122 g, average (of 90) 103 g (Urban et al. 1986). Egg (*coturnix*), average 29.7 × 22.8 mm (Ali and Ripley), estimated weight 8.5 g.

Identification

In the field (7 in.)

Associated with rather heavy grassy cover, such as crop lands, tall pastures, etc. Its small overall size, generally concealing dorsal pattern, and striping on the head allow for separation from other species in most areas. In wintering areas of Africa and southern Asia confusion with other *Coturnix* species is possible. The male's advertising call is a three-syllable 'Wet-my-lips', also described as *whic!, whic-ic* or *whit-whit'tit*. Often difficult to flush, and a generally slow and seemingly weak flier.

In the hand

The small size and barred condition of the outer webs of the primaries separates this species from all but *japonica*, and the latter has an even shorter wing (under 105 mm) and generally more rufous coloration on the throat and underparts, especially in spring.

Adult female. Differs from the male in having no black band down the middle of the throat; the arms of the anchor-shaped mark imperfectly represented, and the chest is variably spotted with black.

Immature. Similar to the adults, but the anchor-shaped mark just indicated in males, and in both sexes the outer two or three remiges persist from the juvenal plumage (Clancey 1967).

General biology and ecology

This is a species of grassland habitats and relatively flat to gently undulating landforms, ranging in elevation up to 1000 m and sometimes considerably higher. It prefers vegetation dense and tall enough to provide concealment but unusually high or wet cover, and avoids both forest edges and hedgerows. Cultivated fields of winter wheat and clover are especially preferred nesting cover, with a variety of other small grain crops, lucerne, rape, and mowing grass or natural grasslands and meadows used to lesser degrees. The birds are relatively omnivorous, consuming primary fallen seeds such as weeds and cereal gleanings, and also a good number of insects, especially beetles, true bugs, ants, earwigs, and orthopterans. The insects are mainly eaten during spring and early summer, when few seeds are available and females presumably need high-protein diets for breeding (Cramp and Simmons 1980).

This and the following species are perhaps the most migratory of all galliform birds, with trans-Mediterranean migrations commonplace for many of the European breeders, and many birds making spring landfalls in Italy and then continuing north. However, there is a substantial circular movement around the western Mediterranean, and apparently at least some birds vary their routes in crossing the Mediterranean between years (Moreau 1963). The Mediterranean passages are also complicated by birds that have bred or been hatched along the African coast and then fly north to Italy. This group includes both post-breeding females and immatures possibly no more than 2 months old, and it has been speculated that perhaps some of these birds from Africa may actually breed in Europe that same summer (Berthet 1949). However, late summer clutches in northern Europe may simply be the result of second nestings or renesting efforts by resident breeders (Cramp and Simmons 1980).

Social behaviour

Males arrive on their breeding areas somewhat in advance of females, and immediately establish singing territories that evidently serve more to attract females than repel other males. The 'song' uttered by such males is the familiar 'wet-my-lips' vocalization, although males also often utter a less loud and guttural 'growl-call' as a preliminary to their primary territorial call (Moreau and Wayre 1968). Upon arrival, the females apparently initially seek out possible nest sites, and only then do they respond to the male's calls by uttering their own 'attraction call'. This call attracts the local territory-holder, who approaches and displays to her. The typical initial male display is the circle-display, which is similar to (and probably homologous with) waltzing inasmuch as the male moves around the female with the throat and breast feathers ruffled, the nearer wing drooped or even trailing on the ground, and uttering soft notes. Males also perform tidbitting while uttering a food-call (Stokes and Williams 1971). Prior to copulation the female crouches and utters an invitation call. Although some persons have reported possible polygynous matings in situations where an extra female establishes herself on the male's territory. The pair-bond is at least sometimes quite strong, and duetting between them is common (Bannerman 1963).

Reproductive biology

Nests are placed in herbaceous, often grassy vegetation, with laying in northern Europe extending from the middle of May to the latter part of August. In southern Africa the laying season is from September until about March (mainly November to January in Natal; September–October in Cape Province), and in Kenya breeding occurs during the wet season of January and February. The clutch-size is quite variable, but usually ranges between 8 and 13 eggs, and in Europe probably averages about 10. In South Africa the clutch reportedly varies from 6 to 12 eggs, with the larger extreme possibly reflecting laying by two females (Clancey 1967). The eggs are laid on a daily basis, and up to at least two renesting attempts may be made following nest failure. Males remain fairly close to the nesting females, and there are cases known of a male sitting on eggs and of accompanying a brood of young, although these appear to be very rare (Bannerman 1963). The incubation period lasts 17–20 days, and is normally by the female alone, as is brood care. The chicks fledge in about 19 days, although short fluttering flights are possible after only about 11 days (Cramp and Simmons 1980).

Evolutionary relationships

Beyond the question of the possible conspecificity of *coturnix* and *japonica*, which is discussed in the next section, the broader question of relationships of *Coturnix* need to be mentioned. As noted in Chapter 1, the *Coturnix*-like forms have at times been separated tribally from the more typical partridges, but I am not convinced that such a distinction is currently warranted. There are to be sure some unusual aspects of its pectoral myology, placing it slightly closer in that respect to some grouse than to most genera of perdicines, but none the less seemingly having the closest myological affinities with *Francolinus* and *Alectoris* (Hudson *et al.* 1986). Similarly, DNA hybridization data suggest a fairly close relationship of *Coturnix* (including '*Synoicus*') with *Alectoris* and *Francolinus* (C. G. Sibley, personal communication).

Status and conservation outlook

This species exhibits marked fluctuations in abundance, which mask long-term trends. Most evidence from Europe suggest that some local declines in population have occurred as a result of habitat changes and hunting pressures, but no clear pattern emerges (Cramp and Simmons 1980).

ASIAN MIGRATORY QUAIL (Plate 92)

Coturnix japonica Temminck and Schlegel 1849
Other vernacular names: eastern common quail; Japanese grey quail; Japanese rice quail; Manchurian golden quail; red-throated quail; caille Japonaise (French); Japanwachtel (German); uzura (Japanese).

Distribution of species (see Map 30)

Seasonal resident of eastern Asian grasslands, from northern Mongolia and Transbaicalia north to the Vitim Plateau and east through Amurland to Ussuriland, Sakhalin, the Kuriles, Japan, Manchuria, Korea, and China (to Hopeh). Winters south from central China and central Japan to northern Indochina, Burma, and Assam. Introduced into Hawaiian Islands; now resident on main islands east from Kauai, excepting Oahu. Also introduced on Réunion. No subspecies recognized by Vaurie (1965), who synonymized *ussuriensis*. Previously often considered a subspecies of *coturnix*, but now generally considered specifically distinct.

Measurements

Wing, males 92–101 mm, females 93–101 mm; tail, males 35–49 mm, females 36–49 mm (Taka-Tsukasa 1967). Estimated weight, both sexes 90 g. Egg, average 29.8 × 21.5 mm, estimated weight 7.6 g.

Identification

In the field (7 in.)

Found in similar grassy habitats as the last species, which it replaces in eastern Asia. The two differ in their territorial 'song'; males of this species utter a soft and muted *choo-peet-trrr*, and are more rufous below in spring. Field distinction from *coturnix* in autumn or winter may be impossible.

In the hand

Separated from *coturnix* by its smaller size (wing under 105 mm) and in having more sharply pointed upper throat feathers during autumn and winter, forming a small 'beard'. In spring the more rufous colour, especially on the underparts, provides a useful means of separation. Occasionally males may exhibit a partial or complete 'anchor' or 'collar' on the throat (Lyons 1962). Hybrids (or at least morphologically intermediate specimens) may rarely occur, and so exact identification might not always be possible. Avicultural varieties may also be encountered with plumages that greatly diverge from wild-type *japonica*, varying from uniformly brown ('British range') to piebald ('tuxedo') and albinistic ('English white'). Females differ from *coturnix* in that the feathers of the chin and throat are elongated and lanceolate, and are usually margined on the outer web with rufous; the margins of the chest, flank, and side feathers are mostly rufous, and much less spotted with black.

Adult female. Differs from males in normally lacking any rufous on the breast, and also usually lacking black markings on the throat, although these traits are not invariable criteria of sex (Lyons 1962).

Immature. Young males have the elongated throat feathers of the adult female, and the middle of the throat is suffused with dull brick-red.

General biology and ecology

Although much less studied than *coturnix*, it is highly probable that the general biology and habitat characteristics of these two species are nearly identical. It has been suggested that in the USSR the birds may prefer somewhat damper meadows than the European species, but they also extend to steppes and dry mountain slopes, and in Japan generally seem to prefer areas that are fairly dry but close to water. Foods include a wide variety of plant materials, supplemented in summer by terrestrial insects and probably other invertebrates (Taka-Tsukasa 1967). Like *coturnix*, the birds are highly mobile, and migrate annually, with some birds wintering as far north as the USSR, but these are also apparently from more northerly populations (Dementiev and Gladkov 1952). Of more than a thousand recoveries from birds ringed in Japan, only three were from outside the Japanese archipelago, suggesting that considerable north–south movement occurs between the Japanese islands, supplemented by some flow across the southern Japan Sea to and from Korea (McClure 1974).

Social behaviour

Pairing may at times be non-monogamous, perhaps (as has been suggested for the European species) depending on the availability of excess females, and is facilitated by the males uttering advertisement calls shortly after their arrival on breeding areas. The male's 'song' has been variously described as being rather muted, resembling a hum, thunder, or a snore at a distance, and as sounding like a repeated *chzhu-chzhir* or *chut-pit-trr*. The only sonagraphic comparisons between *coturnix* and *japonica* are those of Moreau and Wayre (1968), who note that whereas *coturnix* has a series of three clear, brief, and sharp notes with a dactylic rhythm, males of *japonica* (from Japan) do not, and the latter form's notes are so much harsher and more blurred that they sound something like the crowing of a francolin. However, a recording of a male in Amurland was very low-pitched, soft, and apparently more drone-like or humming, but having a rhythm more like the Japanese than the European samples.

The vocalizations and displays of captive-bred *japonica* have been analysed in some detail by Eynon (1968). He recognized seven distinct male displays, of which the full strut is the most elaborate. It is used to assert dominance over other males, and always occurs after copulation and sometimes also before it. Seven male vocalizations were recognized, three of which also were observed in females. Eynon noted that the male's advertisement or crowing call usually consisted of three notes, with the third note emphasized, but rarely had two or four syllables. The female's sexual call was termed the 'long call', the hearing of which typically excites males.

Reproductive biology

Breeding in the USSR is fairly prolonged, with eggs observed from late April at least until 10 August, the late clutches presumably being renests rather than second nestings (Dementiev and Gladkov 1952). In Japan nesting occurs from the end of May to August, or rarely into September (Taka-Tsukasa 1967). Clutch-sizes from the USSR have been reported as usually nine to ten, and from Japan of five to eight eggs, suggesting that perhaps clutch-size may increase with latitude and associated increasing photoperiods. At least photoperiodic regulation of sexual behaviour is known for both sexes in laboratory situations (Follett and Farner 1966; Sachs 1966). The incubation period of 18 days is essentially the same as in *coturnix*. Several studies of post-hatching growth and plumage development of *japonica* have been performed (Lyons 1962; Summers 1972) and are partially summarized in Chapter

4. The ontogeny of crowing and copulatory behaviour has been reported by Ottinger and Brinkley (1979).

Evolutionary relationships

Although long regarded as only a subspecies of *coturnix*, Moreau and Wayre (1968) reviewed the comparative evidence on plumage, size, vocalizations, and breeding ranges of *coturnix* and *japonica*. They believed that the two forms are narrowly sympatric in the vicinity of the USSR's Khara River, and near the north-eastern end of Lake Baikal. Although hybrids have been produced in captivity and are at least partially fertile, these authors believed that no natural hybridization occurs in the area of sympatric contact. They judged the two forms to be in an intermediate stage of speciation for which conventional taxonomic categories are inadequate, but that some hybrid fertility data suggest separation of the taxa into two species may be justified.

Status and conservation outlook

The broad breeding range and its generally remote nature would suggest that this species is probably relatively secure.

AFRICAN HARLEQUIN QUAIL (Plate 93)

Coturnix delegorguei Delegorgue 1847
Other vernacular names: harlequin quail; caille arlequine (French); Harlekinwachtel (German).

Distribution of species (see Map 31)

Seasonal or permanent resident of much of sub-Saharan Africa except for the Congo basin, north to the Ivory Coast and Ethiopia, and south to the Cape; Socotra, Pemba, and Zanzibar islands; also (? formerly) Arabia. Migratory at the southern and northernmost portions of the African range; resident in equatorial areas. Comprises a superspecies with *coromandelica*.

Distribution of subspecies

C. d. delegorguei: range as indicated for the species, excepting Arabia.

C. d. arabica: known only from a few early specimens from Arabia.

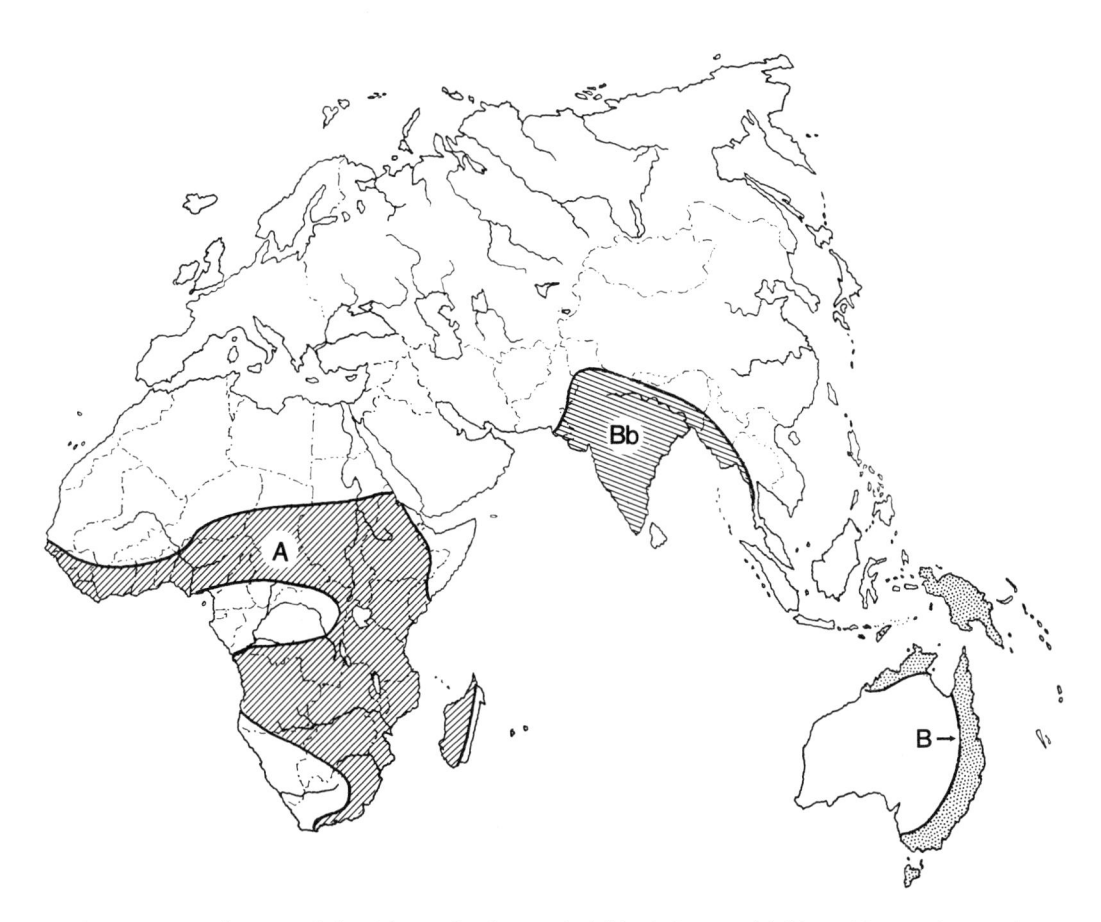

Map 31. Distribution of the African harlequin (A), black-breasted (Bb), and brown (B) quails.

C. d. histrionica: São Thome Island (Gulf of Guinea).

Measurements

Wing (*delegorguei*), males 93–100 mm, females 96–104 mm; tail, males 30–33 mm, females 32–34 mm (Clancey 1967). Weight, males 65–81 g, average (of 11) 72.4 g, females 73–94 g, average (of 16) 78.5 g (Urban *et al.* 1986). Egg, average 29 × 23 mm (Mackworth-Praed and Grant 1952–73), estimated weight 8.5 g.

Identification

In the field (6.5 in.)

Associated with open grasslands, cultivated fields, grassy edges, and similar habitats, especially somewhat damp ones. The males are generally darker than the European migratory quail, but females are probably indistinguishable in the field, although they too are slightly darker dorsally. Males differ from the African blue quail in lacking that species' bluish tones. Females lack the latter's somewhat barred flank pattern, but have darker chest streaking and a rudimentary black throat pattern approaching the male's 'anchor'. The male's advertisement call is a loud, metallic whistling double note *twee-twit*, with the second note higher and shorter than the first. He sometimes utters a treble chirp more like that of *C. coturnix*, but this call is reportedly more metallic, higher pitched, and less rhythmic. Both sexes utter a low *peet* when flushed.

In the hand

Males are rather easily distinguished by their dark, mostly chestnut undersides, while females differ from those of *adansonii* in having the first or second primary the longest and having rufous buff underparts except for a nearly white throat with a vestigial black 'anchor' mark.

Adult female. Differs from *coturnix* in the much darker upperparts, the outer webs of the primaries uniform, and the underparts rufous buff or dull chestnut, washed with dusky on the chest. Most of the feathers have pale but dark-edged shaft-stripes, and the feathers of the sides are mottled and barred with black and buff.

Immature. Similar to the adults of each sex, but more olivaceous above, less greyish, and retaining the outer two or three juvenal primaries (Clancey 1967).

General biology and ecology

Widely distributed in Africa, this species favours grassland habitats, especially rather open grasslands and cultivated areas. However, it also occurs in the grassy borders of rivers, in rank grasses of damp regions, in fallow cultivated areas, and occasionally in bushvelt. They occasionally occur in company with *C. coturnix* and perhaps compete to some extent with this species, but the two species are not known to hybridize. Both species are irruptive, and this one is distinctly migratory in some areas. In favourable situations densities of up to about 10 birds per hectare may develop. In southern Africa the birds are present from about October to April, with year-to-year numbers varying greatly, and apparently undergoing their post-nuptial moult before migrating back north. They feed mostly on seeds of grasses and weeds, and on insects, particularly termites (Clancey 1967; Urban *et al.* 1986).

Social behaviour

During migration these birds are often found in very large numbers, sometimes moving on foot, and during both daytime and night-time hours. When breeding they become more scattered, although females have in favoured areas been reported to nest in compact aggregations. During that period the males can be readily captured in snares, by putting out 'decoy' males whose calls quickly attract the resident territorial male. The male's whistled call can also be easily imitated, and thus also can be used to bring males up to a few yards away. The female's reply call is much lower and reportedly almost inaudible. When calling, the advertising male stands in a nearly upright position, but he also has another display posture with the breast close to the ground, shaking his body from side to side, and uttering low notes. A lateral display posture, with the wing nearest the female lowered, is also present, as is courtship feeding. The pair-bond pattern is likely to be comparable to that of the European migratory quail, or essentially monogamous but facultatively polygynous, with two females possibly even laying at times in the same nest (Clancey 1967; Urban *et al.* 1986).

Reproductive biology

In eastern South Africa breeding occurs mainly during the summer months, peaking in December and January (Clancey 1967), while in Namibia it is associated with the rainy period between July and September. Farther north, as in Zimbabwe, breeding also peaks during the rainy periods. The clutch-size seems to be generally between three and eight eggs, the average of nine being 4.8 eggs. The incubation period is of 14–18 days, and is by the female alone. When the chicks are only 5 days old they can fly short distances, and males may remain with the brood until they fledge. By a month they average about 21 g. During the non-breeding season coveys of 6–20 birds are typical, probably representing one or more family units (Urban et al. 1986).

Evolutionary relationships

This species is certainly part of a superspecies including *coromandelica*, and these two allopatric forms could easily be considered conspecific.

Status and conservation outlook

The broad range and opportunistic movements of this species make it quite resistant to local extirpation or over-exploitation, so it is unlikely to pose any problems for conservationists.

BLACK-BREASTED QUAIL (Plate 94)

Coturnix coromandelica (Gmelin) 1789
Other vernacular names: rain quail; caille natte (French); Regenwachtel (German).

Distribution of species (see Map 31)

Seasonal resident of grassy habitats throughout Pakistan, India, Bangladesh, and Burma, up to about 2000 m elevation in the Himalayas. Winters southwardly; migrants occasionally reach Sri Lanka in winter. No subspecies recognized.

Measurements

Wing, males 93–96 mm, females 90–97 mm; tail, males 29–32 mm, females 28–31 mm. Weight, both sexes c. 64–85 g (Ali and Ripley 1978). Egg, average 27.4 × 20.8 mm (Ali and Ripley), estimated weight 6.5 g.

Identification

In the field (6.5 in.)
Associated with grassy areas, cultivated lands, and dry scrub. Males have a distinctive black patch on the centre of the breast, and a black 'anchor' on the white throat; females are probably inseparable from migratory quail in the field. The male's advertisement call is a repeated loud, metallic, and high-pitched *whit-whit*, often in doublets that are repeated in runs of three to five calls, especially in mornings and evenings but also throughout the day during the nesting period.

In the hand
Separated from other Asian quails by the unbarred condition of the outer webs of the primaries, and a wing that is over 85 mm, with the outermost or adjacent primary the longest.

Adult female. Females are similar to *C. coturnix*, but are smaller and have no buff or rufous bars or mottling on the primaries.

Immature. Young males have a less distinct black anchor-mark on the throat, and the chest-patch is also reduced. Probably the outer juvenal primaries are also retained.

General biology and ecology

Like the previous species, this form occurs in a variety of grassland habitats, including grain crops, stubble fields, meadows, and even gardens. During the dry season they may move into heavier cover, especially near sources of water, but basically they are open-country birds. They are local migrants, nomadic, or residents, depending upon the area, occasionally reaching Sri Lanka during winter, and at times moving as high as about 2500 m in the Himalayas, and breeding there to elevations of about 1800 m. Probably most movements are fairly local in nature. They evidently eat the usual mixture of grass and weed seeds, supplemented by some insects (Baker 1930; Ali and Ripley 1978).

Social behaviour

These birds rarely if ever occur in large flocks, but instead move about in small groups. Reportedly the parents remain with their chicks until the latter are about 8 months old, and then they are driven off and scatter, presumably to mate and breed themselves. The male is reportedly strongly monogamous, to the extent of helping to care for the brood.

Reproductive biology

Nesting occurs during the wet season, or generally from March to October, depending on local rainfall patterns. The nest is usually placed in standing crops or thin grasses, but occasionally is hidden in scrub or low bush jungle. The usual clutch is of four to six eggs, but considerably larger numbers of 12 or even rarely as many as 18 have been found, these higher numbers clearly the result of two or more females laying in the same nest. Incubation is by the female, and probably lasts 18–19 days. No specific information on the development of the young is available (Baker 1930; Ali and Ripley 1978).

Evolutionary relationships

Certainly this is a close relative of the *coturnix–japonica* and *delegorguei* assemblage, especially the latter.

Status and conservation outlook

This small, mobile, and opportunistic species is unlikely to provide any conservation problems in the foreseeable future.

STUBBLE QUAIL (Plate 95)

Coturnix novaezeelandiae Quoy and Gaimard 1830
Other vernacular names: grey quail; pectoral quail (*pectoralis*); caille de Nouvelle-Zélande (French); Schwartzbrustwachtel (German).

Distribution of species (see Map 32)

Seasonal or permanent resident of grassy and cropland habitats of southern Australia (mainly south of Tropic of Capricorn); rare in Tasmania. Migratory or nomadic. Extirpated from New Zealand.

Distribution of subspecies

C. n. novaezeelandiae: previously resident on New Zealand, now extinct.

C. n. pectoralis: Australia as indicated above; introduced in New Zealand and Hawaiian islands, apparently both unsuccessfully (Long 1981). Often considered specifically distinct. Includes *sordidus*.

Map 32. Distribution of the stubble (S) and Snow Mountain (Sn) quails.

Measurements

Wing, adults 104–117 mm; tail 38–46 mm. Weight (*pectoralis*), adults 99–128 g (Long 1981), males c. 114 g, females c. 95 g (Cruise 1966). Egg, average 30.3 × 23.4 mm (Oliver 1955), estimated weight 9.2 g.

Identification

In the field (6.5–7.5 in.)

Occupies nearly all habitats except forest, but most common in well-drained plains and agricultural lands. Easily confused with the brown quail, but more streaked on the flanks, paler on the underparts, and with a more pronounced pale superciliary stripe. The male's usual advertisement call is a whistled, three- or four-noted *chuch-ee-whit* or *chip-a-ter-weet*. A sharp two-syllable *to-weep* is sometimes also uttered. Tends to flush and land abruptly, with a loud whirring of wings.

In the hand

Separable from the other Australian quails by the combination of white axillary feathers and a wing longer than 100 mm. The upperparts are as in *coturnix*, but are somewhat darker, especially on the wing-coverts; the superciliary stripe, sides of head, chin, throat, and foreneck are uniform dull brick-red to bright chestnut, bounded in *novaezeelandiae* with black markings, and the foreneck also black in *novaezeelandiae*.

Adult female. Differs from the male in having no brick colours on the head and throat, no buff stripe down the middle of the crown, and no dark vermiculations on the upperparts.

Immature. Young males resemble the adult female, but the black markings on the underparts are much wider and more completely fused at the shaft. The outer juvenal remiges (8–10) are also retained through the first year of life (Disney 1978), and the outer webs of these primaries are slightly mottled with buff. Breeding can occur in birds as young as 4 months old; such birds may still have their sixth and seventh juvenal primaries.

General biology and ecology

This is a ground-foraging, mainly seed-eating bird, that is associated with natural grassland areas and more recently grain crop lands in Australia. The birds are largely nomadic and probably opportunistic in their distribution and breeding cycles. They occur in all kinds of habitats except for forests, but especially favour well-drained plains and agricultural lands.

They move largely in response to rainfall and food availability, dispersing when food becomes scarce, and, on the other hand, sequentially raising up to three broods in a single area when conditions are favourable. Ringed birds have been known to move as far as 1300 km. Because of this mobility, the range of the species has changed greatly in historical times, having disappeared from both New Zealand and Tasmania, but increasing on the mainland of Australia in response to increased irrigation activities, and gradually spreading into Western Australia (Blakers *et al.* 1984). Their foods consist mainly of the seeds of cultivated cereals, grasses, and weeds, along with leafy materials, and a fairly small incidence of insects (McNally 1956; Frith *et al.* 1977).

Social behaviour

This is a species that occurs in small coveys of up to about 20 birds for most of the year. The breeding season is relatively regular, occurring from August or September to February in much of Australia, but is more affected by rainfall in the interior and parts of that country (Frith 1977). During that period the male persistently utters his advertisement call, usually in mornings and evenings, but sometimes throughout the night on moonlit nights. Observations on captive birds suggest that this occurs even in well-mated birds, and that pairing may be permanent in this species (Cruise 1966).

Reproductive biology

Nests are very well hidden in the cover of grasses or grain fields, and the eggs are laid at the rate of slightly more than a day per egg (Cruise 1966). The clutch ranges from about six to eleven eggs, usually seven to eight, and incubation is by the female alone. Incubation was found by Cruise to require 21 days; other estimates have been as short as 18 days. The young grow rapidly, and are fully mature and grown at 4 months (Disney 1978). Cruise noted that under captive conditions the young birds tended to divide themselves into unisexual groups after 3 months and remained thus until the following spring, but this is unlikely to be true of wild birds.

Evolutionary relationships

This is certainly a close relative of the last species, and perhaps evolved when a pre-*coromandelica* form somehow reached Australia during migration and became isolated there.

Status and conservation outlook

Probably this species has in general been favoured by human activities in Australia such as forest-clearing, cereal agriculture, and irrigation activities, although it was extirpated quite early from New Zealand and has seemingly recently disappeared from Tasmania. It is a highly mobile and adaptable species that is likely to adjust well to present ecological trends.

BROWN QUAIL (Plate 96)

Coturnix ypsilophorus Bose 1792

Other vernacular names: silver quail; swamp partridge; swamp quail; Tasmanian brown quail (*ypsilophorus*); caille Tasmane (French); Australische Sumpfwachtel; Ypsilonwachtel (German).

Distribution of species (see Map 31)

Seasonal or permanent resident of grassy habitats of northern, eastern, and southern Australia, mainly in coastal and adjacent areas, but extending locally inland in less arid habitats. Locally nomadic. Also resident in Tasmania, some of the Lesser Sundas, and New Guinea. Introduced in Fiji and the North Island (and adjoining offshore islands) of New Zealand.

Distribution of subspecies (excluding introductions)

C. y. ypsilophorus: south-eastern Australia and Tasmania.

C. y. australis: south-western Australia and southern Queensland south to Victoria. Sometimes considered specifically distinct.

C. y. queenslandicus: northern Queensland.

C. y. cervinus: north-western Australia.

C. y. dogwa: lowlands of southern New Guinea.

C. y. plumbeus: lowlands of eastern New Guinea.

C. y. saturatior: lowlands of northern New Guinea.

C. y. mafulu: south slopes of south-eastern New Guinea.

C. y. lamonti: mid-mountain grasslands in central highlands of New Guinea.

C. y. monticola: alpine grasslands of south-eastern New Guinea.

C. y. pallidior: Sumba and Savu islands.

C. y. raaltenii: Flores and Timor islands.

Measurements

Wing, males 82–109 mm, females 82–103 mm (Rand and Gilliard 1967); tail, both sexes 43–51 mm. Weight, males 74.5–84 g, average (of two) 79.3 g, females 88–92 g, average (of four) 90.2 g (Goodwin 1975). Egg (*australis*), average 30 × 23.6 mm (Schönwetter 1967), estimated weight 9.2 g.

Identification

In the field (7.5 in.)
Associated with rank, moist, and swamp-like vegetation, but sometimes also found in crop lands. When flushed, the birds tend to fly at a low angle, and usually show completely brownish coloration above and below. At close range the barred flanks and underparts should be visible. The usual call is a whistling crow, *tu-whee* or *gop-warr*, rising in pitch and with the second syllable extended. It often utters a cackling call as it flushes.

In the hand
Identified by the combination of a wing-length from 80 to 105 mm, grey axillaries, and narrow black barring on the flanks and underparts.

Adult female. Differs from the male in having coarser black patches and markings on the upperparts, less black on the underside, and pale buff shaft-stripes that are sometimes wider.

Immature. Similar to the adults, but with more conspicuous shaft-stripe markings, and with yellowish on the lower mandible and tip of the upper mandible (museum specimens).

General biology and ecology

This species is much more associated with rank grasses, moist meadows, and generally heavier cover than is the previous one, although at times the two are found together. It is less mobile than the previous form, and often remains in the vicinity of creeks or swamps, although occasionally being forced to higher ground during periods of heavy rainfall. It is normally to be found as pairs or in small coveys (Frith 1977). In New Guinea it occurs from sea-level to as high as 11 800 ft (3600 m), the latter in alpine grasslands (Rand and Gilliard 1967).

Reproductive biology

Breeding occurs in south-eastern Australia during the late spring period of October to December, while in the northern tropical areas it is associated with the late wet and early dry seasons from January to May (Frith 1977). Few nests have been found, but in Australia the estimated usual clutch-size is from seven to eleven eggs, whereas in New Guinea it has been reported to lay only four to six eggs (Rand and Gilliard 1967). Apparently based on avicultural information, Robbins (1981) reported a clutch of 10–18 eggs and an incubation period of 18 days.

Evolutionary relationships

This is seemingly a somewhat isolated form judging from its very distinctive underpart and flank patterning. In part this plumage pattern is the ultimate in concealing coloration trends and probably is to be expected in a species that is so closely associated with thick grassy cover. However, C. G. Sibley (personal communication) has found that the DNA of *australis* is quite distinct from that of *coturnix*, and that possibly generic distinction of the former (*Synoicus*) is thus warranted.

Status and conservation outlook

Although the brown quail has declined locally in Australia as a result of the draining of wetlands, in other areas it has benefited from the increase in stubble fields. On balance it has probably suffered an overall loss of habitat (Frith 1977). It has become locally established if limited in distribution on the North Island of New Zealand and also on Three Kings, Poor Knights, Alderman, Mayor, Great and Little Barrier islands, and is also locally established on Viti Levu and Nanua Levu, Fiji Islands (Long 1981).

ASIAN BLUE QUAIL (Plate 97)

Coturnix chinensis (L.) 1766
Other vernacular names: blue quail; blue-breasted quail; Chinese painted quail, king quail (Australia); caille peinte (French); Zwergwachtel (German).

Distribution of species (see Map 33)
Seasonal or permanent resident of open and grassland habitats throughout the Indian subcontinent from western India, Sri Lanka, and Bangladesh east to south-eastern China, and south through Indochina, Malaya, Greater Sundas, the Philippines, Indonesia, New Guinea, and northern and eastern Australia. Established on Guam; other introductions have apparently been failures. Locally nomadic or migratory.

Distribution of subspecies
C. c. chinensis: India (east of a line from Bombay to Simla) to Malaya, Indochina, and south-eastern China (Fukien to Kwangtung and Yunnan; Taiwan).
   C. c. trinkutensis: Nicobar Islands.
   C. c. palmeri: Sumatra and Java.
   C. c. lineata: Philippines, Borneo, and Celebes. Includes *minima*. Introduced on Guam.
   C. c. lineatula: Lombok, Sumba, Flores, and Timor islands.
   C. c. lepida: islands of Bismarck Archipelago.

Map 33. Distribution of the Asian (As) and African (Af) blue quails, and of the Madagascar partridge (M).

C. c. novaeguinea: mid-mountain valleys of New Guinea.
C. c. papuensis: south-eastern New Guinea.
C. c. colletti: Northern Territory of Australia.
C. c. australis: Queensland to Victoria.

Measurements

Wing (*chinensis*), males 65–78 mm, females 66–67 mm; tail, males *c.* 25 mm, females no data. Weight, both sexes *c.* 43–57 g (Ali and Ripley 1978). Egg, average 24.5 × 19 mm (Baker 1935), estimated weight 5.0 g.

Identification

In the field (5.5 in.)
Associated with very dense grasslands, swamp edges, and cultivated areas with heavy weedy growth, avoiding all open grassy areas. Birds are prone to squat and hide rather than fly, and if they do flush it is on silent wings, and the flights are short and laboured. The typical call is a high-pitched series of two- or three-noted descending piping notes, *ti-yu* or *quee-kee-kew*, and repeated weak *tir* notes may also be uttered upon flushing. If seen, the distinctive bluish colour of the male provides easy identification, but females resemble some of the small bustard-quail (*Turnix*) species rather closely.

In the hand
The very small size (wing under 80 mm) and presence of a hind toe (eliminating *Turnix* species) separates this from all but the African blue quail, males of which have some chestnut on the underparts and females of which are more distinctly barred with black on the wing-coverts. Various avicultural varieties of this species exist, which range from fawn-coloured to leucistic ('silver') and white.

Adult female. Upper surface generally browner than in the male, but with the forehead, wide superciliary stripe and cheeks rufous buff; the chin, throat, and foreneck are white, shading into rufous buff toward the sides; rest of underparts buff, paler toward the middle of the breast and abdomen; chest, sides, and flanks barred with black (less so in older birds). Iris brown rather than red.

Immature. Young males are similar dorsally to the female, but have larger and more numerous black markings, and more conspicuous white or buffy shaft-stripes. Young females have the feathers of the upper breast and sides spotted.

General biology and ecology

This is another typical *Coturnix* species that is associated with a variety of essentially open, grassland habitats that originally probably consisted of native grasses but now include gardens, roadside edges, crop lands, and deforested woodlands, ranging in altitude from sea-level up to about 2000 m. It is distinctly nomadic or migratory, often moving into an area after rainfall for breeding, and generally occurring in small coveys of about six to seven individuals that perhaps are family groups, although flocks of as many as 40 have been observed. Its foods are primarily grass seeds, with a limited number of insects taken as well (Ali and Ripley 1978).

Social behaviour

Members of a pair remain in contact with soft contact notes that become louder when the birds are separated or excited. Pair-bonds, at least in captive birds, are strong, and are in part maintained by tidbitting behaviour of the male. During the breeding season the male utters a whistled crowing or advertisement call, throwing his head back with each such call, thereby exposing the black and white throat-patch. Another male call, a hoarse and low-pitched 'growl', is performed with fluffed feathers and a nearly horizontal body angle; this call may be used as a separation or rally call for restoring contact between birds separated by some distance. Males also perform a lateral wing-display, or waltzing. This apparently sometimes serves as a precopulatory display, at which time the male persistently follows the female in a rather upright posture, with his abdominal and flank feathers maximally ruffled (Harrison 1956). An ethogram of social and egocentric behaviour patterns of this species has been provided by Schleidt *et al.* (1984); these patterns have been summarized in Chapter 5 and illustrated in Figs 8 and 9.

Reproductive biology

The breeding season in this species is quite variable and probably is related to the rainy period. In southern Australia breeding occurs between September and March, and in Northern Territory from February to April (Frith 1977). In the Malay Peninsula it occurs from January to August (Robinson and Chasen 1963), in Assam from June to August, and in southern India during March and April (Ali and Ripley 1978). In all these areas the normal clutch-size is apparently about four to seven eggs, with occasional reports of larger clutches. Incubation is by the female alone, and requires only 16 days, perhaps the shortest incubation period of any galliform species.

Evolutionary relationships

This is certainly part of a superspecies with *adonsonii* and perhaps is best considered conspecific with it (Snow 1978). These two forms have a counterpart pair of allopatric Indian and African taxa in *C. coromandelica* and *C. delegorguei* that seem to be equally far along in the speciation process.

AFRICAN BLUE QUAIL (Plate 98)

Coturnix adonsonii J. and E. Verreaux 1851
Other vernacular names: African painted quail; blue quail; blue-breasted quail; caille bleue (French); Afrikanische Zwergwachtel (German).

Distribution of species (see Map 33)

Seasonal or permanent resident in wet grassy and marshy habitats of sub-Saharan Africa from Sierra Leone and Ethiopia south to Zambia, Cape Province, and Natal, but absent from both the Congo basin and the driest areas. Migratory and irruptive. No subspecies recognized; but often considered a subspecies of *chinensis* (Snow 1978; Urban *et al.* 1986).

Measurements

Wing, males 78–82 mm, females 80–84 mm; tail, males 26–32 mm, females 29–31 mm (Clancey 1967). Weight, male 43 g, female 44 g (Newman 1983). Egg, average 25 × 18 mm (Mackworth-Praed and Grant 1952–73), estimated weight 4.5 g.

Identification

In the field (6 in.)

This tiny quail is found in most grasslands and fields overgrown with weeds, where the cover is fairly thick. More often heard than seen; the male's advertising call is a series of three notes descending the scale in semitones, the first shrill and loud, and the others softer. Also utters three squeaky notes when flushed. If seen, the male's bluish colour is distinctive; females lack white throats and superciliary stripes.

In the hand

Separated from all quails but *chinensis* by the small wing (under 85 mm), and from the latter by the male's lack of chestnut on the underparts and the female's tendency for heavier black barring on the wing-coverts.

Adult female. Easily distinguished from the male by the brown rather than red iris, and the absence of dark grey underparts or bright chestnut markings on the flanks.

Immature. Like the adult female, but with distinct light shaft-stripes on the greater wing-coverts, rump, and upper tail-coverts.

General biology and ecology

Associated with fairly dense grassy cover, including wet areas near rivers, edges of cultivation, cultivated lands, gardens, and similar habitats. It is found in pairs or small coveys that probably are of family units, and at least in some areas is nomadic or migratory, often moving unpredictably in response to rainfall patterns. It may be fairly sedentary in tropical areas. Its foods are mainly the seeds of grasses and weeds, plus some green materials and invertebrates (Clancey 1967).

Social behaviour

Not studied yet in detail, but evidently nearly identical to *chinensis*. The male's advertising call is seemingly almost identical. Reportedly territorial, but so far there is little evidence of this (Urban *et al.* 1986).

Reproductive biology

In South Africa nesting occurs from December to April, in Zimbabwe from January to April, in Zambia during January and February, in Malawi from February to June, in Kenya in October–November and again from May to July, in Uganda from April to June, and in Cameroon during November and again in May–June. There is thus much regional variation that is probably mainly related to temperature (in temperate Africa) and especially rainfall patterns (elsewhere), with birds mostly breeding during the rainy season or shortly afterward. The clutch-size under natural conditions is usually estimated at six to eight or rarely nine eggs, and the incubation period (in captivity) has been reported as 16 days (Robbins 1981). Both parents care for the chicks, and family groups remain together until the young can fly well (Urban *et al.* 1986).

Evolutionary relationships

As noted in the last account, this is a part of the *chinensis* superspecies, and is often regarded as conspecific with it.

Status and conservation outlook

Widespread in Africa, this species is at least locally common if not conspicuous. Its unpredictable movements make population estimates impossible, but like other *Coturnix* forms its status is unlikely to warrant the concern of conservationists.

23 · Genus *Anurophasis* van Oort 1910

The Snow Mountain quail is a medium-sized alpine species endemic to New Guinea having a short tail comprised of soft rectrices that are indistinguishable from and hidden by the upper tail-coverts. The wing is rounded, with the sixth and seventh primaries the longest. The sexes are similar, with distinctive dorsal and lateral black and buff barring. The tarsi are unspurred in both sexes, but the innermost toe has an unusually elongated claw. A single monotypic species is recognized:

A. monorthonyx van Oort 1910: Snow Mountain quail

SNOW MOUNTAIN QUAIL (Plate 99)

Anurophasis monorthoryx van Oort 1910
Other vernacular names: New Guinea alpine quail; caille de montagne (French); Neuguineahuhn; Schneebergswachtel (German); gimaabut (Danis tribe of New Guinea).

Distribution of species (see Map 32)
Endemic resident of alpine grasslands of the Kemabu Plateau, Mt Wilhelmina, and Mt Carstensz, western New Guinea. No subspecies recognized.

Measurements

Wing, males 157–161 mm, females 158 mm; tail, males 61–70 mm, females 65 mm (Ripley 1964; Rand and Gilliard 1967). Estimated weight, both sexes 401 g. Egg, average 45 × 33 mm, estimated weight 25.4 g.

Identification

In the field (10–11 in.)
Limited to alpine grasslands above 3000 m in the Snow Mountains, where it is the only partridge-like bird. When flushed, it utters a noisy cackling call, then usually flies to heavy cover where it is difficult to flush again. The strongly barred upperparts should provide for easy recognition.

In the hand
The heavily patterned black barring on the upperparts is distinctive; only *Lerwa* has a somewhat similar plumage pattern, and in that species the legs and bill are red rather than yellow. The claw of the inner toe is straight and elongated; no tarsal spur is present.

Adult female. Similar to the male, but the pale areas of underparts are much more whitish or buff, less chestnut, and much more heavily barred with black; upperparts have broader pale shaft-stripes and darker brown, less conspicuous bars.

Immature. Like the adult male, but duller, less rufous above and below, the barring less distinct and more irregular. Downy young are more like those of *coturnix* than of *ypsilophorus*.

General biology and ecology

This very little-studied species occurs on alpine grasslands between 10 300 and 12 850 ft (3100–3900 m) of the Carstensz and Wilhelmina massifs in the Snow Mountains of New Guinea (Rand and Gilliard 1967). It was collected by Ripley (1964) in the Nassau range of the Kemabu Plateau, and more recently has been found on the Carstensz meadow behind the Ertsberg, on tussock grasslands in lower Yellow and Discovery valleys, and on the Kemabu Plateau (3500–4000 m) (Schodde *et al.* 1975). The species' foods consist of flower-heads, leaves, and seeds, and some insects such as caterpillars (Rand and Gilliard 1967).

Social behaviour

These birds are found as solitary individuals, pairs, or sometimes in groups of up to three, but apparently not in larger coveys. They are seen but rarely, usually flushing closely and flying only a short distance before landing and hiding. It is possible that groups of two or three birds may roost together at the base of a shrub or in a dense tussock of grass (Rand and Gilliard 1967).

Reproductive biology

Rand (1942) reported breeding of this species in late September, including a nest with a clutch of three eggs, placed under the edge of a grass tussock.

Evolutionary relationships

Judging from the appearance of its downy young and also its egg pattern, this species is probably quite closely related to typical *coturnix* (Iredale 1956) or '*Synoicus*' *ypsilophorus* (Rand and Gilliard 1967). The adult plumage pattern is unique, lacking the buffy shaft-streaks typical of all *Coturnix* species, and is instead strongly barred and more reminiscent of *Lerwa*, another alpine-adapted partridge.

Status and conservation outlook

Very few specimens of this species have been collected, and it is still known from only a few localities. However, its remote and generally uninhabited habitat presumably places it out of contact with most humans, and it is apparently fairly common in some areas such as the alpine zone of Mt Wilhelmina (Rand and Gilliard 1967).

24 · Genus *Perdicula* Hodgson 1837 (including *Cryptoplectron* Streubel 1842)

The bush-quails are small, tropical grassland- and scrub forest-adapted species of the Indian subcontinent that have short and rounded tails of 10–12 feathers that are about half as long as the wing. The wing is rounded, with the fifth to seventh primaries the longest. The post-juvenal primary moult is incomplete. The sexes are dimorphic and are rather *Coturnix*-like dorsally; males lack sharp tarsal spurs, but may have blunt, wart-like tarsal knobs. Four species are recognized:

P. asiatica (Latham): jungle bush-quail
  P. a. asiatica (Latham) 1790
  P. a. punjabi Whistler 1939
  P. a. vadali Whistler and Kinnear 1936
  P. a. ceylonensis Whistler and Kinnear 1936
P. argoondah (Sykes): rock bush-quail
  P. a. meinertzhageni Whistler 1937
  P. a. argoondah (Sykes) 1832
  P. a. salimali Whistler 1943
P. erythrorhyncha (Sykes): painted bush-quail
  P. e. erythrorhyncha (Sykes) 1832
  P. e. blewitti (Hume) 1874
P. manipurensis Hume: Manipur bush-quail
  P. m. manipurensis Hume 1881
  P. m. inglisi (Ogilvie-Grant) 1909

KEY TO THE SPECIES OF BUSH-QUAILS (*PERDICULA*)

A. Third primary (from inside) equal in length to outermost; feet yellow to dull orange.
  B. Inner webs of primaries brown: jungle bush-quail (*asiatica*)
  BB. Inner webs of primaries mottled or barred with buff: rock bush-quail (*argoondah*)
AA. Innermost primary equal in length to outermost; feet deep red to flesh-coloured
  B. Upperparts brown with black spots: painted bush-quail (*erythrorhyncha*)
  BB. Upperparts slate-grey with black markings: Manipur bush-quail (*manipurensis*)

JUNGLE BUSH-QUAIL (Plate 100)

Perdicula asiatica (Latham) 1790
Other vernacular names: perdicule rousée-gorge (French); Frankolinwachtel (German).

Distribution of species (see Map 34)
Widespread resident of dry scrub and brush habitats of India and Sri Lanka, north to Kashmir and the outer Himalayas, possibly (at least previously) to the lowlands of Nepal. Introduced to Réunion, where it may still occur (Long 1981).

Distribution of subspecies
P. a. asiatica: central and north-eastern India, west to Gujarat, and east to Bihar.
  P. a. punjabi: north-western India, from Kashmir and Punjab to Himachal Pradesh and Uttar Pradesh.
  P. a. vadali: western India, from the Malabar coast south through Kerala.
  P. a. ceylonensis: Sri Lanka.

Measurements

Wing (*asiatica*) males 81–88 mm, females 80–88 mm; tail, males 34–41 mm, females 32–41 mm. Weight, both sexes *c.* 57–82 g (Ali and Ripley 1978). Egg (*asiatica*), average 25.4 × 19.5 mm (Baker 1935), estimated weight 5.3 g.

Identification

In the field (6–7 in.)
Associated with scrub, open deciduous forest, stony grasslands, and similar arid habitats, avoiding very dense and wet areas, and preferring stony, gravelly places with thorny bushes for nesting. Gregarious, and forming tight coveys that clump together. They scatter on flushing, and then utter repeated *tiri* notes as separation calls. The male's advertisement call is a harsh and grating *chee-chee-chuck* that is repeated over and over. Females are almost entirely rufous, and are not separable in the field from *argoondah*, but

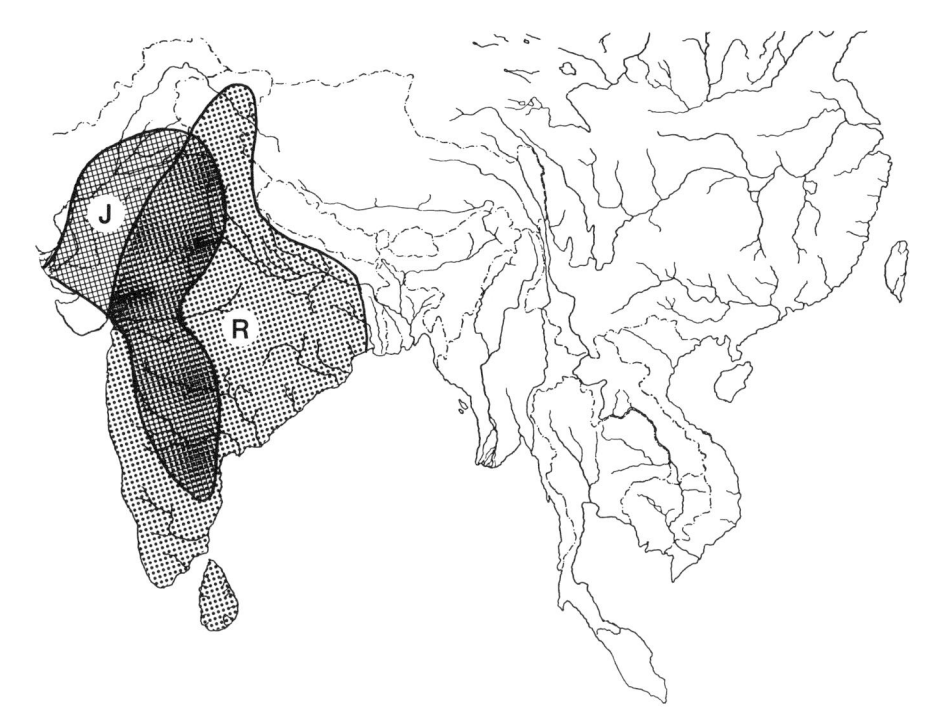

Map 34. Distribution of the jungle (J) and rock (R) bush-quails.

males have more conspicuous head streaking and a brown rather than rufous-toned back.

In the hand
Distinguished from other bush-quails by the combination of the outermost primary being equal in length to the third and longer than the first, and the inner webs of the primaries uniformly brown.

Adult female. Differs from the male in having no buff shaft-streaks or bars on the feathers of the back, rump, and upper tail-coverts; the wing-coverts are much less barred with buff and blotched with black, and the foreneck, chest, breast, and abdomen are uniform vinaceous buff.

Immature. Young birds have wider dorsal buffy shaft-stripes than adults. Young males have vinaceous buff underparts that are barred with black, while young females have the wing-coverts more heavily barred with buff and blotched with black.

General biology and ecology

This species occurs in a rather wide range of generally scrubby to well-wooded habitats, ranging from thin grass to fairly dense deciduous forests, but preferring rather dry and stony substrates, and extending from the plains up to about 4000 ft (1200 m). They occur in small coveys of up to about a dozen birds outside the breeding season, but during nesting the males are highly pugnacious (Baker 1930). They consume a variety of grass and weed seeds, millets, lentils, and small insects (Ali and Ripley 1978).

Social behaviour

The male's advertisement call is a harsh series of grating notes, sounding like *chee-chee-chuck*, that apparently serves as a territorial challenge, as captive 'caller' males quickly attract the local resident males. The birds are believed to be normally monogamous, with males perhaps occasionally taking two mates (Baker 1930).

Reproductive biology

The breeding season is generally prolonged and over most of the species' range it breeds from the end of the rains to the end of cold weather. This includes the period January to March in Travancore and Mysore (Karnatka), from October to February or March in the Deccan plateau, and during March and April in Orissa, West Bengal, and Behar (Baker 1930). The race *vidali* is said to breed nearly throughout the year, but perhaps mainly between July and February, while the Sri Lankan race has been reported to breed in March and April (Ali and Ripley 1978). The nest is placed in cover that affords good protection, but is not too dense nor too moist. The clutch usually numbers five to six eggs, with a range of four to seven. Incubation is by the female alone, although the male remains near by and takes a strong interest in helping brood and feed the young. The incubation period has recently been determined as 21–22 days in captive birds, and the sex of the young could not be determined with certainty until they were 3 months old (Thornhill 1981), although Wennrich (1983) observed some differences among birds only 3 weeks old.

Evolutionary relationships

The relationships of *Perdicula* are probably fairly close to *Coturnix*, judging from the similarities in adult plumage patterns and overall morphological characteristics. The two forms *asiatica* and *argoondah* comprise a superspecies that previously often have been regarded as representing a single species. The juvenal plumages of the two differ in minor details, and the two forms are apparently sympatric in many areas, although they have somewhat different habitat preferences (Ali and Ripley 1978). In a recent phenetic analysis (Crowe and Crowe 1985), *Perdicula* was associated with the typical quails and well separated from the partridges and francolins.

Status and conservation outlook

Evidently rather common and widespread if not conspicuous. The species perhaps benefits from some deforestation and cultivation, since the birds occur around the edges of terraced fields, but do not enter standing crops (Ali and Ripley 1978).

ROCK BUSH-QUAIL (Plate 101)

Perdicula argoondah (Sykes) 1832
Other vernacular names: red bush-quail; perdicule argoondah (French); Madraswachtel (German).

Distribution of species (see Map 34)

Resident of semi-desert habitats of northern and west-central India, from Punjab and Kutch east to Rajasthan and Madhya Pradesh, and south through peninsular India to Madras. This species and *asiatica* comprise a partially sympatric superspecies.

Distribution of subspecies

P. a. meinertzhageni: from Punjab and Uttar Pradesh south to Kutch and Gugerat, and east to western Madhya Pradesh.

P. a. argoondah: peninsular India from about Berar (Maharashtra) south through the Deccan plateau at least to Madras.

P. a. salimali: limited to stony lateritic soils of east-central Mysore (Karnatka).

Measurements

Wing (*argoondah*), males 82–89 mm, females 82–86 mm; tail, males 44–47 mm, females 40–47 mm (Ali and Ripley 1978). Estimated weight, males 62 g, females 59 g. Egg, average 25.6 × 20.1 mm (Baker 1935), estimated weight 5.7 g.

Identification

In the field (6–7 in.)

Associated with open semi-desert and desert scrub, in drier and more sandy or stony areas than the preceding species, but overlapping with it. Like that species it is highly gregarious, and always found in small, tightly packed coveys that roost together and tend to flush as a group. Its calls are evidently almost the same as those of the jungle bush-quail. Males can be recognized by their more rufous tones on the back, but females of the two species are probably not separable in the field.

In the hand

Separated from other bush-quails by the combination of having the inner webs of the primaries mottled or barred with buff, and the outermost primary as long as the third from the inside.

Adult female. Separated from the male by the lack of barring on the breast and flanks.

Immature. Young males have upperparts with narrow buff shaft-stripes edged on either side with black, which with increasing age become isolated buff patches. Immature females have the upperparts more or less distinctly barred and mottled with buff and black.

General biology and ecology

This species is perhaps more of a plains and lowland form than is the last one, rarely ascending mountains above 600 m, and mostly occurring on dry, open, and sandy to rocky substrates that are vegetated with thin, thorny bushes. Apart from having a more arid habitat than the previous form, the two are very similar, with opportunistic breeding and probably a certain capacity for local nomadic movements (Baker 1930; Ali and Ripley 1978).

Social behaviour

These birds occur in coveys of from about 12 to 20 birds, probably consisting of a few family units, and roost in small clusters, with all the members of the group in close contact. The male's advertisement call is apparently not yet well described.

Reproductive biology

Breeding in this species is prolonged or erratic and perhaps associated with the rain cycle. It occurs in general from August to October or November, and from January to February or March (Ali and Ripley 1978). The nest is placed under a rock, shrub, or in a grass tussock, and the average clutch is probably of five to six eggs, ranging from four to seven. The incubation period was reported by Robbins (1981) as 21 days, and it is performed by the female alone. Soon after hatching the chicks are apparently merged with groups containing several adult birds (Baker 1930).

Evolutionary relationships

As noted in the last account, *asiatica* and *argoondah* comprise a closely related and slightly sympatric superspecies. No definite hybrids are known, although Baker (1930) reported that some birds are 'quite intermediate' and might be assigned to either form. He considered these to be simply associated with intermediate habitats and forming a morphological link between these two 'subspecies' rather than as hybrids. Clearly this situation requires additional field study.

Status and conservation outlook

These small and inconspicuous birds are able to exist in dry and rocky habitats that are submarginal for cultivation, and therefore are probably quite secure.

PAINTED BUSH-QUAIL (Plate 102)

Perdicula erythrorhyncha (Sykes) 1832
Other vernacular names: perdicule à bec rouge (French); Buntwachtel; Weisscheitelwachtel (German).

Distribution of species (see Map 35)

Resident of grasslands and scrub habitats of central and south-western India, from Maharashtra east to West Bengal, and south through Kerala.

Distribution of subspecies

P. e. erythrorhyncha: western Ghats, from about Khandala (Maharashtra) south through Kerala, and associated hills.

P. e. blewitti: eastern Maharashtra, eastern Madhya Pradesh, Bihar, Orissa, and West Bengal.

Measurements

Wing (*erythrorhyncha*) males 81–87 mm, females 81–86 mm; tail, males 40–44 mm, females 33–45 mm. Weight, both sexes (*erythrorhyncha*) c. 70–85 g, both sexes (*blewitti*) c. 50–70 g (Ali and Ripley 1978). Egg (*erythrorhyncha*), average 25.4 × 19.5 mm (Baker 1935), estimated weight 5.3 g.

Identification

In the field (7 in.)

Associated with low, hilly land covered with grasses, scanty bush, or light deciduous forest, and often at the edges of gardens, compounds, or other human habitation. Except when breeding always found in coveys. The usual call is a repeated, whistled *tutu* or *tu* note, and soft whistles are used to reassemble scattered coveys.

In the hand

Recognized by the combination of an outermost primary no longer than the innermost one, and the upperparts and flanks brownish, with black spots and blotches.

Adult female. Like the male, but the forehead, lores, superciliary stripe, sides of the head, chin, and throat dull rufous chestnut, and the crown of the head almost entirely brown, with only a few black spots.

Immature. Young males have white shaft-stripes on some of the mantle feathers, and the forehead, superciliary stripes, sides of the head, chin, and throat are white washed with rufous. In young females the crown is mostly black.

General biology and ecology

This species has a similar ecological distribution to the other bush-quails, namely plateaux and foothills areas of grass, thin scrub, forest edges, and interspersed cultivation, mainly between about 600 and 2000 m. The birds especially favour mixtures of cultivated land interspersed with patches of woody cover and open wasteland, particularly where low foothills and ravines occur (Baker 1930). Foods consist of grass and weed seeds, green materials, and insects, especially termites (Ali and Ripley 1978).

Social behaviour

Like the other *Perdicula* this species is found in coveys for most of the year; these coveys number about 6–15 birds, and quickly reunite when scattered. The rally or separation call is a series of very soft whistles similar to those of *asiatica*, that rise and fall in pitch and last for about 2 s. The male's advertisement call is a triple-noted *kirikee, kirikee*.... This note, or an imitation of it, quickly attracts resident territorial males, suggesting that it

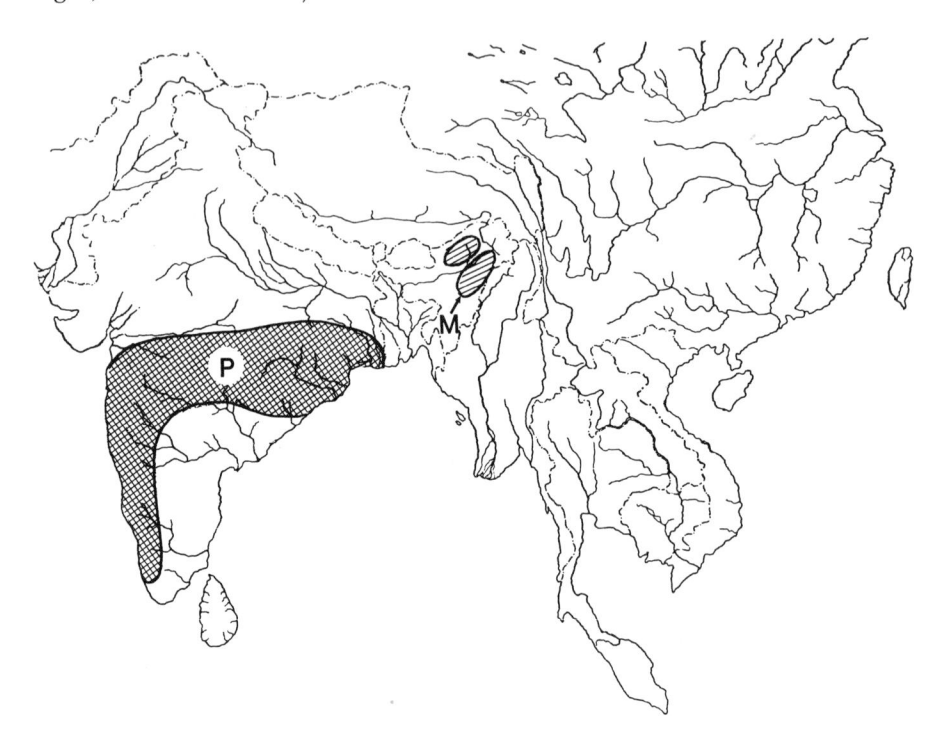

Map 35. Distribution of the Manipur (M) and painted (P) bush-quails.

is used to intimidate other males as well as to attract females.

Reproductive biology

The breeding season is extended and probably related to rainfall, with the birds typically laying after the break of the rains, when insects are most abundant. August to October is the primary breeding season, but in Travancore eggs have been collected from December to February as well as from July to September, and in the Nilgris there is also an apparent double breeding season, from January to March and again from August to November. The nest is usually poorly hidden, and often under a bush, a rock, or in a grass tussock. The normal clutch is probably five to seven eggs, with records of 10 or more probably reflecting the efforts of more than one female (Baker 1930). The incubation period is reportedly 21 days (Robbins 1981).

Evolutionary relationships

This and the following species are sometimes separated as the genus *Cryptoplectron*, based on the male's lack of definite spurs, but this is clearly not a generic-level trait. The female plumage of *erythrorhyncha* closely approaches that of the two previous species, and probably provides a better index of affinities than male tarsal traits.

Status and conservation outlook

Because this species is favoured by interspersed cultivation and natural scrubby habitats that are submarginal for agricultural purposes, and can coexist in close proximity to humans, it is likely that it will survive indefinitely.

MANIPUR BUSH-QUAIL (Plate 103)

Perdicula manipurensis Hume 1881
Other vernacular names: perdicule du Manipur (French); Manipurwachtel (German).

Distribution of species (see Map 35)
Resident of moist grassland and deciduous habitats of north-eastern India, on both sides of the Brahmaputra River.

Distribution of subspecies
P. m. manipurensis: Manipur and the Assam hills (Cachar, Khasi, Naga, etc.), south of the Brahmaputra River.
  P. m. inglisi: northern parts of West Bengal and Assam, north of the Brahmaputra River.

Measurements

Wing (*manipurensis*), both sexes 80–86 mm; tail, both sexes 45–52 mm. Weight, both sexes c. 64–78 g (Ali and Ripley 1978). Egg (*manipurensis*), average

30.5 × 24.1 mm (Schönwetter 1967), estimated weight 9.6 g.

Identification

In the field (7 in.)

Associated with damp and fairly tall grasses, in which the birds are very difficult to see unless they are flushed. They are usually to be found in tightly knit coveys, and fly only short distances. When running on the ground they are said to resemble black rats. Their calls reportedly resemble those of the painted bush-quail.

In the hand

The combination of an outermost primary no longer than the innermost one, and greyish upperparts with black barring and spotting, these markings becoming chevrons on the flanks, provides for identification within *Perdicula*.

Adult female. Differs from the male in having none of the dark chestnut markings on the head, chin, and throat, the two latter areas being white or greyish white, and the breast and abdomen being buff instead of tawny.

Immature. Differs from adults in that the innermost secondaries are black, and more widely barred with tawny buff.

General biology and ecology

This species is associated with taller and denser grasslands than are the other bush-quails, often occurring in jungle-like stands of elephant grass (*Typha elephantina*), usually remaining close to water. In such cover they are almost impossible to observe, and little is known of their biology, except that they remain in coveys of about six to eight birds for much of the year. They feed on the seeds of grasses, pods, or seeds or other herbs such as lentils, and some insects (Baker 1930; Ali and Ripley 1978).

Social behaviour

Presumably the social behaviour is like that of other bush-quails, but little has been written. One probable separation call has been described as a low, soft whistle, and a captive female has been heard uttering a soft, whistled *whit-it-it-it-t-t*, with each of the *it* notes slightly higher in pitch until they run together at the end. This call is repeated several times, becoming louder each time (Ali and Ripley 1978).

Reproductive biology

Apparently only a single nest of this elusive species has ever been described. It was found in North Cachar during May, in rather short sun grass, and contained four eggs. Nothing more is known of the reproduction of this species, although adults of *inglisi* showing signs of breeding have been collected in March, and a fledged youngster was obtained in January (Baker 1930).

Evolutionary relationships

Judging from its plumage, this is the most isolated of the *Perdicula* forms, and it is also ecologically distinctive.

Status and conservation outlook

The extremely elusive nature of this species makes any judgement of its status impossible.

25 · Genus *Ophrysia* Bonaparte 1856

The Indian mountain-quail is a small montane species originally endemic to the western Himalayas having a wedge-shaped tail of 10 rectrices and about 80 per cent as long as the wing. The wing is rounded, with the fifth and sixth primaries the longest. The sexes are dimorphic, but both have unusually long, lanceolate body feathers. The bill and legs of both sexes are bright to dusky red, and tarsal spurs are lacking. A single extinct and monotypic species is recognized:

O. superciliosa (J. E. Gray) 1846: Indian mountain-quail

INDIAN MOUNTAIN-QUAIL (Plate 104)

Ophrysia superciliosa (J. E. Gray) 1846
Other vernacular names: ophrysia de l'Himalaya (French); Hangwachtel (German).

Distribution of species

Apparently now extinct (Ripley 1952); once inhabited the western Himalayas of India, last reliably reported about 1868. No subspecies recognized.

Measurements

Wing, males 85–93 mm, females 87–91 mm; tail, males 76–78 mm, females 64–65 mm (Ali and Ripley 1978). Estimated weight, both sexes 705 g. Egg, no information.

Identification

In the field (10 in.)
Now apparently extinct, and thus field identification is a moot point. Similar in size to a see-see partridge, but with a red bill and leg colour, and originally found in grass and brushy cover on the steep Himalayan slopes around Dehra Dun and Naini Tal.

In the hand
Identified by the 10 fairly long (over 60 mm) rectrices, the bristle-shafted and stiffened forehead feathers, and the rather long and lanceolate body feathers that are relatively lax and soft.

Female. Separated from the male by the brown to tawny body colour, and lack of black on the chin and throat.

Immature. Immature males have the innermost secondaries black, with a narrow buff shaft-stripe and transverse bars and marks like those of the female; a few of the feathers of the middle of the back are also marked with black, and are almost like those of the female.

General biology and ecology

This extinct species apparently occurred in groups of five to ten birds that lived in high grasses, where they fed on fallen seeds and rarely could be seen. In the afternoons they apparently would descend into sheltered hollows, sometimes occupying very steep slopes with patches of brushwood (Baker 1930).

Social behaviour

Beyond the fact that the birds occurred in coveys, and used shrill whistles to reassemble after having been flushed, nothing is known of their social behaviour.

Reproductive biology

No information.

Evolutionary relationships

The most distinctive feature that characterizes the genus *Ophrysia* are the long and somewhat decomposed feathers. It has rather bristly shafted forehead feathers similar to those of *Perdicula*, which I believe to be its nearest living relative, and indeed a rather close one. Although larger than *Perdicula*, its beak is of a similar shape, the tarsus is completely short and stout, and the male's plumage is somewhat similar to that of *P. manipurensis*, while the female is rather more like those of the other *Perdicula* species.

Status and conservation outlook

This is the only full species of perdicine bird known to be extinct. It was last collected in 1876, and only 10 specimens are known to exist in the world's museums. Ripley (1952) mentioned a possible specimen being shot more recently in eastern Kumaon, but this has not been confirmed. Kukherjee (1966) has reviewed the history of this species.

26 · Genus *Arborophila* Hodgson 1837 (including *Tropicoperdix* Blyth 1959)

The hill-partridges are medium-sized, forest-adapted species of the tropical to temperate areas of eastern and south-eastern Asia, Taiwan, and the Greater Sundas, with rounded tails that are less than half as long as the wing. The wing is rounded, with the sixth and seventh primaries the longest. The sexes are alike or very similar, usually having black and white facial markings and small areas of bare red skin around the eyes. Tarsal colour varies but usually is reddish to yellowish; spurs are lacking in both sexes. Sixteen species are recognized here (taxonomy mainly after Davison 1982); at least three forms (*rufipectus*, *rufogularis*, and *ardens*) generally regarded as species and considered here as allospecies; might perhaps best be regarded as subspecies:

1. Grey-breasted group
    A. torqueola Valenciennes: necklaced (common) hill-partridge
        A. t. millardi (Baker) 1921
        A. t. torqueola Valenciennes 1826
        A. t. batemani (Ogilvie-Grant) 1906
        A. t. griseata Delacour and Jabouille 1930
    A. [torqueola] rufipectus Boulton 1932: Sichuan hill-partridge
    A. [gingica] mandellii Hume 1874: red-breasted hill-partridge
    A. gingica (Gmelin) 1789: collared hill-partridge
    A. [gingica] rufogularis (Blyth): rufous-throated hill-partridge
        A. r. rufogularis (Blyth) 1850
        A. r. intermedia (Blyth) 1856
        A. r. tickelli (Hume) 1880
        A. r. euroa (Bangs and Phillips) 1914
        A. r. guttata Delacour and Jabouille 1928
        A. r. laotiana Delacour 1926
        A. r. annamensis (Robinson and Kloss) 1919
    A. atrogularis (Blyth) 1850: white-cheeked hill-partridge
    A. [atrogularis] crudigularis (Swinhoe) 1864: white-throated hill-partridge
    A. [atrogularis] ardens (Styan) 1892: white-eared hill-partridge
    A. javanica (Gmelin): Javan hill-partridge
        A. j. javanica (Gmelin) 1789
        A. j. bartelsi Siebers 1929
        A. j. lawuana Bartels 1938
    A. orientalis (Horsefield): bare-throated hill-partridge
        A. o. campbelli (Robinson) 1904
        A. o. rolli (Rothschild) 1909
        A. o. sumatrana Ogilvie-Grant 1891
        A. o. orientalis (Horsefield) 1821
2. Brown-breasted group
    A. [orientalis] brunneopectus (Blyth): brown-breasted hill-partridge
        A. b. brunneopectus (Blyth) 1855
        A. b. henrici (Oustelet) 1896
        A. b. albigula (Robinson and Kloss) 1919
    A. davidi Delacour 1927: orange-necked hill-partridge
    A. cambodiana Delacour and Jabouille: chestnut-headed hill-partridge
        A. c. cambodiana Delacour and Jabouille 1928
        A. c. diversa Riley 1930
    A. hyperythra (Sharpe): Borneo hill-partridge
        A. h. hyperythra (Sharpe) 1879
        A. h. erythrophrys (Sharpe) 1890
    A. rubrirostris (Salvadori) 1879: red-billed hill-partridge
3. Scaly-breasted group (= *Tropicoperdix* Blyth)
    A. charltonii (Eyton): scaly-breasted hill-partridge
        A. c. chloropus (Blyth) 1859
        A. c. peninsularis de Schauensee 1941
        A. c. olivacea (Delacour and Jabouille) 1924
        A. c. cognacqi (Delacour and Jabouille) 1924
        A. c. merlini (Delacour and Jabouille) 1924
        A. c. vivida (Delacour) 1926
        A. c. tonkinensis (Delacour) 1927
        A. c. atjehensis de Schauensee and Ripley 1940
        A. c. charltonii (Eyton) 1845
        A. c. graydoni (Sharpe and Chubb) 1906

KEY TO THE SPECIES OF HILL-PARTRIDGES (*ARBOROPHILA*)

A. Flanks not distinctly spotted or barred with black or white, nor uniformly chestnut-coloured

B. Chest chestnut, banded above with black and white
　C. Throat with a white gorget, banded narrowly below with black: red-breasted hill-partridge (*mandellii*)
　CC. Throat with a black gorget, banded below with white: collared hill-partridge (*gingica*)
BB. Chest grey or duller brown, not bounded above by black and white as described
　C. Chest brown vermiculated with black; throat spotted with black, legs olive to yellowish: scaly-breasted hill-partridge (*charltonii*)
　CC. Chest not vermiculated with black; throat mostly black or streaked with black, legs reddish
　　D. Forehead white or greyish
　　　E. Throat white above and black below; chest grey: white-throated hill-partridge (*crudigularis*)
　　　EE. Throat streaked with black; chest grey (males) or russet (females): Sichuan hill-partridge (*rufipectus*)
　　DD. Forehead brown or olive-brown
　　　E. Crown bright chestnut; no white on face: necklaced hill-partridge (*torqueola*)
　　　EE. Crown not chestnut; face partly white
　　　　F. Crown black; cheeks white: white-cheeked hill-partridge (*atrogularis*)
　　　　FF. Crown dull olive; cheeks not pure white
　　　　　G. Throat mostly rufous with black spots; cheeks white with black spots: rufous-throated hill-partridge (*rufogularis*)
　　　　　GG. Throat black; cheeks black with white ear-coverts: white-eared hill-partridge (*ardens*)
AA. Flanks variously barred or spotted with black and white (or buffy); rarely entirely chestnut
　B. Bill entirely red, crown black: red-billed hill-partridge (*rubrirostris*)
　BB. Bill not entirely red; crown not usually entirely black
　　C. Chest and head mostly bright brown
　　　D. Head, neck, and chest uniformly chestnut: chestnut-headed hill-partridge (*cambodiana*)
　　　DD. Head and neck distinctly paler than chest: red-breasted hill-partridge (*hyperythra*)
　　CC. Chest and head not bright chestnut-brown
　　　D. Chest grey; chin black: bare-throated hill-partridge (*orientalis*)
　　　DD. Chest brownish; chin white or rufous
　　　　E. Lower throat rufous, bounded by a black band below: orange-necked hill-partridge (*davidi*)
　　　　EE. Lower throat buffy or white, with variable black spotting
　　　　　F. Flanks spotted with black and white; underparts brown to buffy: brown-breasted hill-partridge (*brunneopectus*)
　　　　　FF. Flanks and underparts entirely rufous: Javan hill-partridge (*javanica*)

NECKLACED HILL-PARTRIDGE (Plate 105)

Arborophila torqueola Valenciennes 1826
Other vernacular names: common hill-partridge; Indian hill-partridge (*torqueola*); torqueola à collier (French); Hugelhuhn (German).

Distribution of species (see Map 36)

Resident of evergreen forests and heavy undergrowth from north-western India (Himachal Pradesh and

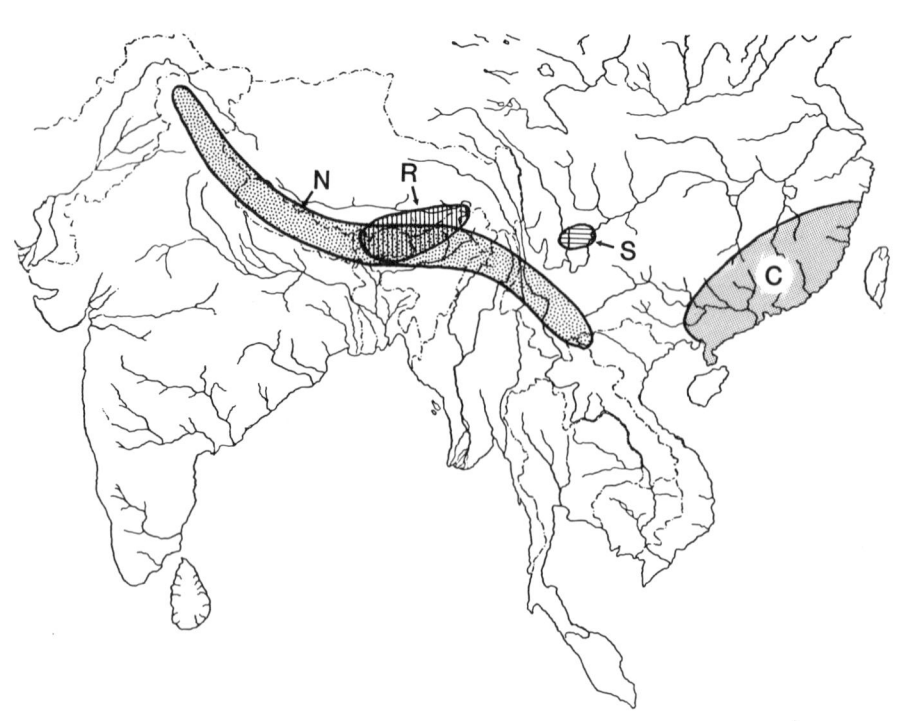

Map 36. Distribution of the collared (C), necklaced (N), red-breasted (R), and Sichuan (S) hill-partridges.

Uttar Pradesh) east through Assam to Burma, southern China, and north-western Tonkin.

Distribution of subspecies

A. t. millardi: western Himalayas from Chamba (Himanchal Pradesh) and Garwhal (Uttar Pradesh) east to western Nepal, intergrading there with *torqueola*.

A. t. torqueola: Himalayas, from Nepal east to southern Xinzang (Tibet), northern Burma, and north-western Yunnan.

A. t. batemani: Kachin and Chin hills of upper Burma to south-western Sichuan and north-western Yunnan.

A. t. griseata: north-western Tonkin.

Measurements

Wing (*torqueola*), males 144–160 mm, females c. 140–150 mm; tail, both sexes c. 60–80 mm (Ali and Ripley 1978). Weight (*torqueola*), males 325–430 g, females 261–386 g (Cheng *et al.* 1978). Egg (*torqueola*), average 40.6 × 31.9 mm (Baker 1935), estimated weight 22.8 g.

Identification (see Fig. 20)

In the field (11.5 in.)

Associated with evergreen forest and scrub habitats, especially along mountain streams or valleys. More often seen than heard, the male's advertisement call consists of a repeated melancholy whistle, *peeor*, descending in tone and uttered at frequent intervals with a quick head-toss. The necklace-like streaked black throat pattern of the male is a distinctive field mark, and females have similar but reduced black streaking on the neck and upper chest.

In the hand

The combination of a chestnut crown and a black chin and throat that becomes broken and is terminated by a white breast-band identifies the male, while females have the crown streaked with black, a tawny throat, and no white breast-band.

Adult female. Differs from the male in having the crown brown or olive-brown; the wing-coverts more or less mottled with black; the ear-coverts olive-brown; the supercilium, sides of the face, chin, throat, and neck rust-coloured or rufous buff and spotted terminally with black; a rust-red band separates the neck and chest, which is brownish grey tinged with rust, and the sides and flanks have subterminal white spots and blotches.

Immature. Young males have white-spotted chest feathers, almost lack chestnut margins on the side and flank feathers, and the outer webs of the primaries are margined with rufous buff. The eye-ring is blue-grey, tinged with red. Young females also have rufous buff margins on the primaries and white-spotted chest feathers.

General biology and ecology

This species is found in woodlands mostly between 5000 and 9000 ft (1500–2700 m), rarely either higher (questionably to 5000 m) or lower. It is mostly associated with heavy forests in hilly areas with ravines, rocks, and torrents, and with a dense undergrowth of ferns and associated plants (Baker 1930). In China the birds especially favour dense evergreen forests of oaks (*Quercus*) and laurels (*Cinnamomum*) along mountain streams or in valleys (Cheng *et al.* 1978).

Social behaviour

These birds are usually found in pairs or small coveys of up to about five to ten individuals, which probably are primarily family groups. They forage on the forest floor for seeds, green shoots, berries, and various invertebrates. While foraging they commonly call with hen-like calls, and probable pairs perform a duet, in which one bird (possibly the female) begins with a series of *kwikwikwikwikwik*... notes that are soon supplemented by repeated *du-eat* calls (the second syllable more prolonged than the first) from a second bird, leading to a duet crescendo and an abrupt ending (Ali and Ripley 1978).

Reproductive biology

In India the breeding season is mostly from April to June or July, or earlier at lower elevations. An early nesting season of February to May has also been reported for the race *batemani*. The nest apparently is a rather simple scrape rather than a domed structure, and usually at the base of a tree or under a bush. However, a nest of *millardi* has been described as being in a scraped out and rounded hole in a bank, and domed over with grass. The usual clutch-size is four to five eggs, with an extreme range of four to nine (Baker 1930). An incubation period of 24 days has been reported, apparently from captivity (Robbins 1981).

Evolutionary relationships

Davison (1982) regarded this as an apparently central species, perhaps most closely approaching the ancestral tree-partridge species within his 'grey-breasted group' of *Arborophila*. He considered the

218 The Quails, Partridges, and Francolins of the World

Fig. 20. Adult plumage variation in the: (A) Sichuan hill-partridge; necklaced hill-partridge, subspecies (B) *batemani* and (C) *torqueola*; and the rufous-throated hill-partridge, subspecies (D) *rufigularis*, (E) *intermedia*, and (F) *annamensis*.

plumages of the two sexes to be precursors of the sexually monomorphic plumages of the two assemblages of remaining species in this group. I believe that the phylogeny shown by Cheng *et al.* (1978), showing *rufipectus* and *torqueola* as very close allopatric relatives, to be a likely assessment of relationships, and believed the two might even be considered as subspecies.

SICHUAN HILL-PARTRIDGE (Plate 106)

Arborophila rufipectus Boulton 1932
Other vernacular names: Boulton's hill-partridge; torqueole de Boulton (French); Boulton-Buschwachtel (German).

Distribution of species (see Map 36)

Resident of forests and bamboo thicket hills of southeastern Sichuan, and presumably also in northeastern Yunnan. No subspecies recognized.

Measurements

Wing, males 148–158 mm, females 143–145 mm; tail, males 68–74 mm, females 64–71 mm. Weight, males 410–470 g, females 350–380 g (Cheng *et al.* 1978). Egg, no information.

Identification (see Fig. 20)

In the field (11–12 in.)

Found in dense bamboo thickets and deciduous forests of Sichuan, where it is rare and little studied, living in pairs or small groups. Its call is a double-noted *ho-wo, ho-wo*. Very similar to the preceding species but evidently wholly allopatric with it, and with a mostly white rather than mostly black throat, and hazel to russet rather than grey on the chest of males.

In the hand

See the comments above for distinction of males from *torqueola*; distinction of females is questionable.

Adult female. Forehead, eye-stripe, and median line black; crown, back, rump, and upper tail-coverts dark brown, the feathers with dark stripes and lighter edges; remiges black; rectrices dark brown, with four to five black stripes; chin and throat light yellow; chest greyish brown, breast light greyish brown; abdomen white, flanks and sides grey, with greyish white stripes.

Immature. Undescribed.

General biology and ecology

This species occurs in dense bamboo thickets and associated shrubbery (*Elaeagnus*, *Rubus*, *Prunus*) of broad-leaved forests containing oaks, chestnuts, rhododendrons, and camellias. Its foods have been found to include fruits and seeds of oaks, *Elaeagnus*, *Euonymus*, and *Rubus*, plus various invertebrates (Cheng *et al.* 1978).

Social behaviour

Coveys of this species are usually of five to six birds during autumn and winter, and pairs become solitary during the breeding season. The usual (advertising?) call is a *ho-wo, ho-wo*, and other notes are used by startled birds (Cheng *et al.* 1978).

Reproductive biology

Breeding occurs from April to June, with some chicks hatched by mid-May. The nests are depressions, often concealed among roots of trees, and typically have five to six eggs (Cheng *et al.* 1978). The incubation period is unreported.

Evolutionary relationships

This form comprises a superspecies of allospecies with *torqueola* as noted in the last account, and perhaps the two should even be considered as conspecific.

Status and conservation outlook

According to Cheng *et al.* (1978), this species requires special attention for its conservation, as it is a Chinese endemic, with a restricted distribution and a low population density.

RED-BREASTED HILL-PARTRIDGE (Plate 107)

Arborophila mandellii Hume 1874
Other vernacular names: Mandell's hill-partridge; torqueole de Mandell (French); Mandellibuschwachtel (German).

Distribution of species (see Map 36)

Resident of undergrowth in wet evergreen forests of Arunachal Pradesh (hills north of Brahmaputra River), Sikkim, Bhutan, and south-eastern Xinzang. No subspecies recognized.

Measurements

Wing, both sexes 133–145 mm; tail, both sexes 56–58 mm (Ali and Ripley 1978). Estimated weight, both sexes 268 g. Egg, average *c.* 43 × 33 mm (Ali and Ripley), estimated weight 25.8 g.

Identification (see Fig. 21)

In the field (11 in.)

Associated with dense, deep forests, especially those with streams or rivers, and mainly between 900 and 1800 m. The call is a series of ascending double notes leading to a climax. Probably best identified by the chestnut chest and flank feathers, the chest bordered above with a black-edged white gorget.

In the hand

This is the only *Arborophila* species with a narrow white crescent on the lower foreneck that is bounded above by a rusty red neck and below by a bright chestnut chest, the white separated from the chestnut by a narrow black band.

220 *The Quails, Partridges, and Francolins of the World*

Fig. 21. Adult plumage variation in the (A) white-cheeked, (B) white-eared, (C) white-throated, (D) collared, (E) red-breasted, and (F) Javan hill-partridges.

Adult female. Not noticeably different from the male.

Immature. Not described.

General biology and ecology

This species occurs in evergreen forests having dense undergrowth at elevations of 350–2450 m, especially those having streams or rivers running through them (Baker 1930; Ali and Ripley 1978).

Social behaviour

Nothing is known of this, except that the birds reportedly call with a deep and prolonged note, followed immediately by a series of ascending double notes that end on the highest one, sounding quite similar to the call of *torqueola* (Cheng *et al.* 1978).

Reproductive biology

The breeding season is reportedly from March to June (Cheng *et al.* 1978), but the apparent only record is of

Plate 96. Brown quail (*australis*), pair. Painting by H. Jones.

Plate 97. Asian blue quail, pair. Painting by H. Jones.

Plate 98. African blue quail, pair. Painting by H. Jones.

Plate 99. Snow Mountain quail, adult. Painting by Mark Marcuson.

Plate 100. Jungle bush-quail, pair. Painting by H. Jones.

Plate 101. Rock bush-quail, pair. Painting by H. Jones.

Plate 102. Painted bush-quail, pair. Painting by H. Jones.

Plate 103. Manipur bush-quail, pair. Painting by H. Jones.

Plate 104. Indian mountain-quail, pair. Painting by H. Jones.

Plate 105. Necklaced hill-partridge, pair. Painting by H. Jones.

Plate 106. Sichuan (left), orange-necked (middle), and chestnut-headed (right) hill-partridges, adults. Painting by Timothy Greenwood.

Plate 107. Red-breasted hill-partridge, pair. Painting by H. Jones.

Plate 108. Collared hill-partridge, pair. Painting by H. Jones.

Plate 109. Rufous-throated hill-partridge (*intermedia*), pair. Painting by H. Jones.

Plate 110. White-cheeked hill-partridge, pair. Painting by H. Jones.

Plate 111. White-throated hill-partridge, pair. Painting by H. Jones.

Plate 112. White-eared hill-partridge, pair. Painting by H. Jones.

Plate 113. Javan hill-partridge, pair. Painting by H. Jones.

Plate 114. Bare-throated hill-partridge (*campbelli*), pair. Painting by H. Jones.

Plate 115. Brown-breasted hill-partridge (*brunneopectus*), pair. Painting by H. Jones.

Plate 116. Borneo hill-partridge, pair and presumed immature. Painting by H. Jones.

Plate 117. Red-billed hill-partridge, pair. Painting by H. Jones.

Plate 118. Scaly-breasted hill-partridge (*chloropus*), pair. Painting by H. Jones.

Plate 119. Ferruginous wood-partridge, pair. Painting by H. Jones.

Plate 120. Crimson-headed wood-partridge, pair. Painting by H. Jones.

Plate 121. Crested wood-partridge, pair. Painting by H. Jones.

Plate 122. Stone partridge, pair. Painting by H. Jones.

Plate 123. Mountain bamboo-partridge, pair. Painting by H. Jones.

Plate 124. Chinese bamboo-partridge, pair. Painting by H. Jones.

Plate 125. Red spurfowl, pair. Painting by H. Jones.

Plate 126. Painted spurfowl, pair. Painting by H. Jones.

Plate 127. Ceylon spurfowl, pair. Painting by H. Jones.

a nest with four eggs taken in early June, in an evergreen forest of oaks and rhododendrons, at about 8000 ft (2400 m) in the Chambi Valley (Baker 1930).

Evolutionary relationships

This is obviously a very close relative of *gingica*, and was placed by Davison (1982) in a species group that additionally included *rufogularis* and *javanica*. I generally concur with this assignment.

COLLARED HILL-PARTRIDGE (Plate 108)

Arborophila gingica (Gmelin) 1789
Other vernacular names: Chinese hill-partridge; Rickett's hill-partridge; Sonnerat's hill-partridge; torqueole de Sonnerat (French); Chinabuschwachtel (German).

Distribution of species (see Map 36)

Resident of dense forests of Kujian, Guangdong, and Guangxi, south-eastern China. Recently collected in Taiwan, where it may have been introduced (Kang 1969). No subspecies recognized.

Measurements

Wing, males 146–147 mm, females 137–143 mm; tail, both sexes 50–55 mm (Cheng *et al.* 1978 and personal observations). Weight, adult 253 g (Kang 1969). Egg, no information.

Identification (see Fig. 21)

In the field (12 in.)
Associated with densely wooded hilly areas of south-eastern China between about 700 and 900 m. It is the only *Arborophila* of that region having a throat that is banded with black and white, and a chestnut breast contrasting with grey underparts. The call consists of seven to eight hoarse whistles that rapidly rise in pitch.

In the hand
Similar to *mandellii* in having a chestnut breast and chestnut-spotted flanks with grey lower underparts and black and white lower throat banding, but in this species the white is below rather than above the black band, and there is a definite whitish superciliary stripe rather than a diffuse grey area above the eye.

Adult female. Similar to the male, but the base of the nape maroon-orange, and the under tail-coverts chestnut and white.

Immature. Similar to adults, but the superciliary stripe greyish white rather than white.

General biology and ecology

This species occurs in densely wooded habitats between 700 and 900 m, and reportedly feeds on seeds, berries, insects, and the like (Cheng *et al.* 1978).

Social behaviour

Little information is available. At dusk the birds reportedly utter seven to eight hoarse whistles, rising in pitch. Shortly after making this call they are said to arrive quickly at their roosting site (Cheng *et al.* 1978).

Reproductive biology

The breeding season reportedly extends from April to May, and the clutch-size is said normally to be of five to seven eggs (Cheng *et al.* 1978).

Evolutionary relationships

The close similarities and allopatric distributions of this species and *rufogularis* suggest that they are very close relatives, and it additionally is very similar to *mandellii*, the three forms possibly comprising a superspecies.

Status and conservation outlook

According to Cheng *et al.* (1978), this species is more numerous than are the other Chinese hill-partridges, and the birds have potential for economic development as domesticated or cage birds.

RUFOUS-THROATED HILL-PARTRIDGE (Plate 109)

Arborophila rufogularis (Blyth) 1850
Other vernacular names: torqueole à gorge rousse (French); Rotkehl-Buschwachtel (German).

Distribution of species (see Map 37)

Resident of thick undergrowth and secondary montane forest from north-western Indian (Uttar Pradesh) east through Nepal, Sikkim, and Bhutan to Assam and the hills of upper Burma, Laos, and Vietnam.

Distribution of subspecies

A. r. rufogularis: Uttar Pradesh (Garhwal) to Mishmi hills of Assam.

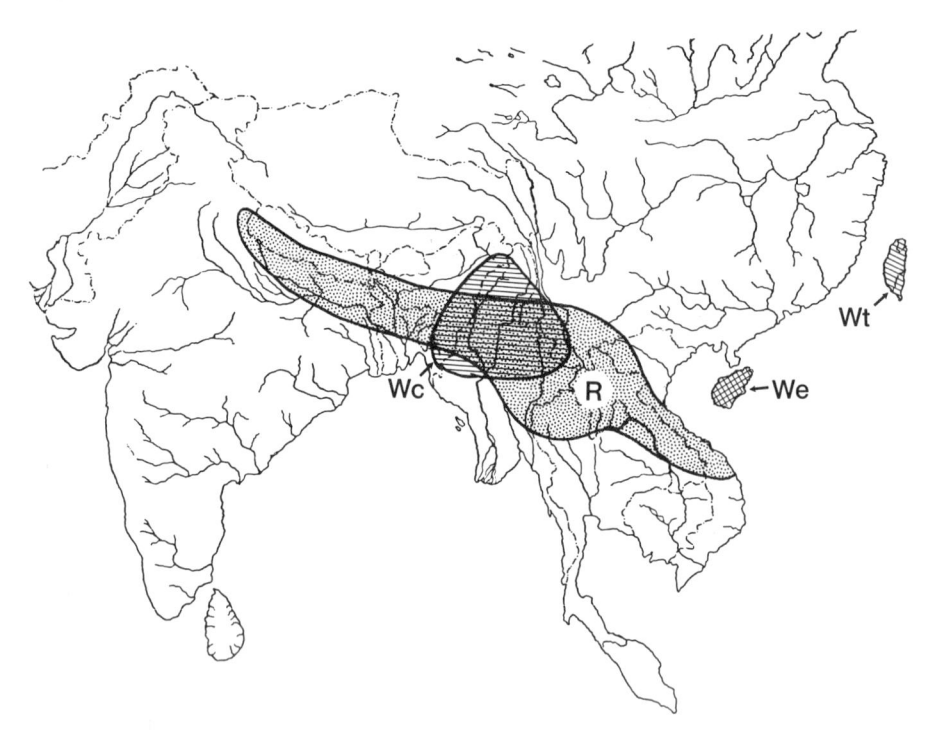

Map 37. Distribution of the rufous-throated (R), white-cheeked (Wc), white-eared (We), and white-throated (Wt) hill-partridges.

A. r. intermedia: Arakan Yomas, Chin, and Kachin hills of Burna and north-western Yunnan.

A. r. tickelli: southern Shan states and Tenasserim of Burma, east to Thailand and south-western Laos.

A. r. eurosa: south-eastern Yunnan to northern Laos.

A. r. laotiana: northern Laos and northern Vietnam (Tonkin).

A. r. guttata: central Vietnam (Annam).

A. r. annamenis: southern Vietnam (Langbien area).

Measurements

Wing (*rufogularis*), males 138–149 mm, females 131–142 mm; tail, males (of *intermedia*) 52–60 mm, females (of *rufogularis*) 50–56 mm. Weight, both sexes (*rufogularis*) c. 200–300 g, also (*intermedia*) c. 340–370 g (Ali and Ripley 1978). Egg (*rufogularis*), average 39.9 × 30.3 mm (Ali and Ripley 1978), estimated weight 20.2 g.

Identification (see Fig. 20)

In the field (12 in.)

Associated with tall grasses, dense secondary scrub, and undergrowth in evergreen forests. Its call is a loud, sweet, and clear double whistle, repeated each time on a higher pitch and with less pause between calls. Similar in appearance to *torqueola*, but lacking dark barring on the upperparts and the throat variably orange-rufous and black. No white is present on the throat, but it has a white supercilium that is lacking in *torqueola*.

In the hand

Identified by the combination of a grey breast, a rufous or black and rufous throat lacking in white, and black spotting extending upward along the sides of the neck to include the whitish cheeks.

Adult female. Like the male, but with less black on the chin and throat, and more white spots on the breast and abdomen (Baker 1928).

Immature. Underparts, abdomen, and flanks with numerous white spots, and (at least in *rufogularis*) the throat immaculate rufous brown (Baker 1928).

General biology and ecology

Associated with evergreen forests above 1000 ft (300 m) up to at least 8500 ft (2600 m), and probably mostly found under 5000 ft (1500 m) in heavy hungle having a fairly thick undergrowth. Less frequently occurs in bamboo jungle, scrub, heavy second growth, and along forest edges. Birds sometimes occur in the same general area as *torqueola* but are generally at lower elevations (Baker 1930). They feed on the usual assortment of seeds, berries, greens, and terrestrial invertebrates (Ali and Ripley 1978).

Social behaviour

The usual covey size is of about five to twelve birds, which move through the forest keeping in contact with low, whistled calls. At least two types of vocalizations are known. One is a loud double whistle, *wheea-whu*, the first note prolonged and the second short and sharp, often uttered at dawn and dusk, and perhaps the male's advertisement call. This call may be repeated two or three times and then followed by a series of ascending notes. A presumably different call, probably a duet between rival birds, consists of a series of double descending whistles, starting very soft and low, but gradually becoming louder, higher in pitch, and more and more rapid until it ends rather breathlessly (Baker 1930; Ali and Ripley 1978).

Reproductive biology

Breeding in the race *rufogularis* reportedly occurs from April to July, and that of *intermedia* from April to August, but mainly May and June. Nests are apparently of two types, either rather simple scrapes hidden in grass, or rather large domed structures in tufts of grass with the roof domed over with grass and only a small entrance hole as an opening, the particular type of nest that is built apparently varying with location or available substrate. The clutch is most often of four eggs, less frequently of five, and rarely up to about eight (Baker 1930). The incubation period from (presumably captive) birds has been reported as 20–21 days (Robbins 1981). Baker (1930) reported a case of a female rearing two broods of four young in a single year, with all 10 birds remaining constantly together the following winter.

Evolutionary relationships

Davison (1982) includes this species in a species-group assemblage that also includes *javanica*, *mandellii*, and *gingica*. Cheng *et al.* (1978) regard it as a close relative of *gingica*, and I agree that these species are probably all close relatives.

WHITE-CHEEKED HILL-PARTRIDGE
(Plate 110)

Arborophila atrogularis (Blyth) 1850
Other vernacular names: white-necked hill-partridge; torqueole à joues blanches (French); Weisswangen-Buschwachtel (German).

Distribution of species (see Map 37)
Resident of lowland and lower montane forest undergrowth from Arunachal Pradesh and the Chittagong region of Bangladesh east to northern Burma and westernmost Yunnan. No subspecies recognized.

Measurements

Wing, males 135–147 mm, females 126–130 mm; tail, males no data, females *c.* 60–65 mm. Weight, males 256 g, females 220 g (Davison 1985). Egg, average 37 × 28.3 mm (Ali and Ripley 1978), estimated weight 16.3 g.

Identification (see Fig. 21)

In the field (11 in.)
Associated with forest, bamboo, scrub jungle, or even grassland, thus not as forest-adapted as most *Arborophila*. The advertising call is a repeated loud, clear double-whistled note, *whew, whew*, often ending with a sharper single note. Separated from the other mainland *Arborophila* by its white ear-patch, surrounded by black, but otherwise very similar to *torqueola*.

In the hand
The combination of grey breast, a black throat that breaks up into scattered black spots, and a large white patch on the cheeks and ear-coverts identifies this species.

Adult female. Like the male, but the tarsus more yellowish, rarely tinged with red, and the bill brownish at the base (based on an immature specimen?).

Immature. Undescribed.

General biology and ecology

This species occurs from the lower foothills to about 1500 m, but is more common below 750 m, where it usually is found in the undergrowth of less dense evergreen forests than are used by *torqueola* or *rufogularis*, as well as in bamboo jungles, scrub-covered hills, and even grasslands or cotton cultivations so long as they are near forest or jungle cover. Their foods are the same as for others of their genus (Baker 1930).

Social behaviour

Coveys that probably consist of a pair and their last brood are typical, and usually number about five to six birds. The male's call during the breeding season is a loud and clear double whistle, and additionally a series of rolling whistles that are repeated many times and wind up with a sharper note may be heard, especially toward dusk (Ali and Ripley 1978).

Reproductive biology

The breeding season is from March to July, being earliest at lower levels and latest at higher ones. The nest is often quite simple when nesting in scrub, but when laying in tall grass of bracken habitats it is often more complex and domed-over, with an entrance tunnel that may be a couple of feet in length. The clutch-size is usually of four to five eggs, but rarely as many as seven have been found. The incubation period is unknown (Baker 1930; Ali and Ripley 1978).

Evolutionary relationships

Davison (1982) considered this to be part of a species group that also included *ardens, orientalis, crudigularis,* and *rufipectus*. Cheng *et al.* (1978) regarded it as a rather isolated offshoot from early stock that gave rise to *rufipectus* and *torqueola* and perhaps is related to *ardens* and *crudigularis*. I believe its closest relatives are *ardens* and *crudigularis*, especially the former, with which it might even be considered conspecific (Moroika 1957).

Status and conservation outlook

In China this is considered a rare species of little economic significance, and is apparently only locally common in India, such as in Assam. Probably it is less sensitive than most *Arborophila* to deforestation, as it seems to be quite tolerant of scrub forest and cultivated forest edge habitats.

WHITE-THROATED HILL-PARTRIDGE (Plate 111)

Arborophila crudigularis (Swinhoe) 1864
Other vernacular names: Formosa hill-partridge; Taiwan hill-partridge; torqueole de Formose (French); Taiwan-Hugelwachtel; Formosabuschwachtel (German).

Distribution of species (see Map 37)
Endemic resident of montane forests of Taiwan. No subspecies recognized.

Measurements

Wing, males 141–143 mm, females 123–134 mm; tail, males 58–62 mm, females 47–51 mm (various sources). Weight, males 311 g (Museum of Vertebrate Zoology specimen), estimated female weight, 212 g. Egg, average 40.6 × 29.5 mm (Schönwetter 1967), estimated weight 19.5 g.

Identification (see Fig. 21)

In the field (11 in.)
Associated with mountain forests of Taiwan, where it occurs in small flocks but is little studied. The male is said to utter a soft whistle. It is the only *Arborophila* native to that island, and has a distinctive white throat, with a black spotting or banding, and a grey chest and breast.

In the hand
Separated from other *Arborophila* by the combination of a grey breast and a white throat variably bisected by and sometimes terminated with black.

Adult female. Similar to the male, but with more white spotting on the underparts, and fewer black spots on the throat, making the throat-band narrower or incomplete.

Immature. Undescribed.

General biology and ecology

This species inhabits dense forest from 1000 to 3000 m, in the same general area used by the Swinhoe's pheasant (*Lophura swinhoei*). It reportedly feeds on earthworms, seeds, berries, seedlings, leaves, and insects (Cheng *et al.* 1978).

Social behaviour

Other than during the breeding season, these birds are found in small groups of two to three birds that forage on the forest floor during the day and roost in tree branches at night. The male is reported to utter a soft whistle in early summer (Cheng *et al.* 1978).

Reproductive biology

The breeding season extends from March (or possibly as early as the end of February) to August. The nest may be placed in a crevice among boulders, or hidden at the base of a tree. The clutch varies from six to eight eggs, and the reported incubation period is 24 days (Cheng *et al.* 1978). Robbins (1984) indicated a clutch of four to six eggs and a probable incubation period of 20–21 days, presumably from captive birds.

Evolutionary relationships

Cheng *et al.* (1978) indicated a close relationship between *ardens* and *crudigularis*, and Davison (1982) listed these two species plus *rufipectus, atrogularis,* and *orientalis* as part of an interrelated species group. Moroika (1957) believed that possibly this species should be considered conspecific with *atrogularis*, and I agree that like *ardens, crudigularis* can probably be considered an insular derivative of early *atrogularis* stock.

Status and conservation outlook

This species is currently protected on Taiwan, and efforts to propagate and release the birds have begun (Cheng *et al.* 1978).

WHITE-EARED HILL-PARTRIDGE (Plate 112)

Arborophila ardens (Styan) 1892
Other vernacular names: Hainan hill-partridge; torqueole scintillante (French); Hainan-Buschwachtel (German).

Distribution of species (see Map 37)
Endemic in forests of Hainan Island. No subspecies recognized.

Measurements

Wing, males 122–144 mm, females 112–120 mm; tail, males 41.5–55 mm, females 36 mm (Moroika 1957; Cheng *et al.* 1978). Weight, males 300 g (Davison 1985), females 237 g (Cheng *et al.* 1978). Egg, no information.

Identification (see Fig. 21)

In the field (11 in.)
Limited to the woodlands of Hainan Island, where it is the only *Arborophila* species present. It is distinctive in having a white ear-patch and white supercilium contrasting with an otherwise black face.

In the hand
Identified by the combination of a grey chest, a white ear-patch surrounded by black, and a patch of scarlet to orange hair-like feathers on the chest (which may be faded on museum specimens).

Adult female. Like the male but the upper breast has a light orange rather than scarlet colour, and the abdomen is white, with a slight amount of red.

Immature. On a young (male?) specimen the upperparts are similar to those of the adult male, but the tail is brown, the breast white, and the chest grey (Cheng *et al.* 1978).

General biology and ecology

This little-known species is rarely seen but is said to have a range and habitat similar to that of the silver pheasant (*Lophura nycthemera*), and to feed on tree seeds and snails (Cheng *et al.* 1978).

Social behaviour

Cheng *et al.* (1978) stated the the birds occur as solitary individuals, pairs, and coveys of three to five birds during winter. Their vocalizations are unreported.

Reproductive biology

No information.

Evolutionary relationships

Cheng *et al.* (1978) indicate diagrammatically that *ardens* and *crudigularis* are a closely related species pair, perhaps less closely related to *atrogularis*. I believe that the mainland form *atrogularis* is a close relative of both of these insular forms, and probably approximates the common ancestral type that independently produced each of these two isolates. Moroika (1957) suggested that *ardens* possibly should be considered conspecific with *atrogularis*.

Status and conservation outlook

Cheng *et al.* (1978) stated that this species is rarely seen and highly valued, and hoped that appropriate governmental departments might pay attention to its conservation.

JAVAN HILL-PARTRIDGE (Plate 113)

Arborophila javanica (Gmelin) 1789
Other vernacular names: brown-breasted tree-partridge; chestnut-bellied hill-partridge; torqueole de Java (French); Java Hugelhuhn; Javabuschwachtel (German).

Distribution of species (see Map 38)
Endemic on central and western Java. This form is sometimes considered to be conspecific with *brunneopectus* and/or *orientalis*.

Distribution of subspecies
A. j. javanica: mountains of western Java.
   A. j. bartelsi: mountains of central Java.
   A. j. lawuana: mountains of central Java.

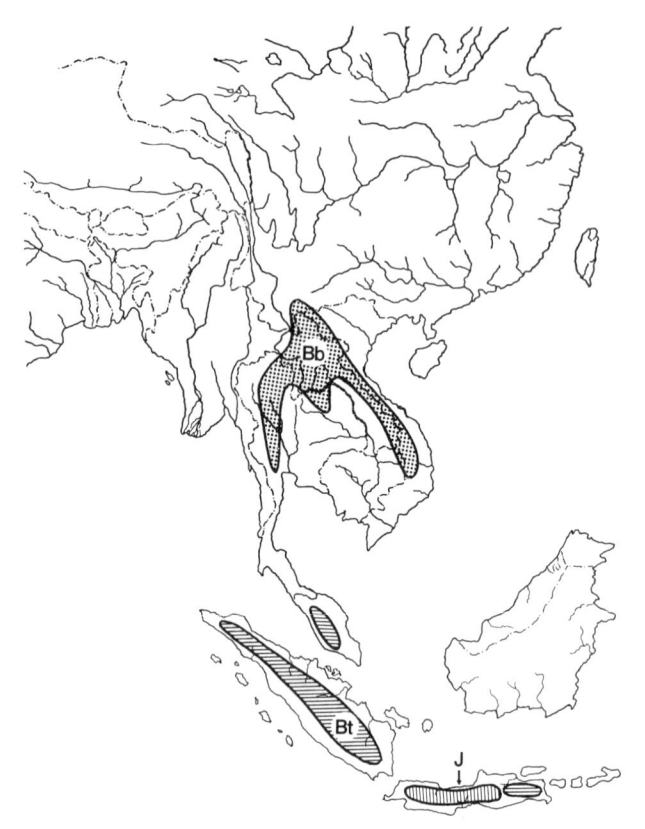

Map 38. Distribution of the bare-throated (Bt), brown-breasted (Bb), and Javan (J) hill-partridges.

Measurements

Wing, male 142 m, female 137 mm; tail, male 56 mm, females 51 mm (Ogilvie-Grant 1893). Estimated weight, males 286 g, females 257 g. Egg, average 40.7 × 31.6 mm (Schönwetter 1967), estimated weight 22.4 g.

Identification (see Fig. 21)

In the field (11 in.)
Endemic to the mountain forests of central and western Java, where it replaces *orientalis* of the eastern parts of that island, and is the only *Arborophila* species. Nothing is known of its ecology, but its calls consist of several monosyllabic spaced whistles, followed by a series of whistled couplets that rise in volume (Robinson 1982).

In the hand
Separated from other *Arborophila* by the combination of a grey breast, uniformly chestnut-coloured flanks, and rufous on the cheeks and throat.

Adult female. Apparently identical to the male.

Immature. Juveniles have a few small white shaft-spots on the scapulars, which are reddish brown, and the chin and throat are white, with no rusty or black tones present. The bill is reddish brown (in dried skins), and the breast and chest are dark brown.

General biology and ecology

In Java this species occurs in montane forests above 3500 ft (1100 m), but nothing specific is known of its ecology.

Social behaviour

No specific information.

Reproductive biology

The breeding season is apparently quite extended, with records for January through to April, and from July through to November. Clutches probably consist of up to four eggs (Hellebrekers and Hoogerwerf 1967).

Evolutionary history

Davison (1982) regards this species and *orientalis* as only convergently similar, and believes that *brunneopectus* is likewise not a close relative, in spite of the obvious zoogeographic problems that such proposed relationships pose. I am of the belief that these three species are too similar in their plumages and zoogeographic relationships to be the result of convergent evolution.

Status and conservation outlook

No information on the status of this species is available.

BARE-THROATED HILL-PARTRIDGE
(Plate 114)

Arborophila orientalis (Horsefield) 1821
Other vernacular names: Campbell's hill-partridge (*campbelli*); Sumatran hill-partridge (*sumatrana*); torqueole de Sumatra (French); Sumatrabuschwachtel (German); san serok gunong (Malaysia)

Distribution of species (see Map 38)
Resident in montane forests of Malaysia, Sumatra, and Java. Sometimes considered part of *brunneopectus*.

Distribution of subspecies
A. o. campbelli: mountains of Malay Peninsula.
   A. o. rolli: mountains of north-western Sumatra.

A. o. sumatrana: mountains of central Sumatra.
A. o. orientalis: mountains of eastern Java.

Measurements

Wing (*campbelli*), both sexes 138–147 mm; tail, both sexes 48–53 mm (Robinson and Chasen 1936). Estimated weight, males 274 g, females 263 g. Egg (*campbelli*), 41 × 31.7 mm (Robinson and Chasen 1936); estimated weight 21.4 g.

Identification (see Fig. 22)

In the field (11 in.)

Associated with montane woods and forests, including dense jungle and swampy forests or river valleys. Vocalizations include a series of monosyllabic spaced whistles. Probably best identified visually by the white ear-coverts.

In the hand

Identified by the combination of a grey to brownish

Fig. 22. Adult plumage variation in: the brown-breasted hill-partridge, subspecies (A) *brunneopectus*, (B) *albigula*, and (C) *henrici*; the bare-throated hill-partridge, subspecies (D) *campbelli*, (E) *rolli*, and (F) *sumatrana*.

grey breast, black and whitish spotted flanks, and a mostly black head with white ear-coverts. There usually is a white supraloral stripe (absent in *sumatrana*), but not such a long superciliary stripe as in *atrogularis*. (The amount of white on the head varies markedly individually and perhaps with age.) The chin and upper throat are white (in *sumatrana* and *orientalis*) or black (*campbelli* and *rolli*), the feathers of the lower throat arranged in vertical rows and sometimes showing bare neck skin between them. The flanks are greyish black to olive-brown, with white (*orientalis*, *rolli*, and *sumatrana*) to rufous buff (*campbelli*) spots and black bands. Similar to some races of *brunneopectus*, which also have exposed throat skin, but which do not have their whitish ear-coverts so completely enclosed by darker feathers.

Adult female. Similar to the male, but with the black of the head less intense, somewhat tinged with olive.

Immature. Resembles the adult, with with pale shaft-stripes, more rufous below, and the breast slightly barred.

General biology and ecology

Very little known, this species occurs between about 3500 and 5000 ft (1100–1500 m) in Malaysia, and has also been reported at about 3000 ft (900 m) in Sumatra. In Malaysia it occupies gullies and river valleys having overgrown rotan palms, which it largely feeds upon, and damp, swampy flatlands where a creeping plant *Pratia* produces small red berries that it eats. It also eats invertebrates, especially termites and snails (Robinson and Chasen 1936).

Social behaviour

The birds occur in pairs or groups of up to about five birds. They apparently utter a low whistle that is softer and more melodious than that of the long-billed partridge, often at dusk (Robinson and Chasen 1936). Davison (1982) stated that the male calls of this and several related species are a series of whistled couplets that increase in volume or pitch, and are preceded by several monosyllabic spaced whistles.

Reproductive biology

Nesting in Selangor, Malaysia, has been reported to occur in March. A nest with two eggs was found below small stemless palms. In all likelihood the normal clutch is somewhat larger than this, and an undated and unlocalized clutch of four eggs has also been reported (Medway and Wells 1976).

Evolutionary relationships

Davison (1982) concluded that *orientalis* is a full species that is not closely related to the similar *javanica*, and that their outward similarities may be the result of convergent evolution. He also placed it in a different species group from *brunneopectus*, with which it (or at least the race *campbelli*) has at times been considered conspecific.

Status and conservation outlook

No specific information on the status of this species is available.

BROWN-BREASTED HILL-PARTRIDGE (Plate 115)

Arborophila brunneopectus (Blyth) 1855
Other vernacular names: bar-backed hill-partridge; torqueole à pointrine brune (French); Rotbauch-Buschwachtel (German).

Distribution of species (see Map 38)
Resident of tropical forests from south-western Yunnan south through eastern and southern Burma, Thailand, Laos, and Vietnam. Sometimes included in *javanica*, and often expanded to include part of *orientalis*.

Distribution of subspecies
A. b. brunneopectus: south-western Yunnan south to southern Thailand.
   A. b. henrici: northern and central Vietnam (Tonkin, Annam); probably also in Guangxi.
   A. b. albigula: southern Vietnam (southern Annam).

Measurements

Wing (*albigula* and *henrici*), males 135–159 mm, females 125–153 mm; tail, males 55–72 mm, females 50–68 mm (Delacour and Jabouille 1931). Estimated weight, males 317 g, females 268 g. Egg, average 37.2 × 28.4 mm (Schönwetter 1967), estimated weight 16.6 g.

Identification

In the field (11 in.)
Associated with evergreen forests at moderate elevations. Its calls are said to include a three-noted whistle that descends in scale, preceded by spaced

monosyllabic whistles, during which the bare reddish throat skin is distended. The conspicuous white chin and throat, and the white spotting on the sides of the flanks, probably are the best field marks.

In the hand

The combination of strong transverse barring on the back, a greyish brown breast, white-spotted flank feathers with black edges, and a sparsely feathered whitish or buffy throat that is spotted with black serves to identify this species of *Arborophila*. The chin, throat, cheeks, and ear-coverts vary from pale buff (*brunneopectus*) to white (*albigularis* and *henrici*), and the sides and flank feathers are black and brownish buff, with subterminal black-bordered white spots, or mostly white with black spots (*henrici*).

Adult female. Similar to the male, but probably with less colourful soft-parts.

Immature. With pink to orange tarsi and toes; other differences undescribed.

General biology and ecology

This species is associated with evergreen forests, and occurs in northern Thailand at lower levels than does *rufogularis* (Deignan 1945), or up to 4500 ft (1350 m), but to about 5000 ft (1500 m) in Burma, and about 6000 ft (1800 m) in Malaysia. Similarly, it apparently does not overlap with *chloropus* in the Pegu Hills, but instead the two species seem to occupy mutually exclusive areas in similar habitats (Baker 1928).

Social behaviour

Apparently this species is usually found in small groups of from four to nine birds, probably representing one or two family groups. Their call is said to be a musical two- or three-noted whistle, repeated on a descending scale. This sequence is preceded by a series of spaced monosyllabic whistles (Davison 1982).

Reproductive biology

The only information on this subject is that four eggs, presumably but not definitely of this species, were collected in open bamboo jungle during June (Baker 1928; Etchécopar and Hüe 1978).

Evolutionary relationships

As noted earlier, this species has received highly variable taxonomic treatment as to its species limits, and has been taxonomically associated with both *javanica* and *orientalis*. I have followed Davison's (1982) recommendation that *brunneopectus* be separated from both of these, and instead be considered part of his 'brown-breasted' group, which includes the four following species. Actually, I think that *orientalis* may serve to connect these two groups, and thus might in fact be fairly closely related to *brunneopectus*.

Status and conservation outlook

No specific information available.

ORANGE-NECKED HILL-PARTRIDGE
(Plate 106)

Arborophila davidi Delacour 1927
Other vernacular names: David's hill-partridge; torqueole de David (French); Davidbuschwachtel (German).

Distribution of species (see Map 39)

Endemic to forests of southern Vietnam. No subspecies recognized.

Map 39. Distribution of the Borneo (B) chestnut-headed (C), orange-necked (O), and red-billed (R) hill partridges.

Measurements

Wing (sex?) 134 mm; tail, 49 mm (Delacour and Jabouille 1931). Estimated weight, 241 g. Egg, no information.

Identification

In the field (11 in.)

A little-known species found in the lower hills of the Annamitic Chain at elevations of less than 300 m, and characterized by strong black and white flank barring and a broad black facial stripe extending from the lores through the eyes and down the sides of the neck, joining on the lower foreneck to enclose a mostly tawny orange throat. Vocalizations are undescribed.

In the hand

Identified by the combination of a barred back, brown chest, barred black and white flanks, and a tawny orange tint on the sides and front of the neck, where it is mostly bounded by black.

Female and immature plumages. Undescribed.

General biology and ecology

No information.

Social behaviour

No information.

Reproductive biology

No information.

Evolutionary relationships

Davison (1982) includes this as part of his 'brown-breasted' species group, but offers no opinions on its nearest relative.

Status and conservation outlook

Collected only once, in 1927, and evidently not seen since. This species is presumably endangered if not already extinct.

CHESTNUT-HEADED HILL-PARTRIDGE
(Plate 106)

Arborophila cambodiana Delacour and Jabouille 1928

Other vernacular names: Cambodian hill-partridge; rufous-faced hill-partridge; torqueole du Cambodge (French); Kambodschabuschwachtel (German).

Distribution of species (see Map 39)

Endemic to tropical forests of Cambodia (Kampuchea) and south-eastern Thailand.

Distribution of subspecies

A. c. cambodiana: south-western Cambodia.
A. c. diversa: south-eastern Thailand.

Measurements

Wing, males 144–150 mm, females 136–138 mm; tail, males 59–65 mm, females 50–63 mm (Delacour and Jabouille 1931). Estimated weight, males 318 g, females 257 g. Egg, no information.

Identification

In the field (11.5 in.)

Associated with forests of Thailand and Cambodia between about 600 and 1500 m. Calls of the species are still undescribed, but the birds are extensively hazel (in *cambodiana*) to russet (in *diversa*) throughout the head, neck, chest, and breast; on the abdomen they become paler, and white flank spotting provides the only distinctive markings.

In the hand

The combination of a strongly barred back, white-spotted flanks, and an almost uniformly russet to chestnut colour on the head, neck, and underparts provides for identification of this species of *Arborophila*.

Adult female. Similar to the male, but smaller. At least in *diversa* the females vary considerably in patterning, with some females having heavier barring on the upperparts than do males (Riley 1938).

Immature. A probable immature specimen (US National Museum) had less spotting below and the lower mandible became pale toward the tip.

General biology and ecology

No specific information.

Social behaviour

No information.

Reproductive biology

No information.

Evolutionary relationships

Davison (1982) included this form in his 'brown-breasted' species group, with no suggested nearest relatives.

Status and conservation outlook

According to Lekagula and Cronin (1974), this species is a locally common if highly geographically restricted resident of south-eastern Thailand, occurring in the Kao Sabap mountains. Otherwise it is known only from the area around Bokor, southern Cambodia.

BORNEO HILL-PARTRIDGE (Plate 116)

Arborophila hyperythra (Sharpe) 1890
Other vernacular names: red-breasted tree-partridge; torqueole de Borneo (French); Borneobuschwachtel (German).

Distribution of species (see Map 39)
Endemic on north-western Borneo highlands. Sometimes considered a race of *orientalis*.

Distribution of subspecies
A. h. *hyperythra*: Sarawak to north Borneo border.
A. h. *erythrophrys*: endemic on Mt Kinabalu.

Measurements

Wing, both sexes 132–147 mm; tail, both sexes 45–56 mm. Estimated weight, both sexes 270 g. Egg, no information.

Identification

In the field (10 in.)
Limited to the dense montane forests of northern Borneo, where the only other *Arborophila* is the scaly-breasted hill-partridge. If observed, the large white flank spots and the rufous throat and cinnamon-coloured breast should serve to identify the species. Its calls are similar to those of other *Arborophila*.

In the hand
The combination of a heavily barred back, flanks that are spotted with black and white, and a mostly cinnamon head, neck, and chest colour, with the head and neck distinctly paler than the chest, should serve to identify this *Arborophila* species.

Adult female. Apparently similar to the male, but lacking the black chin and throat in *erythrophrys*, and probably with paler soft-part colours.

Immature. Undescribed.

General biology and ecology

Associated with a variety of habitats, but especially river flats in old or secondary forests, and rarely seen in trees except during roosting. It eats acorns, seeds, fruits, and various insects (Smythies 1968).

Social behaviour

No specific information. The species' full call is a series of whistled couplets that ascend in pitch (Davison 1982), but it also utters a single ringing call repeated three times per second, that is answered by a double-noted call that drops in pitch and is uttered about once per second (Smythies 1968).

Reproductive biology

No information.

Evolutionary history

At times this form has been considered a subspecies of a complex including *orientalis* and/or *brunneopectus*, but I have followed Davison (1982) in considering it a full species within his 'brown-breasted' species group, with *cambodiana* its probable nearest relative.

Status and conservation outlook

No information is available on the status of this species.

RED-BILLED HILL-PARTRIDGE (Plate 117)

Arborophila rubrirostris (Salvadori) 1879
Other vernacular names: torqueole à bec rouge (French); Rotschnabel-Buschwachtel (German).

Distribution of species (see Map 39)
Endemic to montane forests of Sumatra. No subspecies recognized.

Measurements

Wing, males 132–137 mm, females 124–132 mm; tail, males 43 mm, females 41 mm (various sources). Estimated weight, males 243 g, females 209 g. Egg, no information.

Identification

In the field (11 in.)
Limited to the forests of Sumatra, where it is the only red-billed partridge. Its calls include a series of whistled couplets that ascend in pitch (Robinson and Kloss 1924).

In the hand
Easily separated from the other hill-partridges by its coral-red bill.

Adult female. Like the male, but with more white on the lores and chin.

Immature. Undescribed.

General biology and ecology

No information.

Social behaviour

No information.

Reproductive biology

No information.

Evolutionary relationships

This species is considered by Davison (1982) a part of the 'brown-breasted' species group. No other speculations on its relationships seem to have been made.

Status and conservation outlook

No information.

SCALY-BREASTED HILL-PARTRIDGE
(Plate 118)

Arborophila charltonii (Eyton) 1845
Other vernacular names: Annamese hill-partridge (*merlini* group); chestnut-breasted hill-partridge; Eyton's hill-partridge (*charltonii* group); green-legged hill-partridge (*chloropus* group); torqueole à poitrine chataine (*charltonii* group); torqueole des bois (*chloropus* group); torqueole de Merlin (*merlini* group) (French); Grünfuss-Buschwachtel (*chloropus* group); Gelbfuss-Buschwachtel (*merlini* group) (German).

Distribution of species (see Map 40)
Resident of forests from southern Yunnan and Burma south through Malaysia and Indochina to northern Sumatra and northern Borneo. Up to three species are sometimes recognized in this group.

Map 40. Distribution of the scaly-breasted hill partridge (broken line) and ferruginous (dotted line) wood-partridge.

Distribution of subspecies
A. c. chloropus: Yunnan, Burma south to northern Tenasserim, and western Thailand.
   A. c. peninsularis: south-western Thailand.
   A. c. olivacea: Laos and Cambodia (Kampuchea).
   A. c. cognacqi: southern Vietnam. (These first four races are often considered specifically distinct, as *A. chloropus*, or green-legged hill-partridge.)
   A. c. merlini: central interior Vietnam (vicinity of Quantri).
   A. c. vivida: central coastal Vietnam (vicinity of Hue). (These last two races are often considered specifically distinct, as *A. merlini*, or Annamese hill-partridge.)
   A. c. tonkinensis: northern Vietnam (Tonkin and northern Annam).
   A. c. charltonii: southern Tenasserim, southern Thailand, and Malaysian Peninsula.
   A. c. atjehensis: northern Sumatra (Atjeh).
   A. c. graydoni: northern Borneo (Sabah). (These last four races are sometimes considered specifically distinct.)

Measurements

Wing (*chloropus*), males 140–146 mm, females 146 mm; tail, males 70–84 mm, females no data. Weight, males 280–300 g, females 250 g (Cheng *et al.* 1978). Wing (*merlini* and *vivida*), males 150–162 mm, females 141–150 mm; tail, males 70–80 mm, females 70–75 mm. Wing (*cognacqi*, *olivacea*, and *tonkinensis*), males 145–168 mm, females 140–166 mm; tail, males 73–90 mm, females 75–78 mm (Delacour and Jabouille 1931). Wing (*charltonii*), males 160 mm, females 155 mm (Davison 1985). Weight (*chloropus*), males 290 g, females 250 g (Davison 1985). Egg (*charltonii*), average 39.7 × 30.1 mm (Schönwetter 1967), estimated weight 19.8 g.

Identification (see Fig. 23)

In the field (12 in.)

Associated with forests, woody second growth, and edges of cultivated lands, from lowlands up to about 1400 m. The yellowish to greenish legs are distinctive, as are the absence of white flank markings and generally obscured head and breast patterning. The

Fig. 23. Adult plumage variation in the scaly-breasted hill-partridge, subspecies (A) *chloropus*, (B) *vivida*, and (C) *charltonii*.

usual call is a low and soft double whistle; the full call is preceded by a series of monosyllabic whistles, followed by an increasingly wild crescendo of double or treble notes; in *charltonii* these latter notes are of steady pitch and volume, and sometimes appear to be triplets, perhaps because of duetting (Davison 1982).

In the hand
Identified by the combination of greenish (in the *chloropus* group) to yellowish (in the *charltonii* group) legs, a vermiculated or barred chest, and a hidden patch of downy white feathers behind the axilla. The lower foreneck and upper breast are separated by a bright unmarked rufous chest-band in the *charltonii* group.

Adult female. Like the male, but probably with less bright soft-part colours.

Immature. With white shafts and tips on the breast and side feathers, dark barring on the upperparts, and less colourful soft-parts (Diegnan 1945).

General biology and ecology

This is a generally lowland species, preferentially occupying moist jungles and thick evergreen forests, both in flat areas and sometimes in steep valleys or ravines, and more rarely found in drier scrub jungle habitats. It reportedly feeds on seeds, berries, and termites.

Social behaviour

These birds occur in small groups like other hill-partridges, and apparently communicately mostly by calls. Beyond the calls mentioned in the identification section above, the race *charltonii* is said to utter a series of monotone notes increasing in tempo, followed by a second series of gradually ascending notes, and finally one of wildly ascending and descending notes that end abruptly (Lekagula and Cronin 1974). This contradicts Davison's (1982) view that in the scaly-breasted forms the couplet notes ascend neither in pitch nor volume.

Reproductive biology

No specific information. Riley (1938) reported the finding of three eggs, obtained in May, and apparently of *olivacea*, but averaging somewhat smaller than those reported by Schönwetter (1967) for *charltonii* (and of unstated origin).

Evolutionary relationships

Davison (1982) considers this form to be a part of *Arborophila*, but comprised of two species, *charltonii* and *chloropus*. A third species, *merlini*, has at times been recognized within the group, but was included by Davison within *chloropus*. I see no compelling reason for accepting more than a single species, considering the variability of various other *Arborophila* forms and the complex geographic problems posed by separating the generally southern *charltonii* (with one race occurring in northern Vietnam) from the more northern *chloropus* (with one race occurring in southern Vietnam). Unless these forms can be proven partially sympatric, it seems unwise to recognize more than one species in my view.

Status and conservation outlook

No specific information. In Thailand this species is considered a common and widespread resident (Lekagula and Cronin 1974).

27 · Genus *Caloperdix* Blyth 1861

The ferruginous wood-partridge is a medium-sized, tropical forest-adapted species of the Malaysian Peninsula and the Greater Sundas having a tail of 14 rectrices and less than half as long as the wing. The wing is rounded, with the fifth to the seventh primaries the longest. The sexes are alike, uncrested, and mostly chestnut-brown. There are one to two tarsal spurs in males and sometimes one in females. The claw of the hallux is rudimentary. A single polytypic species is recognized:

C. oculea (Temminck): ferruginous wood-partridge
- *C. o. oculea* (Temminck) 1815
- *C. o. sumatrana* Ogilvie-Grant 1892
- *C. o. borneensis* Ogilvie-Grant 1892

FERRUGINOUS WOOD-PARTRIDGE (Plate 119)

Caloperdix oculea (Temminck) 1815

Other vernacular names: ocellated partridge; rouloul ocelle (French); Augfenwachtel (German).

Distribution of species (see Map 40)

Resident of tropical forests of Thailand, southern Malaysia, Sumatra, and Borneo.

Distribution of subspecies

C. o. oculea: south-eastern Thailand and southern Malaysian Peninsula.
- *C. o. sumatrana*: Sumatra.
- *C. o. borneensis*: Borneo.

Measurements

Wing, males 141–144 mm, females 134–139 mm (Ogilvie-Grant 1893; Davison 1985); tail, male 63 mm, female 58 mm. Weight, male 190.5 g, females no data (Davison 1982). Egg, no mensural information.

Identification

In the field (10.5 in.)

Associated with lowland forests up to about 900 mm, and differing from all other partridges of the area in its bright rusty brown head, breast, and flank colour, with black spotting on the flanks. Similar to the long-billed partridge, but with a much shorter bill. Its call is a duet, in which the male utters a series of notes ascending the scale, repeated eight to nine times, followed by several *e'-terang* or *e'terang-e'* calls, and the female replies with a long series of up to 20 more whistled and faster notes that also ascend the scale.

In the hand

Identified by its generally rusty body coloration and the rudimentary claw on the hind toe. The male has one or rarely two tarsal spurs.

Adult female. Like the male, but with a single short tarsal spur.

Immature. Like the adults, but with the occipital area having crescentic bars of black, and the chest area sparsely and irregularly spotted or barred with black.

General biology and ecology

This species occurs in low-country jungles, including bamboo and secondary forests as well as swampy jungles, tall forests on hillsides, and dry sandy forests of valley bottoms. It occurs in groups of a few birds that forage on the forest floor, sometimes on fallen wild figs. It also sometimes eats berries, grasses, seeds, and insects (Smythies 1968).

Social behaviour

Apparently similar to *Arborophila* in its social life and vocalization, and with very similar calls that begin with monosyllables and then break into couplets (Davison 1982).

Reproductive biology

One nest was apparently found in Malaysia at the end of May. It resembled an *Arborophila* nest, and was hidden under a scrubby bush. On the other hand, Smythies (1968) stated that the nest is domed, with a lateral entry, and with eight to ten eggs being laid in December and January. The incubation period, presumably based on captive birds, was reported by Robbins (1984) as 18–20 days.

Evolutionary relationships

I believe that this is a very close relative of *Arborophila* and might well be included in that genus. Its fairly long bill and spurred tarsi are suggestive of *Francolinus* affinities, although the phyletic value of these traits is questionable. Davison (1982) considered the similarities of *Caloperdix* and some of the *Arborophila* species to be the result of convergent evolution.

Status and conservation outlook

No specific information.

28 · Genus *Haematortyx* Sharpe 1879

The crimson-headed wood-partridge is a medium-sized, tropical forest-adapted species endemic to Borneo, having a rounded tail of 12 rectrices and less than half as long as the wing. The wing is rounded, with the sixth primary the longest. The sexes are dimorphic, but both are distinctly crimson-rufous on the head and neck. Males have from one to three tarsal spurs; females have none. The bill is relatively weak, and the nail on the hallux is unusually small. A single monotypic species is recognized:

H. sanguiniceps Sharpe 1879: crimson-headed wood-partridge

CRIMSON-HEADED WOOD-PARTRIDGE
(Plate 120)

Haematortyx sanguiniceps Sharpe 1879
Other vernacular names: rouloul sanglant (French); Rotkopfwachtel (German).

Distribution of species (see Map 41)
Endemic to montane forests of northern Borneo. No subspecies recognized.

Map 41. Distribution of the crested (broked line) and crimson-headed (stippled) wood-partridges.

Measurements

Wing, both sexes 149–168 mm; tail, both sexes 40–50 mm (various sources). Weight, male 300 g, female no data (Davison 1985). Egg, 44? mm (Smythies 1968) (only one measurement available).

Identification

In the field (10 in.)
Confined to the mountain forests of Borneo, including both moist and fairly dry variants. The brilliant crimson head colour is distinctive. The calls include harsh, repeated clucking notes, and a *pom-prang* that is repeated several times in a weak, metallic voice.

In the hand
The unique crimson head colour easily identifies this species. The nail of the hallux is very small, and males have from one to three tarsal spurs.

Adult female. Similar to the male, but the chin and throat pale rufous, washed with crimson, the fore-neck and chest reddish chestnut; the longer under tail-coverts widely tipped with crimson. Bill greenish horn, becoming darker above the nostrils, rather than ivory, and the tarsus unspurred.

Immature. Similar to adults, but with duller and less extensive red feathering. Most of the wing-coverts have a rust-coloured spot at the tip of the shaft, and some of the feathers of the middle of the breast are mottled with reddish chestnut.

General biology and ecology

This species is most common in montane moss forests, but unlike other partridges it also occurs on leached-out sandy forests of the valley bottoms and in primary forests. It apparently eats berries and insects (Smythies 1968).

Social behaviour

Little specific information, but the birds utter repeated, harsh clucking notes, and double-noted metallic calls. They evidently lack the complex whistled sequences typical of most or all *Arborophila*.

Reproductive biology

A nest of dry leaves on a tussock of grasses in lichens is placed in *kerangus* forests. The earliest reported egg date is 12 January (Smythies 1968). The incubation period has been reported as 18–19 days, presumably from captivity (Robbins 1984).

Evolutionary relationships

This unusual species is seemingly quite distinctive, but probably was derived from early forest-adapted partridges not very different from present-day *Arborophila* types.

Status and conservation outlook

No specific information, but generally considered fairly common if of local distribution in Borneo.

29 · Genus *Rollulus* Bonnaterre 1791

The crested wood-partridge is a medium-sized, tropical forest-adapted species of the Malaysian Peninsula and the Greater Sundas, having a rounded, soft tail of 12 rectrices and about 40 per cent as long as the wing. The wing is rounded, with the sixth primary the longest. The sexes are dimorphic, the male mostly dark glossy bluish black with a dense maroon occipital crest, and the female dark olive-green. Both sexes have red eye-rings, are sparsely crested on the forehead, lack tarsal spurs, and the hind toe is clawless. A single monotypic species is recognized:

R. rouloul (Scopoli) 1786: crested wood-partridge

CRESTED WOOD-PARTRIDGE (Plate 121)

Rollulus rouloul (Scopoli) 1876

Other vernacular names: crowned wood-partridge; green wood-partridge; red-crowned wood-partridge; roulroul; rouloul couronne (French); Strausswachtel (German); burong siul (Malaysia).

Distribution of species (see Map 41)

Resident of tropical forests of southern Burma and Thailand (from 13°N latitude) south through Malaysia, Sumatra, and Borneo, including Banka and Belitung islands. No subspecies recognized.

Measurements

Wing, males 140–144 mm, female 132 mm; tail, male 63.5 mm, female 61 mm (Ogilvie-Grant 1893; Robinson and Chasen, 1936). Average weight, seven males 232 g, six females 202 g (David Rimlinger, personal communication). Egg, 39 × 32 mm (Smythies 1968), estimated weight 22.0 g.

Identification

In the field (10 in.)

Associated with lowland forests, up to about 1200 m. The dark glossy, almost black appearance of the male, with a bushy crest, and the bright green colour of the female and its more diffuse crest, provides for easy recognition. The calls include a series of low, melancholy glissading whistles as well as a shrill plaintive whistle.

In the hand

Easily recognized by the absence of a claw on the hind toe and the unusual crest on the forehead of both sexes.

Adult female. Entire head blackish grey, with the occipital crest poorly developed. Neck, upperparts, chest, and breast bright grass-green, shading posteriorly into greyish green. Remiges and their coverts as in the male, but the scapulars and smaller wing-coverts mostly chestnut rather than maroon and only slightly glossed with purple.

Immature. Young females (and probably also males) have their wing-coverts broadly tipped with buffy white, and the remiges are more mottled than in adults; young males resemble females, but the abdomen is partly greyish green.

General biology and ecology

This species is broadly distributed, occurring in woodlands or forests almost everywhere except on mangroves and sandy littoral areas, but especially frequent in fairly open, dry jungle where stemless palms and bamboos are common, or in cleared areas of old forests. In Malaysia the species extends at least to 4000 ft (1200 m). The birds feed on seeds, fruits, insects, and small molluscs (Robinson and Chasen 1936), and perhaps eat a larger proportion of animal foods than do most partridges, at least in captivity.

Social behaviour

Groups of from four to twelve or more birds are typical, which communicate in low, mellow whistles. A typical call is a glissading whistle, *su-il*, uttered most commonly at dawn (Medway and Wells 1976).

Reproductive biology

On the Malaysian Peninsula nests or chicks have been found in most months, and the usual clutch appears to be of five to six eggs (Medway and Wells 1976). The nest is typically large and domed, with a small lateral entrance, although nests have also been found that were mere depressions in the substrate. In captivity the incubation period has been reported to be 17–18 days (Whitley 1927) and also as 22 days

(Muller 1969). One captive female was observed to lead her young back to the nest each evening and to close the entrance with twigs, until the young were about 25 days old. Muller noted that both sexes care equally for the young, and that by 2 weeks of age the birds could be sexed by differences in wing colour. Whitley stated that by that age the birds were able to perch on low branches, and by about 3 months they had attained their adult plumages.

Evolutionary relationships

This seems to be a quite distinctive genus, with no obvious near relatives.

Status and conservation outlook

The rather wide ecological range of habitats used by this species no doubt favours its survival, and so far it has not been mentioned as a threatened species.

30 · Genus *Ptilopachus* Swainson 1837

The stone partridge is a medium-sized, arid-adapted sub-Saharan African species, having a long and somewhat vaulted tail of 14 rectrices and about 75 per cent as long as the wing. The wing is rounded, with the sixth primary the longest. The sexes are alike; adults have bare red facial skin, mostly dull greyish brown plumage, and lack tarsal spurs. A single polytypic species is recognized:

P. petrosus (Gmelin): stone partridge
  P. p. petrosus (Gmelin) 1789
  P. p. saturatior Bannerman 1930
  P. p. brehmi Neumann 1908
  P. p. major Neumann 1908
  P. p. florentiae Ogilvie-Grant 1900

STONE PARTRIDGE (Plate 122)

Ptilopachus petrosus (Gmelin) 1789
Other vernacular names: poule de rocher (French); Felsenrebhuhn (German).

Distribution of species (see Map 42)
Resident of arid habitats of sub-Saharan Africa from Gambia to western Ethiopia, and south to Cameroon and northern Kenya.

Distribution of subspecies
P. p. petrosus: Gambia to Nigeria and Cameroon. Urban *et al.* (1986) include all the following races excepting *major* in *petrosus*.
  P. p. saturatior: north-central Cameroon.
  P. p. brehmi: Lake Chad east to the Sudan. Includes *butleri*.
  P. p. major: northern Ethiopia (Eritrea).
  P. p. florentiae: southern Sudan to north-eastern Zaire, northern Uganda, and northern Kenya. Includes *emini*.

Measurements

Wing, males 115–128 mm, females 116–126 mm (Urban *et al.* 1986); tail, males 76–91 mm, female 79 mm (various sources). Weight, two males average 190 g. Egg, average 33 × 25 mm (Mackworth-Praed and Grant 1952–73), estimated weight 11.4 g.

Identification

In the field (10 in.)
Distinctive in its uniformly brownish colour and its habit of cocking its longish tail in a bantam-like manner. Its call is a *ouit,ouit,ouit* that rises in pitch. Usually found on areas of rocky outcrops, but sometimes also on scrub-covered plains.

In the hand
The relatively long (75–91 mm) and somewhat vaulted tail, together with an unspurred tarsus in both sexes, is unique among African gallinaceous birds.

Adult female. Like the male, but the middle breast feathers paler buff (Bannerman 1963).

Immature. Similar to adults, but with distinct barring on innermost secondaries, back, rump, and tail (Mackworth-Praed and Grant 1952–73). Probably also with reduced facial skin.

General biology and ecology

Habitats used by this species consist mainly of densely vegetated areas among boulders associated with rocky hills between 600 and 1500 m, but also sometimes extend to flat-topped laterite hills, steep and wooded dry watercourses, broken and eroded woody country, and adjacent grassy and cultivated areas. In the sub-Saharan Sahel zone it even occurs in sandy substrates where low and shady spots exist. Probably few if any other galliform birds are present in most of these areas as possible competitors; it is allopatric with both *Alectoris* and *Ammoperdix*. The birds eat a variety of plant materials, such as grass and herb seeds, green leaves, fruits, and buds, with some insect materials as well. They are highly sedentary, a covey often being confined to a single rocky hill (Urban *et al.* 1986).

Social behaviour

These birds are usually to be found in pairs, small coveys of three to four, or occasionally as many as

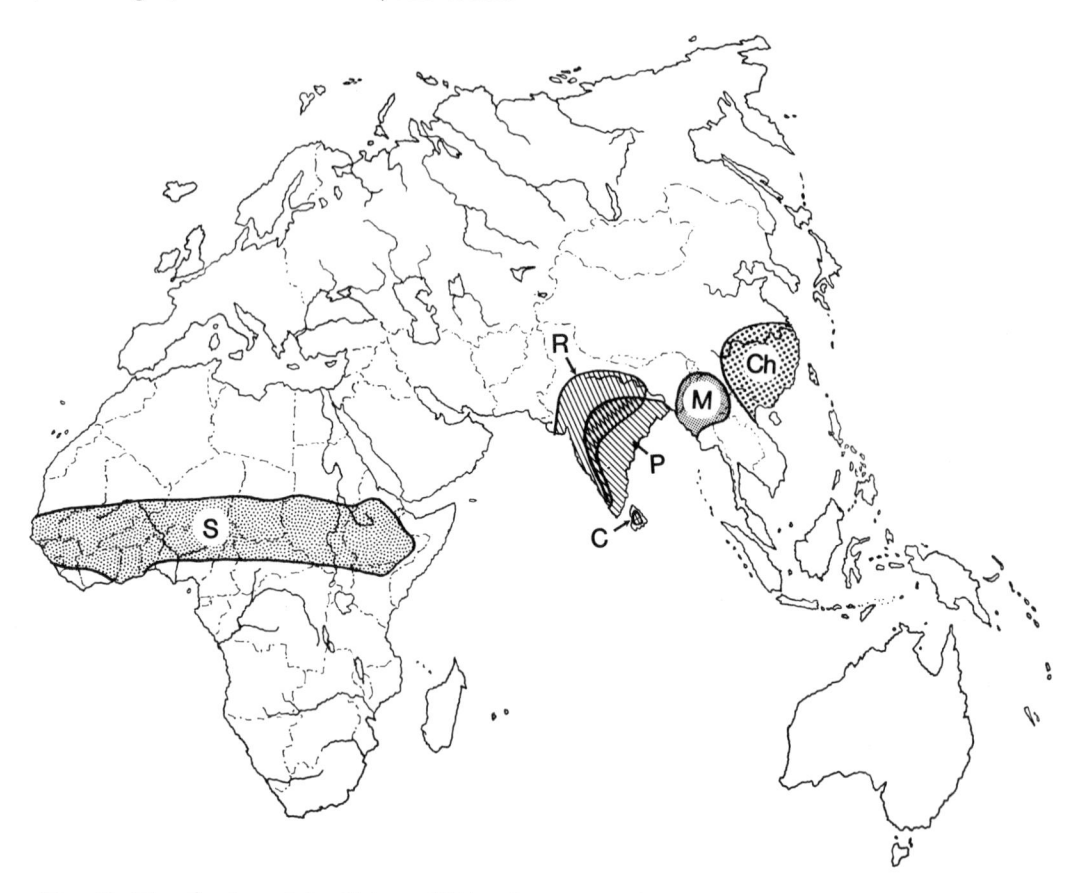

242 *The Quails, Partridges, and Francolins of the World*

Map 42. Distribution of the Chinese (Ch) and mountain (M) bamboo-partridges, the Ceylon (C), painted (P), and red (R) spurfowl, and the stone partridge (S).

15–20 in a covey, the latter groupings being typical of non-breeding periods. At least in captivity the birds are monogamous, and considering the probably erratic breeding cycles associated with infrequent moisture, it is likely that pair-bonds are maintained continuously. According to Ogilvie-Grant (1896–7), the males have been observed going through a series of dancing antics in a cleared space in the bush, with the females watching from near by, and Morris (1951) made a similar comment about a group of males observed strutting and dancing before the females, who watched from near-by cover. This would seem to be a questionable kind of social behaviour for an apparently monogamous species, and needs confirmation. More typical display would seem to be of adults chasing one another with tails cocked, and calling continuously in duets between pairs. Certainly the birds are highly vocal (and presumably territorial) during the breeding season, with one pair's calling often stimulating responses by others. Most calling occurs at dawn and dusk, but may be heard throughout the day during the rainy season. Additionally, males are known to courtship-feed their mates, and to display by hopping before the female with the tail fanned, the neck ruffled, and the wings fanned and trailing on the ground (Urban *et al.* 1986). The lack of tarsal spurs would suggest that aggression associated with social dominance and possible fighting behaviour are not highly developed in this species.

Reproductive biology

The breeding period of this species is highly variable, and in the drier portions of the range is associated with the wet season, while in the moister parts of its range breeding occurs during the dry season. In some areas (Central African Republic) breeding has been reported for all months. The nest is a scrape that is well concealed at the base of a rock, tree, or grass tuft, and the clutch-size is of four to six eggs. The incubation period is apparently still unreported. The families probably remain together until the start of the nest breeding season (Urban *et al.* 1986).

Evolutionary relationships

Relationships of this species pose some problems in zoogeography, for its seeming most likely nearest relatives (*Bambusicola* and *Galloperdix*) are both Asian in distribution. Nevertheless, these seem to be the most likely relatives of *Ptilopachus*; Ogilvie-Grant (1986) included these three genera within his subfamily Phasianinae.

Status and conservation outlook

Although locally distributed, this species is often common to abundant where it occurs, and there is no indication that it presents a conservation problem at the present time. The species has only rarely been maintained in captivity, but Clarke (1966) has described its aviculture.

31 · Genus *Bambusicola* Gould 1862

The bamboo-partridges are medium-sized, temperate woodland and bamboo forest-adapted species of eastern and south-east Asia with somewhat wedge-shaped tails of 14 rectrices that are nearly 70 per cent as long as the wing. The wing is rounded, with the sixth primary the longest. The sexes are very similar, with black or brown flank spots and chestnut-brown remiges and outer rectrices. Males have single tarsal spurs and females are usually unspurred. Two allopatric species are recognized:

B. fytchii Anderson: mountain bamboo-partridge
   B. f. fytchii Anderson 1871
   B. f. hopkinsoni Godwin-Austen 1874
B. thoracica (Temminck): Chinese bamboo-partridge
   B. t. thoracica (Temminck) 1815
   B. t. sonorivox Gould 1862

KEY TO THE SPECIES OF BAMBOO-PARTRIDGES (*BAMBUSICOLA*)

A. Throat grey, face reddish brown: Chinese bamboo-partridge (*thoracica*)
AA. Throat and face tawny buff, with a chestnut collar: mountain bamboo-partridge (*fytchii*)

MOUNTAIN BAMBOO-PARTRIDGE
(Plate 123)

Bambusicola fytchii Anderson 1871
Other vernacular names: Anderson's bamboo-partridge; Fytch's bamboo-partridge; bambusicole de Fytch (French); Gelbbrauen-Bambushuhn (German).

Distribution of species (see Map 42)
Resident of moist temperate forests of south-western Sichuan and Yunnan south to north-eastern India and Bangladesh, east to Burma and northern Vietnam.

Distribution of subspecies
B. f. fytchii: Sichuan and Yunnan south to the Kachin Hills, southern Shan State of Burma, and to northern Vietnam.
   B. f. hopkinsoni: India from Assam (Lakhimpur and hills south of the Brahmaputra River) south to the Chittagong area and the adjoining hill country of northern Burma (Chin Hills, northern Arakan Yomas).

Measurements

Wing (both races), males 140–149 mm, females 129–143 mm; tail, males 105–130 mm, females 90–115 mm. Weight (*fytchii*), males 278–360 g, females 256–300 g (Cheng *et al.* 1978). Egg (*hopkinsoni*), average 40.2 × 29.6 mm (Ali and Ripley 1978), estimated weight 19.4 g. Egg (one of *fytchii*), 38.6 × 28.6 mm (Cheng *et al.* 1978), estimated weight 17.4 g.

Identification

In the field (11–12 in.)
Associated with bamboo thickets, forests, and weedy thickets near cultivated lands, often near flowing water, and up to about 1900 m. It is relatively long-tailed, and in flight the chestnut outer tail feathers are visible. Otherwise, the black spotting on the flanks and the rusty to yellow throat colour assist in field identification. The call is a chattering, rather shrill series of *chirree* notes that are continued indefinitely.

In the hand
The combination of chestnut outer tail feathers, black flank spots and eye-stripe, and a rusty to yellowish throat and neck provide for identification.

Adult female. Like the male, but usually unspurred.

Immature. Undescribed in detail, but probably similar to the following species. A specimen attaining juvenal plumage had black-edged mantle feathers with wide buffy shaft-stripes, and paler underparts, also with buffy shaft-stripes.

General biology and ecology

This species is associated with open scrub jungle in foothills, mixed elephant grassland and willow, oak scrub along streams, bamboo groves, and relatively precipitous gorges having flowing water. It occurs over elevations of about 500–3000 m, but in spite of its vernacular name is not especially restricted to bamboo forests. It consumes a variety of plant materials (seeds, buds, shoots, grain, berries) and various insects (Ali and Ripley 1978; Cheng *et al.* 1978).

Social behaviour

The coveys of these birds usually number five to six birds, which dissolve in March to form breeding pairs. The advertising call of males during the breeding season consists of a *che-chirree-che-chiree, chirree, chiree, chiree* or repeated *pishup* notes, usually uttered at dawn (Ali and Ripley 1978).

Reproductive biology

Nesting in China reportedly usually occurs from April to July, but may start as early as March and extend to September. The season in India is mainly from March to May. The nest is placed in scrub, bamboo jungle, or grasslands, and the clutch is usually of four to five eggs, but ranges from three to seven. The incubation period is of 18–19 days, and is by the female alone (Ali and Ripley 1978; Cheng et al. 1978).

Evolutionary relationships

The genus *Bambusicola* seems to be somewhat intermediate between *Arborophila* and *Galloperdix* in general morphology, and appears to be closely related to both.

Status and conservation outlook

In China this species is of limited geographic distribution and small economic significance (Cheng et al. 1978). In India it has become a rather rare bird in recent years, apparently owing to excessive hunting pressures and habitat losses.

CHINESE BAMBOO-PARTRIDGE (Plate 124)

Bambusicola thoracica (Temminck) 1815
Other vernacular names: chu-chi (Chinese); bambusicole à plastron (French); Graubrauen-Bambushuhn (German); kujukei (Japanese).

Distribution of species (see Map 42)
Resident of temperate hill forests from Zhejiang north-west to Sichuan, and south to Guangdong and Guizhou; also on Taiwan. Introduced in southern Japan (Honshu southward) and locally on the Hawaiian islands (Maui).

Distribution of subspecies
B. t. thoracica: range as indicated above excepting Taiwan.
 B. t. sonorivox: endemic on Taiwan.

Measurements

Wing, males 124–140 mm, females 125–134 mm; tail, males 91–106 mm, females 90–97 mm. Weight, males 242–297 g, females 200–342 g. Egg, average 31.6 × 26.1 mm (Cheng et al. 1978), estimated weight 11.9 g.

Identification

In the field (12–13 in.)
Associated with shrubby vegetation, grassy openings, and bamboo groves on hilly habitats up to about 2000 m. Calls are still undescribed in detail, but probably are similar to those of the preceding species. In flight the outer chestnut-coloured tail feathers are visible, and otherwise the large chestnut spotting on the flanks should serve for field identification.

In the hand
The combination of chestnut outer tail feathers, a bright rufous throat, and chestnut spotting on the flanks should provide for identification.

Adult female. Like the male, but usually without tarsal spurs.

Immature. Differs from adults in that the head is mostly dull brown, with a paler post-ocular stripe, the chin and throat are mixed with buff, and the lower back, rump, and upper tail-coverts are spotted with dark chestnut and black. The feathers of the mantle and breast have buffy shaft-stripes and lack rufous tones.

General biology and ecology

In China this species occurs mainly in hilly and flat terrain having shrubbery, bamboo forests, and grassy areas below 1000 m, but locally extending to 2000 m. Weedy thickets close to cultivated areas are also frequented, and at times the birds will mix with domestic fowl feeding near human habitations. The birds have a mixed diet including seeds, shoots, leaves, nuts, grains, and invertebrates. They also undertake short seasonal migrations, moving downward in winter and to hilltops in summer. In winter they often roost in clusters on tree branches, but in spring they begin to scatter (Cheng et al. 1978).

Social behaviour

The coveys or flocks of this species vary in size from two to twenty birds, with each flock having a relatively fixed home range, foraging area, and roosting site that do not overlap with those of neighbouring flocks. In spring the males become quite aggressive, and can be readily trapped with decoys. Their territorial call is a loud *gi-gi-gi-gi-gi-gi-gigeroi-gigeroi*.

Members of a pair are said to engage in duets, with the male calling first and the female answering (Cheng *et al.* 1978).

Reproductive biology

Nests are built beneath shrubs, at the bases of trees, or in grass clumps, usually in forests or bamboo thickets. The clutch is said usually to be three to seven, although in captivity clutches of seven to nine eggs are common, and some reports of nests with up to 12 or even more eggs have been made. The incubation period is of 17–18 days (Cheng *et al.* 1978; Robbins 1984).

Evolutionary relationships

The two species of *Bambusicola* constitute a closely allopatric pair of taxa, or allospecies. In a recent phenetic study (Crowe and Crowe 1985) this genus nearly fell within the limits of typical *Francolinus*.

Status and conservation outlook

In China this is a locally common species in the south, and is a popular game bird. It is also sometimes raised as a cage bird or for fighting purposes. Its ability to live near humans would seem to favour its long-term survival prospects.

32 · Genus *Galloperdix* Blyth 1844

The spurfowls are medium-sized, tropical scrub and woodland species of the Indian subcontinent having long tails of 14 rectrices that are about 80 per cent as long as the wing. The wing is rounded, with the fifth and sixth primaries the longest. The sexes are dimorphic, but in two of the three species both sexes have a considerable area of bare red skin around the eyes. The tail is somewhat vaulted, and in males of one species is slightly iridescent. Males have one to three tarsal spurs, and females usually few or none. Three species are recognized.

G. spadicea (Gmelin): red spurfowl
    G. s. spadicea (Gmelin) 1789
    G. s. caurina Blanford 1898
    G. s. stewarti Stuart Baker 1919
G. lunulata (Valenciennes) 1825: painted spurfowl
G. bicalcarata (J. R. Forster) 1781: Ceylon spurfowl

KEY TO THE SPECIES OF SPURFOWLS (*GALLOPERDIX*)

A. Only a narrow eye-ring of bare skin present on head; legs brownish: painted spurfowl (*lunulata*)
AA. A fairly large area of bare red skin present on head; legs red
  B. Auricular area greyish, grading to a whitish throat: red spurfowl (*spadicea*)
  BB. Auricular area brownish (females) or black and white (males) contrasting with a white throat: Ceylon spurfowl (*bicalcarata*)

RED SPURFOWL (Plate 125)

Galloperdix spadicea (Gmelin) 1789
Other vernacular names: galloperdix rouge (French); Rotes-Spornhuhn; Rot-Zwergfasen (German).

Distribution of species (see Map 42)
Widespread resident in brushy and forest-edge habitats of the Indian subcontinent from Gujarat, Uttar Pradesh, western Nepal, and Bihar south to Karnataka, Tamil Nadu, and Kerala.

Distribution of subspecies
G. s. spadicea: Uttar Pradesh and the terai of western Nepal south to Mysore and Madras.
  G. s. caurina: endemic in the Aravalli Hills of southern Rajasthan.
  G. s. stewarti: endemic in Kerala.

Measurements

Wing (*spadicea*), males 153–165 mm, females 140–160 mm; tail, males 120–143 mm, females 105–123 mm. Weight, both sexes c. 284–454 g (Ali and Ripley 1978). Egg, average 40.4 × 29.5 mm (Baker 1935), estimated weight 19.4 g.

Identification

In the field (14–15 in.)
Associated with moist-deciduous and dry-deciduous scrub, usually below 1000 m elevation, and often found close to cultivated areas. Both sexes are distinctly reddish brown, with longer tails than those of typical partridges and rather hen-like cackling calls. Males also utter a chuckling crow, a quickly repeated and rattling *k-r-r-r-kwek*.

In the hand
Identified by the combination of a relatively long (over 100 mm) tail, a rather uniformly rufous plumage, with either concolorous or black-tipped chestnut feathers on the breast, and somewhat more silvery brown cheeks than are found in the other spurfowl.

Adult female. Differs from the male in that all the upperparts, especially the mantle and wing-coverts, are mottled and barred with black. The feathers of the neck and underparts are tipped with black, and all the rectrices are mottled with rufous or buff. Orbital skin duller red, the tarsus, and toes also less bright than in the male. Tarsus sometimes unspurred, but up to two spurs may be present (compared with up to three in males).

Immature. Young males resemble adult females but are more richly coloured, with proportionately more black, and have small spurs or none.

General biology and ecology

This species is associated with stony foothills and bamboo jungle having brushy cover, interspersed watercourses, and dry to moist-deciduous woodlands. It occurs from foothills of under 1000 ft (300 m) locally to 7500 ft (2300 m), but with most breeding occurring between 2000 and 4000 ft (600–1200 m). It accepts a wide variety of habitat types, the primary criteria apparently being that ample cover is present and that the topography be broken and hilly. Its foods are a broad mixture of plant and animal materials, with seeds, berries, figs, and other fruits important components (Baker 1930).

Social behaviour

The usual social unit is from two to six birds, which typically are of family groups, and it has been suggested that mating is permanent. The male's advertisement call is a chuckle-like crowing, a quickly repeated and rattling *k-r-r-r-kwek*, somewhat like the calls of a guineafowl (Ali and Ripley 1978).

Reproductive biology

Breeding in *spadicea* occurs nearly throughout the year, but especially from January until June, while in *caurina* breeding has been reported in May and June, and in *stewarti* it seems to be irregular in most months except those of the heaviest monsoon rains (June to August). In all the clutch-size is probably of two to five eggs, most commonly three, and the nest is usually placed in dense bamboo or scrub jungle (Baker 1930; Ali and Ripley 1978). The incubation period is still uncertain.

Evolutionary relationships

The genus *Galloperdix* seems particularly well named, as it certainly helps to bridge the gap between the perdicine and pheasant-like groups, and might readily be placed in either. The species *spadicea* seems to occupy a central place in the genus *Galloperdix*. In a recent phenetic analysis of skeletal traits (Crowe and Crowe 1985), *Galloperdix* 'fell' closest to the genus *Francolinus*.

Status and conservation outlook

This widespread and adaptable species is probably more numerous now than it was in earlier times, as it has adapted well to coffee plantations, marginal agriculture near scrub jungle, and similarly altered ecosystems.

PAINTED SPURFOWL (Plate 126)

Galloperdix lunulata (Valenciennes) 1825
Other vernacular names: galloperdrix lunulée (French); Perlspornhuhn (German).

Distribution of species
Resident of semi-arid steppes and rocky foothills of the Indian subcontinent, south of the Gangetic Plain, except for Rajasthan, Gujarat, and the coastal strip west of the Western Ghats, and north to Gwalior (Madhya Pradesh) and Bengal. No subspecies recognized.

Measurements

Wing, males 148–167 mm, females 132–157 mm; tail, males 111–129 mm, females 99–128 mm. Weight, males c. 255–285 g, females c. 226–255 g (Ali and Ripley 1978). Egg, average 40.9 × 29.3 mm (Baker 1935), estimated weight 19.4 g.

Identification

In the field (12.5 in.)
Associated with dense thorn scrub, bamboo jungle, and generally drier habitats than the previous species, with which it is in local contact. Males can be readily separated from the red spurfowl by their almost entirely black head and more blackish tails, but females are very similar. Calls include scolding, clucking notes when flushed, a fowl-like cackling call (both being male calls), and a peculiar and loud repeated *chur* note.

In the hand
Males are easily told by their glossy blackish brown tail and almost entirely black head; females closely resemble those of the previous species but have no bare red eye-patch, their breast feathers are not tipped with black, and their cheeks are brighter chestnut-brown.

Adult female. Crown with chestnut shaft-stripes; neck and upperparts of the body dull olive-brown, most of the feathers narrowly margined with dusky; lores, superciliary stripes, and cheeks mostly dark chestnut; chin and throat yellowish buff, mottled with chestnut; chest and breast ochraceous brown, shading into dull olive-brown on the lower parts, most of the feathers with a small blackish terminal spot or margin. Remiges brown, as in the male, but the rectrices rich brown, with little blackish. Usually one but often two tarsal spurs present.

Immature. Young males have the wing-coverts and scapulars mostly brownish black, glossed with purplish, and the chest and breast are more heavily spotted with black. Immature females have the upperparts finely mottled and barred with chestnut and black, and the outer secondaries and rectrices are more distinctly barred.

General biology and ecology

This species is not so closely associated with dense forest or bamboo cover as is the last one, and instead favours broken ground with numerous rocks and boulders, especially in hilly areas, where the birds favour the hilltops, with their grassy and thorny scrub cover rather than the more densely vegetated areas below. They occupy lower elevations than does *spadicea*, and probably rarely occur above 3000 ft (900 m) (Baker 1930).

Social behaviour

Like the other spurfowl, this species occurs in pairs and family groups of up to about six birds, with the males having strong pair-bonds and possibly pairing permanently. They are strongly territorial, although the advertising call of the male is still only very poorly described, being called both fowl-like and anything but fowl-like (Baker 1930; Ali and Ripley 1978).

Reproductive biology

Nesting occurs from about February to June, with a peak in April and early May. The nests are hidden under the shelter of rocks or vegetation, and the clutch-size is apparently normally only three eggs, with observed ranges of two to five (Baker 1930). The incubation period is unknown.

Evolutionary relationships

This is a very close relative of *spadicea*, and the two forms probably speciated fairly recently. There is as yet no indication of hybridization in spite of limited sympatry.

CEYLON SPURFOWL (Plate 127)

Galloperdix bicalcarata (J. R. Forster) 1781
Other vernacular names: galloperdrix de Ceylan (French); Ceylon-Zwergfasen; Weisskehlspornhuhn (German).

Distribution of species (see Map 42)
Endemic resident of the wetter forests of southern Sri Lanka (Ceylon), up to about 1500 m. No subspecies recognized.

Measurements

Wing, males 157–174 mm, females 143–150 mm; tail, males 121–130 mm, females no data. Weight, males *c.* 312–368 g, females *c.* 200–312 g (Ali and Ripley 1978). Egg, average 40.6 × 29.7 mm (Baker 1935), estimated weight 19.8 g.

Identification

In the field (13–14 in.)
Found in well-forested habitats of Sri Lanka, especially in the moister areas of the south-west. There it is the only partridge-like bird with a fairly long tail (but shorter than those of junglefowl), and the brightly spangled appearance of the male is unmistakable. Females are almost entirely rufous brown, with a whitish throat. Calls include a ringing, cackling series of three-syllabled whistles, each series on a higher pitch than the one before, and the final one dropping to the starting point. This call evidently serves as a male advertisement call, and may be answered by other near-by males.

In the hand
Males are easily separated from the other spurfowl by the distinctive black and white pattern on the sides, upperparts, and breast. Females are much like those of both other species, but have chestnut breast feathers that lack black tips, an area of red skin around each eye, and a blackish crown with little or no chestnut present.

Adult female. Crown blackish; sides of head dull chestnut, the feathers black-edged; chin and throat white; remiges and rectrices very similar to the male, but the central rectrices slightly mottled with rufous. Rest of the plumage chestnut, brightest on the breast and finely vermiculated with black. Bill, legs, and toes paler red than in male. Tarsi with or without a pair of spurs, as compared with two or three in males.

Immature. Similar to the adult of each sex, but with the white spots and streaks of the young male relatively fewer and larger than in adults.

General biology and ecology

Like the red spurfowl, this species is associated with rather well-forested and moister habitats having ample cover and occurring from the foothills up to about 5000 ft (1500 m). It favours tangled brakes, forests near rivers, jungle-covered hillsides, and virtually any other habitat offering dense protective cover. Their foods are a mixture of plant materials and invertebrates, but they especially favour the ripe berries of lantana (Baker 1930).

Social behaviour

Pairs remain together permanently, and family groups of up to about six birds may sometimes be found. During the breeding season males call in early morning hours with an ascending series of notes, producing a cackling sequence of three-noted whistles (Ali and Ripley 1978). The birds are reportedly highly territorial, and readily trapped by using captive decoy birds (Baker 1930).

Reproductive biology

The breeding season occurs during the north-east monsoon, from November to March, with some breeding also occurring from July to September or October. However, most eggs are probably laid in February and March. The usual clutch-size is of only two eggs, but uncommonly three and rarely four may be laid (Baker 1930). It thus has a somewhat smaller clutch than do the Indian species of spurfowls, and indeed one of the smallest clutches of all perdicine birds. The incubation period is unknown.

Evolutionary relationships

This form is clearly an insular derivative that seems to be especially similar to *lunulata*, its nearest likely relative from a zoogeographic standpoint.

Status and conservation outlook

This is a very shy and elusive species that has the ability to survive in any heavy cover, and will probably survive as long as such habitats persist in Sri Lanka.

Bibliography

This bibliography includes some references, especially those of a bibliographic, taxonomic, or distributional nature, that are not specifically cited in the text but were used during its preparation.

Akande, M. (1977). The biology and domestication of the African bushfowl *Francolinus bicalcaratus bicalcaratus*. Proceedings of the 13th Congress of Game Biology, pp. 474–80.

Aldrich, J. (1946). The United States races of the bobwhite. *Auk* **63**, 493–508.

Ali, S. and Ripley, S. D. (1978). *Handbook of the birds of India and Pakistan* (revised edn) Vol. 2. Oxford University Press, Oxford.

Amadon, D. (1966). The superspecies concept. *Syst. Zool.* **15**, 245.

American Ornithologists' Union (1983). *Check-list of North American birds* (6th edn). Allen Press, Lawrence.

Anderson, W. (1978). Vocalizations of scaled quail. *Condor* **80**, 49–63.

Anthony, R. (1970). Ecology and reproduction of California quail in southeastern Washington. *Condor* **72**, 276–87.

Ash, J. S. (1978). The undescribed female of Harwood's francolin *Francolinus hardwoodi* and other observations on the species. *Bull. Br. Ornithol. Club* **98**, 50–5.

Austin, O. L., Jr., and Kuroda, N. (1953). The birds of Japan, their status and distribution. *Bull. Mus. Comp. Zool. Harv. Univ.* **109**, 277–637.

Bailey, E. D. and Baker, E. (1982). Recognition characteristics in covey dialects of bobwhite quail. *Condor* **84**, 317–20.

Baker, E. C. S. (1928). *The fauna of British India, including Ceylon and Burma*, Vol. 5, *Birds*. Taylor and Francis, London.

—— (1930). *Game birds of India, Burma and Ceylon* Vol. 3. John Bale and Son, London.

—— (1935). *The nidification of birds of the Indian Empire* Vol. IV. Taylor and Francis, London.

Banks, R. C. and Walker, L. W. (1964). A hybrid scaled × Douglas quail. *Wilson Bull.* **76**, 378–80.

Bannerman, D. A. (1963). *The birds of the British Isles* Vol. 12. Oliver & Boyd, Edinburgh.

Bannikov, A. G., ed. (1978). In (*Red data book for the USSR*) pp. 90–149. USSR Ministry of Agriculture. Lesnaya Promyshlennost, Moscow. (In Russian.)

Barclay, H. J. (1981). Plumage development as an aging technique in juvenile California quail. *Murrelet* **62**, 59–60.

—— and Bergerud, A. T. (1975). Demography and behavioral ecology of California quail on Vancouver Island. *Condor* **77**, 315–23.

Bateson, P. P. G. (1982). Preferences for cousins in Japanese quail. *Nature* **295**, 236–7.

Bazier, D. Kh. (1968). (*The snowcocks of the Caucasus; ecology, morphology, evolution.*) Nauka, Leningrad. (In Russian.)

—— (1972). Evolution of the genus *Tetraogallus* Gray 1834. *Acta Ornithol.* **13**, 173–90.

Beebe, C. W. (1914). Preliminary pheasant studies. *Zoologica* **1**, 261–85.

Bell, P. T. and Summers, A. B. (1982). The ecology of the chukar (*Alectoris chukar*) and the black francolin (*Francolinus francolinus*) in north-west Cyprus. *Annu. Rep. Cyprus Ornithol. Soc.* **29**, 67–79.

Bennitt, R. (1951). Some aspects of Missouri quail and quail hunting 1938–1948. Missouri Conservation Department Technical Bulletin No. 2, Columbia.

Bernard-Laurent, A. (1984). (Natural hybridization between rock partridges and red-legged partridges in the Alps-Maritimes.) *Giber Faune Sauvage* **2**, 79–96. (In French, English summary.)

Berthet, G. (1949). La nidification tardive des Cailles et l'hypothèse de le double nidification de cette espèce sur les deux continents. *Nos Oiseaux* **20**, 34–5.

Bishop, R. A. (1964). The Mearns quail (*Cyrtonyx montezumae mearnsi*) in southern Arizona. M.S. thesis, University of Arizona.

—— (1977). A look at Iowa's Hungarian partridge (pp. 10–30 *in* Kobriger 1977).

Blake, E. (1977). *Manual of Neotropical birds* Vol. 1. University of Chicago Press, Chicago.

Blakers, M., Davies, S. J. J. F., and Reilly, P. N. (1984). *The atlas of Australian birds*. Melbourne University Press, Carlton, Victoria.

Blank, T. H. and Ash, J. S. (1955). A population of partridges (*Perdix p. perdix* and *Alectoris r. rufa*) on a Hampshire estate. Proceedings of the 11th International Ornithological Congress, 1954, pp. 424–7.

—— and —— (1956). The concept of territory in the partridge *Perdix p. perdix*. *Ibis* **98**, 379–89.

—— and —— (1958). Factors controlling brood size in the partridge (*Perdix perdix*) on an estate in south England. *Dan. Rev. Game Biol.* **3**, 39–41.

—— and —— (1960). Some aspects of clutch size in the partridge (*Perdix perdix*). Proceedings of the 12th International Ornithological Congress, 1960, pp. 118–26.

—— and —— (1962). Fluctuations in a partridge population. In *The exploitation of natural animal populations* (ed. E. D. Le Cren and M. W. Holdgate) pp. 118–33. John Wiley & Sons, New York.

——Southwood, T. R. E., and Cross, D. J. (1967). The ecology of the partridge. I. Outline of population processes with particular reference to chick mortality and nest density. *J. Anim. Ecol.* **36**, 549–56.

Blot, J. (1985). (Contribution to our knowledge of the biology and ecology of the pale-billed francolin, *Francolinus ochropectus* Dorst and Jouanin.) *Alauda* **53**, 244–56. (French, with English summary; not seen.)

Borrero, H. J. I. (1972). *Aves de caza Colombianas*. Departmento de Biologia, Universidad del Valle.

Boulton, R. (1932). A new species of tree partridge from Szechuan, China. *Proc. Biol. Soc. Washington* **45**, 235–6.

Brennan, L. A. (1984). Summer habitat ecology of mountain quail in northern California. M.S. thesis, Humboldt State University, Arcata.

—— and Block, W. M. (1985). Sex determination of mountain quail reconsidered. *J. Wildl. Manage.* **49**, 475–6.

Brosset, A. (1974). La nidification des oiseaux en forêt Gabonaise: architecture, situation des nids et predation. *Terre Vie* **28**, 579–610 (not seen).

Brown, D. E. and Ellis, D. H. (1977). Status summary and recovery plan for the masked bobwhite. US Fisheries and Wildlife Service, Office of Endangered Species, Albuquerque.

Bump, G. (1958). Red-legged partridges of Spain. US Fisheries and Wildlife Service, Special Scientific Report, Wildlife No. 39.

—— and Bump, J. W. (1964). A study and review of the black francolin and the gray francolin. US Fisheries and Wildlife Service, Special Scientific Report, Wildlife No. 81.

Campbell, H. (1957). Fall foods of Gambel's quail (*Lophortyx gambelii*) in New Mexico. *Southwest. Nat.* **2**, 122–8.

——Martin, D. K., Ferkovich, P. E., and Harris, B. K. (1973). Effects of hunting and some other environmental factors on scaled quail in New Mexico. *Wildl. Mongr.* **34**, 1–49.

Chapin, J. P. (1932). The birds of the Belgian Congo, Pt. I. *Bull. Am. Mus. Nat. Hist.* **65**, 1–756.

Chapman, F. M. (1929). *My tropical air castle*. D. Appleton, New York.

Cheng, Tso-hsin (1963). *China's economic fauna: birds*. Science Publishing Society, Peking. (Translated by US Department of Commerce, Washington D.C.)

——Yao-Kuang, T., Tai-Chung, L., Chan-zhu, T., Gui-Jun, B., and Fu-Lai, L. (1978). *Fauna Sinica, series Vertebrata. Aves* Vol. 4: Galliformes. Science Press, Peking.

Christensen, G. C. (1954). The chukar partridge in Nevada. Nevada Fish & Game Comm. Biological Bull. No. 1. (77 pp.).

——(1970). The chukar partridge: its introduction, life history and management. Nevada Department of Fish and Game Biological Bulletin No. 4.

Church, K. E., Harris, H. J., and Stiehl, R. B. (1980). Habitat utilization by gray partridge (*Perdix perdix* L.) prenesting pairs in east-central Wisconsin. In *Proceedings of Perdix II* (ed. S. P. Peterson and L. Nelson Jr.) pp. 9–20. Contribution No. 211 to the Gray Partridge Workshop, 18–20 March, Moscow, Idaho. Forest, Wildlife, and Range Experiment Station, Moscow, Idaho.

Cink, C. L. (1971). Comparative behavior and vocalizations of three *Colinus* species and their hybrids. M.S. thesis, University of Nebraska, Lincoln.

Clancey, P. A. (1967). *Gamebirds of southern Africa*. American Elsevier, New York.

——(1985). *The rare birds of southern Africa*. Winchester, New York.

Clarke, E. W. (1966). Mantenimento in cattività e riproduzione della pernice della rocce Abissina *Ptilopachus petrosus major* Neumann. *Riv. Ital. Ornithol.* **36**, 362–6 (not seen).

Collar, N. J. and Stuart, S. N. (1985). *Threatened birds of Africa and related islands*. ICBP, Cambridge.

Coomans de Ruiter, L. (1946). Oölogische en biologische aanteekeningen over eenige hoendervogels in de Westerafdeeling van Borneo. *Limosa* **19**, 129–40. (With English summary.)

Cracraft, J. (1981). Toward a phylogenetic classification of the Recent birds of the world (Class Aves). *Auk* **98**, 681–714.

Craft, B. R. (1966). An ecological study of the black francolin in the Gum Cove area of southwestern Louisiana. M.S. thesis, Louisiana State University.

Cramp, S. and Simmons, K. E. L., eds. (1980). *The birds of the western Palearctic* Vol. 2. Oxford University Press, Oxford.

Crispens, G. G., Jr. (1960). *Quails and partridges of North America: a bibliography*. University of Washington Press, Seattle. (Over 2000 references.)

Crome, F. H. J., Carpenter, S. M., and Rushton, D. K. (1981). Aging stubble quail, *Coturnix pectoralis* Gould by using measurements of lengths and the moulting stages of the primaries. *Aust. Wildl. Res.* **8**, 163–79.

Crowe, T. M. (1978). The evolution of guinea-fowl (Galliformes, Phasianidae, Numidinae), taxonomy, phylogeny, speciation and biogeography. *Ann. S. Afr. Mus.* **76**, 43–136.

—— and Crowe, A. A. (1985). The genus *Francolinus* as a model for avian evolution and biogeography in Africa. I. Relationships among species. In *African vertebrates: systematics, phylogeny and evolutionary ecology* (ed. K. L. Schuchmann) pp. 207–40. Zoologische Forschsung und Museum Alexander Koenig, Bonn.

Cruise, J. (1966). Stubble quail, *Coturnix pectoralis*, and their breeding behaviour. *Emu* **66**, 39–45.

Davison, G. W. H. (1982). Systematics within the genus *Arborophila* Hodgson. *Fed. Mus. J. (Malaya)* **27**, 125–34.

—— Avian spurs (1985). *J. Zool.* **206**, 353–66.

Deignan, H. G. (1945). The birds of Northern Thailand. *US Nat. Mus. Bull.* **186**, 1–616

Delacour, J. (1951). *The pheasants of the world*. Country Life, London.

—— and Jabouille, P. (1931). *Les oiseaux de l'Indochina*

française. (Four volumes.) Impr. du Cantal Republicain, Aurillac.

Dellivers, P. (1976). Project de nomenclature Française des oiseaux du Monde. 2. Anhimides aux Otidides. *Gerfaut* **66**, 391–421.

Dementiev, G. P. and Gladkov, N. A., eds. (1952). *Birds of the Soviet Union* Vol. 4. Translated in 1967 by the Israel Programme for Scientific Translations, Jerusalem.

Demers, D. J. and Garton, E. O. (1980). An evaluation of gray partridge (*Perdix perdix*) aging criteria. In *Proceedings of* Perdix II (ed. S. P. Peterson and L. Nelson Jr.) pp. 21–44. Contribution No. 211 to the Gray Partridge Workshop, 18–20 March, Moscow, Idaho. Forest, Wildlife, and Range Experiment Station, Moscow, Idaho.

Dickey, D. R. and van Rossem, A. J. (1938). *The birds of El Salvador.* Zoological Series, No. 23. Field Museum of Natural History, Chicago

Dimminck, R. W. (1974). Populations and reproductive effort among bobwhites in western Tennessee. *Proc. Ann. Conf. S. E. Ass. Game Fish Comm.* **28**, 594–602.

Disney, H. J. de S. (1978). The age of breeding in the stubble quail and Japanese quail. *Corella* **2**, 81–4.

Dorst, J. and Jouanin, C. (1952). Description d'une espèce nouvelle de francolin d'Afrique orientale. *Oiseau* **22**, 71–4.

—— and —— (1954). Précisions sur le position systématique et l'habitat de *Francolinus ochropectus*. *Oiseau* **24**, 161–70.

Dragoev, P. (1974). On the population of the rock partridge (*Alectoris graeca* Meisner) in Bulgaria and methods of census. *Acta Ornithol.* **14**, 394–8.

Dwenger, R. (1973). *Das Rebhuhn.* Neue Brehm Bucherei No. 447. A. Ziemsen Verlag, Wittenberg Lutherstadt.

Edminster, F. C. (1954). *American game birds of field and forest.* Charles Scribner's Sons, New York.

Ellis, C. R., Jr., and Stokes, A. W. (1966). Vocalizations and behavior in captive Gambel quail. *Condor* **68**, 72–80.

Emfinger, J. W. (1966). Survival, dispersal and reproductive success of the black francolin (*Francolinus francolinus*) in Morehouse Parish. M.S. thesis, Louisiana State University.

Emlen, J. T., Jr. (1939). Seasonal movements of a low-density valley quail population. *J. Wildl. Manage.* **3**, 118–30.

—— (1940). Sex and age ratios in survival of the California quail. *J. Wildl. Manage.* **4**, 92–9.

Erwin, M. J. (1975). Comparison of the reproductive physiology, molt and behavior of the California quail in two years of differing rainfall. M.S. thesis, University of California, Berkeley.

Etchécopar, R. D. and Hüe, F. (1967). *The birds of North Africa.* Oliver & Boyd, Edinburgh.

—— and —— (1978). *Les oiseaux de Chine, non-passereaux.* N. Boubée, Paris.

Eynon, A. E. (1968). The agonistic and sexual behavior of captive Japanese quail (*Coturnix coturnix japonica*). Ph.D. dissertation, University of Wisconsin, Madison.

Faruqi, S. .A., Bump, G., Nanda, P., and Christensen, G. C. (1960). A study of the seasonal foods of the black francolin [*Francolinus francolinus* (Linn.)], the grey francolin [*F. pondicerianus* (Gmelin)] and the common sandgrouse (*Pterocles exustus* Temminck) in India and Pakistan. *J. Bombay Nat. Hist. Soc.* **57**, 354–61.

Fitzpatrick, J. W. and Willard, D. E. (1982). Twenty-one bird species new or little known from the Republic of Colombia. *Bull. Br. Ornithol. Club* **102**, 153–8.

Flieg, G. M. (1970). Observations on the first North American breeding of the spot-winged wood quail (*Odontophorus capuiera*). *Avic. Mag.* **76**, 1–4.

Follett, B. K. and Farner, D. S. (1966). Pituitary gonadotropins in the Japanese quail during photoperiodically induced gonadal growth. *Gen. Comp. Endocrinol.* **7**, 125.

Francis, W. J. (1968). Double broods in California quail. *Condor* **67**, 541–2.

—— (1970). The influence of weather on population fluctuations in California quail. *J. Wildl. Manage.* **34**, 249–66.

Friedmann, H., Griscom, L., and Moore, R. T. (1950). Distributional check-list of the birds of Mexico. *Pac. Coast Avifauna* No. 29, pp. 1–436.

Frisch, O. von (1962). Zur Biologie des Rothuhns (*Alectoris rufa*). *Vogelwelt* **83**, 145–9.

Frith, H. J., ed. (1977). *Reader's Digest complete book of Australian birds.* Reader's Digest Services, Sydney.

—— and Carpenter, S. M. (1980). Breeding of the stubble quail *Coturnix pectoralis*, in south-eastern Australia. *Aust. Wildl. Res.* **7**, 117–37.

—— Brown, B. K., and Morris, A. K. (1977). Food habits of the stubble quail *Coturnix pectoralis*, in south-eastern Australia. CSIRO, Division of Wildlife Research, Technical Paper No. 32.

Frost, P. G. H. (1975). The systematic position of the Madagascan partridge *Margaroperdix madagascariensis*. *Bull. Br. Ornithol. Club* **95**, 64–8.

Frye, O. E., Jr. (1954). Aspects of the ecology of the bobwhite quail in Charlotte County, Florida. Ph.D. dissertation, University of Florida.

Galbreath, D. S. and Moreland, R. (1953). The chukar partridge in Washington. Washington Department of Game, Biological Bulletin No. 11.

Gallagher, M. D. and Woodcock, M. W. (1980). *The birds of Oman.* Quartet Books, London.

Gallizioli, S. and Swank, W. (1958). The effects of hunting on Gambel quail populations. Transactions of the 23rd North American Wildlife Conference, pp. 305–19.

Genelly, R. E. (1955). Annual cycle in a population of California quail. *Condor* **57**, 263–85.

Genung, W. G. and Lewis, R. H. (1982). The black francolin in the Everglades agricultural area. *Fl. Field Nat.* **10**(4), 65–9.

Gines, H. and Aveledo, H. R. (1958). *Aves de casa de Venezuela.* Soc. Cien. Nat. LaSalle, Monogr. No. 4. Editorial Sucre, Caracas.

Glading, B. (1938). Studies on the nesting cycle of the

California valley quail in 1937. *Calif. Fish Game* **24**, 318–40.

——(1941). Valley quail census methods and populations at the San Joaquin Experimental Range. *Calif. Fish Game* **27**, 33–8.

—— and Saarni, R. W. (1944). Effect of hunting on a valley quail population. *Calif. Fish Game* **31**, 139–56.

Glenister, A. G. (1951). *The birds of the Malay Peninsula, Singapore, and Penang: an account of the Malayan species, with a note on their occurrence in Sumatra, Borneo and Java and a list of birds of those islands.* Oxford University Press, Oxford.

Glutz, U. N. von Blotzheim, ed. (1973). *Handbuch der Vogel mitteleuropas*, Vol. 5, *Galliformes und Gruiformes*. Akademische Verlag, Frankfurt.

Goodwin, D. (1953). Observations on voice and behaviour of the red-legged partridge *Alectoris rufa*. *Ibis* **95**, 581–614.

——(1954). Notes on captive red-legged partridges. *Avic. Mag.* **60**, 49–61.

——(1958). Further notes on pairing and submissive behaviour of the red-legged partridge *Alectoris rufa*. *Ibis* **100**, 59–66.

——(1975). Birds of the Harold Hall Australian expeditions 1962–70. Galliformes—fowl-like birds. *Publ. Br. Mus. Nat. Hist.* No. 745, pp. 60–2.

Gorsuch, D. M. (1934). Life history of the Gambel quail in Arizona. University of Arizona, *Biol. Sci. Bull.* **2**, 1–89.

Gould, J. (1850). *A monograph of the Odontophorinae or partridges of North America.* London, published by the author.

Grachev, Yu-N. (1983). (*The rock partridge: biology, use and conservation.*) Nauka, Alma-Ata. (In Russian, not seen.)

Gray, A. P. (1958). *Bird hybrids, a check-list with bibliography.* Commonwealth Agricultural Bureaux, Farnham Royal.

Green, R. E. (1984). Double nesting of the red-legged partridge *Alectoris rufa*. *Ibis* **126**, 332–46.

Grinnell, J., Bryant, H. C., and Storer, T. I. (1918). *The game birds of California.* University of California Press, Berkeley.

Griscom, L. (1932). The distribution of bird-life in Guatemala: a contribution to the study of the origin of Central American bird-life. *Bull. Am. Mus. Nat. Hist.* **64**, 1–439.

Gullion, G. W. (1956). Let's go desert quail hunting. *Nevada Fish Game Comm. Biol. Bull.* No. 2, pp. 1–76.

——(1960). The ecology of Gambel's quail in Nevada and the arid Southwest. *Ecology* **41**, 518–36.

——(1962). Organization and movements of coveys of a Gambel quail population. *Condor* **64**, 402–15.

Gutierrez, R. J. (1975). A literature review and bibliography of the mountain quail, *Oreortyx picta* (Douglas). US Department of Agriculture, Forest Service, California region, San Francisco.

——(1980). Comparative ecology of the mountain and California quail in the Carmel Valley, California. *Living Bird* **18**, 71–93.

——Zink, R. M., and Yang, S. Y. (1983). Genic variation, systematics and biogeographical relationships of some galliform birds. *Auk* **100**, 33–47.

Hachisuka, Marquess (1938). Classification and distribution of the gamebirds. Proceedings of the Ninth International Ornithological Congress, pp. 177–82.

Hall, B. P. (1963). The francolins, a study in speciation. *Bull. Br. Mus. Nat. Hist. (Zool.)* **10**, 105–204.

Hammond, M. C. (1941). Fall and winter mortality among Hungarian partridges in Bottineau and McHenry counties, North Dakota. *J. Wildl. Manage.* **4**, 375–82.

Harper, H. T., Harry, B. H., and Bailey, W. D. (1958). The chukar partridge in California. *Calif. Fish Game* **44**, 5–50.

Harrison, C. J. O. (1965). Plumage patterns and behaviour in the painted quail. *Avic. Mag.* **71**, 176–84.

—— and Walker, C. A. (1977). Birds of the British Lower Eocene. *Tertiary Res.* Special Paper No. 3, pp. 1–61.

Hartert, E. (1921–2). *Die Vögel der Paläarktischen Fauna.* Friedlander & Sohn, Berlin.

Haverschmidt, F. (1968). *Birds of Surinam.* Oliver & Boyd, Edinburgh.

Hellebrekers, W. P. J. and Hoogerwerf, A. (1967). A further contribution to our zoological knowledge of the island of Java (Indonesia). *Zool. Verh. Rijksmus. Nat. Hist. Leiden* **88**, 1–164.

Hellmayr, C. E. and Conover, B. (1942). *Catalogue of birds of the Americas and adjacent islands.* Zoological Series, No. 13. Field Museum of Natural History, Chicago.

Helm-Bychowski, K. M. and Wilson, A. C. (1986). Rates of nuclear DNA evolution in pheasant-like birds: evidence from restriction maps. *Proc. Nat. Acad. Sci. USA* **83**, 688–92.

Henry, C. M. (1955). *A guide to the birds of Ceylon.* Oxford University Press, Oxford.

Hickey, J. J. (1955). Some American population research on gallinaceous birds. In *Recent studies in avian biology* (ed. A. Wolfson) pp. 326–95. University of Illinois Press, Urbana.

Hoffman, D. M. (1965). The scaled quail in Colorado. Colorado Department of Game, Fish, and Parks, Technical Bulletin No. 18, pp. 1–47.

Holman, J. A. (1961). Osteology of living and fossil New World quails (Aves, Galliformes). *Fl. Univ. State Mus. Biol. Bull.* **6**, 131–233.

Howard, R. and Moore, A. (1980). *A complete checklist of the birds of the world.* Oxford University Press, Oxford.

Howard, W. E. and Emlen, J. T. (1942). Intercovey social relationships in the valley quail. *Wilson Bull.* **54**, 162–70.

Hubbard, J. P. (1974). Avian evolution in the aridlands of North America. *Living Bird* **12**, 155–96.

Hungerford, C. R. (1962). Adaptations shown in selection of food by Gambel quail. *Condor* **64**, 213–19.

Hudson, G. T., Parker, R. A., Berge, J. V., and Lanzillotti, P. J. (1966). A numerical analysis of the modification of the appendicular muscles in various genera of gallinaceous birds. *Am. Midl. Nat.* **76**, 1–73.

Hupp, J. W., Smith, L. M., and Ratti, J. T. (1980). Gray partridge nesting biology in eastern South Dakota. In *Proceedings of Perdix II* (ed. S. P. Peterson and L. Nelson Jr.) pp. 55–69. Contribution No. 211 to the Gray Partridge Workshop, 18–20 March, Moscow, Idaho. Forest, Wildlife, and Range Experiment Station, Moscow, Idaho.

Huxley, T. H. (1868). On the classification and distribution of the Alecteromorphae and Heteromorphae. *Proc. Zool. Soc.* pp. 294–319.

Iredale, T. (1956). *Birds of New Guinea* Vol. 1. Georgian House, Melbourne.

Irwin, M. P. S. (1981). *The birds of Zimbabwe.* Quest Publications, Harare.

Jenkins, D. (1956). Factors affecting population density in the partridge. Ph.D. dissertation, Oxford University.

——(1957). The breeding of the red-legged partridge. *Bird Study* **4**, 97–100.

——(1961a). Social behaviour in the partridge *Perdix perdix. Ibis* **103a**, 157–88.

——(1961b). Population control in protected partridges (*Perdix perdix*). *J. Anim. Ecol.* **39**, 235–58.

Johansen, H. (1961). Die Vogelfauna Westsibiriens—Otides bis *Gallus. J. f. Ornithol.* **102**, 237–69.

Johnsgard, P. A. (1971). Experimental hybridization of the New World quail (Odontophorinae). *Auk* **88**, 264–75.

——(1973). *Grouse and quails of North America.* University of Nebraska Press, Lincoln.

——(1979). The American wood quails. *Wld Pheasant Ass. J.* **4**, 93–9.

——(1986). *The pheasants of the world.* Oxford University Press, Oxford.

Johnson, B. C. (1969). Home range, movements and ecology of the gray partridge, *Perdix perdix* L., in a selected area of eastern Kings county, Nova Scotia. M.S. thesis, Acadia University (not seen).

Johnson, M. D. (1964). Feathers from the prairie: a short history of upland game birds. North Dakota Game & Fish Department Report W-67-R-5 (239 pp.).

Judd, S. (1905). The bob-white and other quails of the United States in their economic relations. *US Biol. Survey Bull.* **21**, 1–66.

Kang, K. (1969). A note on Rickett's hill partridge, *Arborophila gingica* (Gmelin) in Taiwan. *Q. J. Taiwan Mus.* **22**, 121–3.

Karr, J. R. (1971). Ecological, behavioral and distributional notes on some central Panama birds. *Condor* **73**, 107–11.

King, W. B., ed. (1981). *Endangered birds of the world: the ICBP red data book.* Smithsonian Institute Press and International Council for Bird Preservation, Washington D.C.

Kirkpatrick, C. M. (1955). Factors of photoperiodism of bobwhite quail. *Physiol. Zool.* **28**, 255–64.

Klimstra, W. D. (1950). Bobwhite quail nesting and production in southeastern Iowa. *Iowa State J. Science* **24**, 385–95.

Kobriger, G. D., ed. (1977). *Proceedings of Perdix I,* Hungarian partridge workshop, Minot, N. Dakota, N. Dakota Chapter Wildlife Society & N. Dakota Game and Fish Department, Bismarck.

Kochenderfer, C. (1971). The ontogeny of vocalizations in the bobwhite quail. M.S. thesis, University of Nebraska, Lincoln.

Komen, J. (1986). A comparison of behaviour and vocalization in some African francolins (Phasianidae). Abstract of paper presented at the XIXth International Ornithological Congress, Ottawa, 22–29 June 1986.

—— and Meyer, E. (1984). Hartlaub's francolin, a formal introduction to one of Namibia's 'specials'. In *1984 SWA Annual* pp. 39–43. SWA Publications, Windhoek.

Kruijt, J. P. (1962). Notes on wing display in the courtship of pheasants. *Avic. Mag.* **69**, 11–20.

Kumerloeve, H. (1963). Zur Brutverbreitung des Frankolin *Francolinus francolinus* (L.) im Vorderen Orient. *Vogelwelt* **84**, 129–37.

Kuroda, N. (1970). *Odontophorinae and Perdicinae of the world.* Ornithological Society of Japan, Tokyo.

Lack, D. (1947). The significance of clutch-size in the partridge. *J. Anim. Ecol.* **16**, 19–25.

——(1951). *Population studies of birds.* Oxford University Press, Oxford.

——(1968). *Ecological adaptations for breeding in birds.* Oxford University Press, Oxford.

LeFebvre, E. A. Z. and LeFebvre, J. H. (1958). Notes on the ecology of *Dactylortyx thoracicus. Wilson Bull.* **70**, 372–7.

Lehmann, V. W. (1984). *Bobwhites in the Rio Grande plain of Texas.* Texas A. & M. University Press, College Station.

Lekagula, B. and Cronin, E. W., Jr. (1974). *Bird guide of Thailand.* Association for the Conservation of Wildlife, Bangkok.

Leopold, A. S. (1959). *Wildlife of Mexico: the game birds and mammals.* University of California Press, Berkeley.

——(1977). *The California quail.* University of California Press, Berkeley.

—— and McCabe, R. A. (1957). Natural history of the Montezuma quail in Mexico. *Condor* **59**, 3–26.

Lepper, M. G. (1978). Covey behaviour in California quail (*Lophortyx californicus* Shaw) in Nevada. *Sociobiology* **3**, 107–24.

Lewin, V. (1963). Reproduction and development of young in a population of California quail. *Condor* **65**, 249–78.

Li, K.-y., K;iu, L-t., Chang, J.-y., and Chang, C.-m. (1974). Discovery of the female of the Szechwan hill-partridge (*Arborophila rufipectus*). *Acta Zool. Sin.* **20**, 421–2.

Litun, V. I. (1983). (Daurian partridge ecology in southern Transbaikal.) *Byull. Mosk. Ova Ispyt. Prir. Otd. Biol.* **88**, 25–32. (In Russian, with English summary; not seen.)

Long, J. L. (1981). *Introduced birds of the world.* A. H. and A. W. Reed, Sydney.

Ludlow, F. (1944). The birds of south-eastern Tibet (part). *Ibis* **86**, 348–89.

Lyon, D. L. (1962). Comparative growth and plumage development in coturnix and bobwhite. *Wilson Bull.* **74**, 5–27.

McCabe, R. A. and Hawkins, A. S. (1946). The Hungarian partridge in Wisconsin. *Am. Midl. Nat.* **36**, 1–75.

McClure, H. E. (1974). *Migration and survival of the birds of Asia*. US Army Medical Component, SEATO Medical Project, Bangkok.

McCrow, V. P. (1977). Movements and habitat use of gray partridge in north central Iowa. In *Proceedings of Perdix I, Hungarian partridge workshop*, Minot, N. Dakota, pp. 89–98. N. Dakota Chapter Wildlife Society and N. Dakota Game and Fish Department, Bismarck.

Mackie, J. R. and Buechner, H. K. (1963). The reproductive cycle of the chukar. *J. Wildl. Manage.* **27**, 246–60.

Mackworth-Praed, C. W. and Grant, C. H. B. (1952–73). *African handbook of birds* Series 1–3. Longmans, London.

McMillan, I. I. (1964). Annual population changes in California quail. *J. Wildl. Manage.* **28**, 702–11.

McNally, J. (1956). A preliminary investigation on the food of the stubble quail in Victoria. *Emu* **56**, 367–400.

Marien, D. (1951). Notes on some pheasants from south-western Asia, with remarks on molt. *Am. Mus. Novitates* **1518**, 1–25.

Marsden, H. M. and Baskett, T. S. (1958). Annual mortality in a banded bobwhite population. *J. Wildl. Manage.* **22**, 414–19.

Medway, Lord and Wells, D. R. (1976). *The birds of the Malay Peninsula* Vol. 5. Witherby, London.

Meinertzhagen, R. (1954). *Birds of Arabia*. Oliver & Boyd, Edinburgh.

Meise, W., Schönwetter, M., and Stresemann, E. (1938). Aves Beickianae. Beiträger zur Ornithologie von Nord-west-Kansu nach den Forschungen von Walter Beick in den Jahren 1926–1933. *Journal Ornithol. (Sonderheft)* **88**, 171–221.

Mendel, G. and Peterson, S. (1980). Gray partridge population structure and densities on the Palouse Prairie (pp. 118–29 in Peterson & Nelson 1980).

Mendez, E. (1979). *Las aves de casa de Panama*. Published by the author.

Mentis, M. T. (1973). A comparative ecological study of greywing and redwing francolins in the Natal Drakensberg. M.S. thesis, University of Stellenbosch.

—— and Bigalke, R. C. (1973). Management for greywing and redwing francolins in Natal. *J. S. Afr. Wildl. Manage. Ass.* **3**, 41–7.

—— and —— (1979). Some effects of fire on two grassland francolins in the Natal Drakensberg. *S. Afr. J. Wildl. Res.* **9**, 1–8.

—— and —— (1980). Breeding, social behaviour and management of greywing and redwing francolins. *S. Afr. J. Wildl. Res.* **10**, 133–9.

—— and —— (1981). Ecological isolation in greywing and redwing francolins. *Ostrich* **52**, 84–97.

—— and —— (1985). Counting francolins in grassland. *S. Afr. J. Wildl. Res.* **15**, 7–11.

Menzdorf, A. (1975a). Beitrag zum Balzverhalten des Steinhuhns (*Alectoris graeca graeca*). *J. Ornithol.* **116**, 202–6.

——(1975b). Zum Vorkommen von Doppelbruten bei Hühnern der Gattung *Alectoris*. *Vogelwelt* **96**, 135–9.

(1975c). Zur Brut und Jungenaufzucht beim Steinhuhn *Alectoris graeca graeca* Meisner 1804. *Zool. Gart.* **45**, 491–9.

——(1975d). (On the variability of the behaviour in some gallinaceous birds with special reference to the rock partridge, *Alectoris graeca graeca* Meisner 1804.) *Zool. Anz.* **195**, 64–88. (In German, English summary.)

——(1975e). (On the juvenile development of the rock partridge, *Alectoris graeca graeca* Meisner 1804.) *Osterr. Akad. Wiss. Vienna* **183**(10), 287–305. (In German, English summary.)

——(1975f). Bemerkungen zur Kunsbrut under Gefangenschaftshaltung des Steinhuns—*Alectoris graeca graeca* Meisner 1804—under besonderer Berücksichtigung der ökologischen Verhältnisse. *Zool. Anz.* **195**, 145–54.

——(1976a). Zur Ontogenese einiger Rufe beim Steinhuhn. *Zool. Anz.* **196**, 221–36.

——(1976b). Ausdrucksbewegungen beim Steinhun (*Alectoris graeca*). *Ornithol. Mitt.* **28**, 29–36.

——(1976c). Bemerkungen zun Siedlungdichte und Riviergrösse von Steinhuhnern, *Alectoris graeca*, im Freilauf. *Zool. Gart.* **46**, 389–400.

——(1976d). Zur Möglichkeit der Einbürgerung verschiedener Feldhuhnarten in der Bundesrepublik, Deutschland. *Ornithol. Mitt.* **28**, 221–5.

——(1976e). Komforthandlungen beim Hühnern der Gattung *Alectoris*. *Zool. Anz.* **197**, 238–50.

——(1976f). Bemerkungen zum Hahrungserwerb und zur Futterzusammensetzung bei Feldhühnern. *Vogelwelt* **97**, 99–107.

——(1977). (Contribution to the study of the vocalizations of the chukar.) *Beitr. Vogelkd* **23**, 85–100. (In German, English summary.)

——(1982). Social behaviour of rock partridges (*Alectoris graeca*). *Wld Pheasant Ass. J.* **7**, 70–89.

——(1984). [Regarding the knowledge of social behaviour and calling patterns of diverse partridge species (Phasianidae, Perdicinae: *Alectoris* spp.).] *Vogelwelt* **105**, 9–21. (In German, English summary.)

Middleton, A. D. (1936). The population of partridge in Great Britain during 1936. *J. Anim. Ecol.* **5**, 252–61.

——(1950). Partridge populations. II. *The Field* (London) 2 pp.

Miller, A. H. and Stebbins, R. C. (1964). *The lives of desert animals in Joshua Tree National Monument*. University of California Press, Berkeley.

Milon, P., Petter, J.-J., and Randriansola, G. (1973). Oiseaux. *Faune Madagascar* **35**, 1–263.

Monroe, B. L. J. (1968). A distributional survey of the birds of Honduras. American Ornithologists' Union, *Ornithol. Monogr.* No. 7.

Moreau, R. E. (1963). The migration of the quail. In *The birds of the British Isles* (ed. D. A. Bannerman) Vol. 12, pp. 374–5. Oliver & Boyd, Edinburgh.

—— and Wayre, P. (1968). On the Palearctic quails. *Ardea* **56**, 209–27.

Moroika, H. (1957). The Hainan tree-partridge *Abrorophila ardens*. *Ibis* **99**, 334–6.

Morris, D. (1951). The courtship of pheasants. *Zoo Life* **12**, 8–13.

Morrison, J. A. and Lewis, J. C., eds. (1972). *Proceedings of the First National Bobwhite Quail Symposium*. Oklahoma State University, Stillwater.

Mukherjee, A. K. (1966). The extinct and vanishing birds and mammals of India. *Bull. Ind. Mus.* **1**, 7–41.

Muller, K. (1966). A further note on the black or common francolin (*Francolinus francolinus*). *Avic. Mag.* **72**, 130.

—— (1969). Raising the crested wood partridge at the National Zoological Park (*Rollulus roulroul*). *Avic. Mag.* **75**, 9–11.

Murphy, D. A. and Baskett, T. S. (1952). Bobwhite mobility in central Missouri. *J. Wildl. Manage.* **16**, 498–510.

Newman, K. (1983). *Newman's birds of southern Africa*. Macmillan South Africa, Johannesburg.

Niekerk, J. H. van (1983). Observations on courtship in Swainson's francolin. *Bokmakierie* **35**, 90–2.

Nygren, L. R. (1963). A contribution toward the bibliography of *Alectoris* partridges. Utah Cooperative Wildlife Research Unit, Logan; Special Research Report No. 9, pp. 1–16.

Ogilvie-Grant, W. R. (1893). *Catalogue of the game-birds ... in the collection of the British Museum* Vol. 22. British Museum (Natural History), London.

—— (1896–7). *A hand-book to the game-birds* (2 vols). Lloyds Natural History, London.

Ohmart, R. B. (1967). Comparative molt and pterylography in the quail genera *Callipepla* and *Lophortyx*. *Condor* **69**, 535–48.

Oliver, W. R. B. (1955). *New Zealand birds* (2nd ed.). A. H. & A. H. Reed, Wellington.

Olson, S. L. (1974). *Telecrex* restudied: a small Eocene guineafowl. *Wilson Bull.* **86**, 246–50.

Ormiston, J. H. (1966). The food habits, habitat, and movements of mountain quail in Idaho. M.S. thesis, University of Idaho.

Ottinger, M. A. and Brinkley, H. J. (1979). The ontogeny of crowing and copulatory behavior in Japanese quail (*Coturnix c. japonica*). *Behav. Process.* **4**, 43–51.

Paludan, K. (1954). Agerhønens ynglesaeson 1953. *Dansk Vildt.* **3**, 1–20.

—— (1963). Partridge markings in Denmark. *Danish Rev. Game Biol.* **4**, 25–60.

Paynter, R. J., Jr. (1955). The ornithogeography of the Yucatan Peninsula. *Bull. Peabody Mus.* (New Haven) **9**, 1–347.

Pepin, D. (1984). (Partner-changing in the red-legged partridge *Alectoris rufa*.) *Oiseau* **54**, 293–304. (In French.)

—— (1985). Morphological characteristics and sex classification of red-legged partridges. *J. Wildl. Manage.* **49**, 228–37.

Peters, J. L. (1934). *Check-list of birds of the world* Vol. 2. Harvard University Press, Cambridge.

Peterson, S. R. and Nelson, L., Jr., eds. (1980). *Proceedings of Perdix II*. Contribution No. 211 to the Gray Partridge Workshop, 18–20 March, Moscow, Idaho. Forest, Wildlife, and Range Experiment Station, Moscow, Idaho.

Petrides, G. A. (1942). Age determination in American gallinaceous game birds. *Trans. 7th N. Am. Wildl. Conf.* **7**, 308–28.

—— and Nestler, R. B. (1943). Age determination in juvenal bobwhite quail. *Am. Midl. Nat.* **30**, 774–82.

Phelps, J. E. (1955). The adaptability of the Turkish chukar partridge (*Alectoris graeca* Meisner) in central Utah. M.S. thesis, Utah State Agricultural College, Logan.

Pinshow, B., Degen, A. A., and Alkon, P. U. (1984). Water intake, existence energy, and responses to water deprivation in the sand partridge *Ammoperdix heyi* and the chukar *Alectoris chukar*: two phasianids of the Nagev Desert. *Physiol. Zool.* **56**, 128–36.

Pittman-Robertson Reports. Quarterly reports of P. R.-funded research, US Fisheries and Wildlife Service, Washington. (mimeo.)

Podor, M. (1984). (The male red-legged partridge *Alectoris rufa* helping with incubation.) *Alauda* **52**, 70. (In French.)

Portal, M. (1924). Incubation by male red-legged partridge. *Br. Birds* **17**, 31–16.

Potts, G. R. (1973). Factors governing the chick survival rate of the grey partridge. Proceedings of the 10th International Congress on Game Biology, pp. 85–96.

—— (1986). *The partridge*. Collins, London.

Priolo, A. (1984). Variabilita in *Alectoris graeca* e descrizione di *A. graeca orlandoi*, subsp. nova degli Appenini. *Rev. Ital. Ornithol.* **54**, 45–76.

Prososki, A. E. (1970). Social behavior and adult vocalizations of some *Colinus* and *Lophortyx* hybrids. M.S. thesis, University of Nebraska, Lincoln.

Raitt, R. J., Jr. (1960). Breeding behavior in a population of California quail. *Condor* **62**, 284–92.

—— (1961). Plumage development and molts of California quail. *Condor* **63**, 294–303.

—— and Genelly, R. E. (1964). Dynamics of a population of California quail. *J. Wildl. Manage.* **28**, 127–41.

—— and Ohmart, R. D. (1966). Annual cycle of reproduction and molt in Gambel quail of the Rio Grande valley, southern New Mexico. *Condor* **68**, 541–61.

—— and —— (1968). Sex and age ratios in Gambel quail of the Rio Grande valley, southern New Mexico. *Southwest. Nat.* **13**, 27–33.

Rand, A. L. (1936). The distribution and habits of Madagascan birds. *Bull. Am. Mus. Nat. Hist.* **72**, 143–499.

—— (1942). Results of the Archbold Expeditions. No. 43. Birds of the 1938–1939 New Guinea expedition. *Bull. Am. Mus. Nat. Hist.* **79**, 425–515.

—— and Gilliard, E. T. (1967). *Handbook of New Guinea birds*. Weidenfeld and Nicolson, London.

Ratti, J. T., Smith, L. M., Hupp, J. W., and Laake, J. L. (1983). Line transect estimates of density and winter mortality of gray partridge. *J. Wildl. Manage.* **47**, 1088–95.

Rea, A. M. (1973). The scaled quail (*Callipepla squamata*) of the Southwest: systematic and historical considerations. *Condor* **97**, 322–9.

Reinhard, R. (1981). Bemerkungen zur Schwarzwachtel *Melanoperdix nigra* (Vigors 1829). *Gefiederte Welt* **105**, 201–3.

Ricci, J.-C. (1983). (Two cases of male red-legged partridges *Alectoris rufa* helping with incubation.) *Alauda* **51**, 64–5. (In French.)

Ridgway, R. and Friedmann, H. (1946). The birds of North and Middle America. *US National Museum Bulletin* Vol. 50, Part X. Smithsonian Institution, Washington D.C.

Riley, J. H. (1938). Birds from Siam and the Malay Peninsula in the United States National Museum collected by Drs Hugh M. Smith and William L. Abbott. *Bull. US Nat. Mus.* **172**, 1–581.

Ripley, S. D. (1952). Vanishing and extinct bird species of India. *J. Bombay Nat. Hist. Soc.* **50**, 902–6.

—— (1964). A systematic and ecological study of birds of New Guinea. *Bull. Peabody Mus.* (New Haven) **19**, 1–87.

Ripley, T. H. (1958). Ecology, population dynamics and management of the bobwhite quail, *Coturnix virginianus marilandicus* (L.) in Massachusetts, Ph.D. dissertation, Virginia Polytechnic Institute.

Robbins, G. E. S. (1981). *Quail: their breeding and management*. World Pheasant Association and Payn Essex Printers, Sudbury.

—— (1984). *Partridges: their breeding and management*. Boydell & Brewer, London.

Robinson, H. C. and Kloss, C. B. (1924). On a large collection of birds chiefly from West Sumatra, made by Mr E. Jacobson. *J. Fed. Malay States Mus.* **11**, 89–347.

—— and Chasen, F. N. (1936). *The birds of the Malay Peninsula*, Vol. III, *Sporting birds: birds of the shore and estuaries*. Witherby, London.

Robinson, J. E. (1980). The photoperiodic control of seasonal reproduction in Japanese quail. Ph.D. dissertation, University of Washington.

Robinson, T. S. (1957). The ecology of bobwhites in south-central Kansas. University of Kansas Museum of Natural History and State Biological Survey, Miscellaneous Publication No. 15, pp. 1–84.

Romero-Zambrano, H. (1983). Revision del status zoogeographico y redescription de *Odontophorus strophium* (Gould) (Aves: Phasianidae). *Caldasia* **13**, 777–86.

Roseberry, J. L. and Klimstra, W. D. (1977). Some aspects of the dynamics of a hunted bobwhite population. In *Proceedings of the First National Bobwhite Quail Symposium* (ed. J. A. Morrison and J. C. Lewis) pp. 268–81. Oklahoma State University, Stillwater.

—— and —— (1983). *Population ecology of the bobwhite*. Southern Illinois University Press, Carbondale.

Rosene, W. (1969). *The bobwhite quail: its life and management*. Rutgers University Press, New Brunswick.

Rowley, J. S. (1966). Breeding records of birds of the Sierra Madre del Sur, Oaxaca, Mexico. *Proc. Western Found. Vert. Zool.* **1**, 107–204.

Russell, S. M. (1964). A distributional survey of the birds of Honduras. *Ornithol. Monogr.* No. 7.

Rutgers, A. and Norris, K. A., eds. (1970). *Encyclopedia of aviculture* Vol. 1. Blandford, London.

Sachs, B. D. (1966). Photoperiodic control of sexual behavior and physiology in the male quail (*Coturnix coturnix japonica*). Ph.D. dissertation, University of California, Berkeley.

Saunders, G. B. (1950). The game birds and shorebirds of Guatemala. US Fisheries and Wildlife Service, Special Scientific Report (Wildlife) No. 5, pp. 3–98.

Schäfer, E. (1934). Zur Lebenweise der Fasanen des chinesisch-tibetischen Grenzlandes. *J. f. Ornith.* **82**, 487–509.

Schauensee, R. M. de (1984). *The birds of China*. Smithsonian Institution Press, Washington D.C.

Schemnitz, S. D. (1961). Ecology of the scaled quail in the Oklahoma panhandle. *Wildl. Monogr.* No. 8.

—— (1964). Comparative ecology of bobwhite and scaled quail in the Oklahoma panhandle. *Am. Midl. Nat.* **71**, 429–33.

Schitoskey, F., Jr., Schitoskey, E. C., and Talent, L. G., eds. (1982). *Proceedings of the Second National Bobwhite Quail Symposium*. Oklahoma State University, Stillwater.

Schleidt, W. M., Yakalis, G., Donnelly, M., and McGarry, J. (1984). A proposal for a standard ethogram, exemplified by an ethogram of the bluebreasted quail (*Coturnix chinensis*). *Z. Tierpsychol.* **64**, 193–220.

Schodde, R., van Tets, G. F., Champion, C. R., and Hope, G. S. (1975). Observations on birds at glacial altitudes on the Carstenz Massif, western New Guinea. *Emu* **75**, 65–72.

Schönwetter, M. (1967). *Handbuch der Oologie* Lieferung 8. Akademie-Verlag, Berlin.

Schulz, J. W. (1977). Population dynamics of Hungarian partridge in north central North Dakota: 1946–1975 (pp. 133–45 in Kobriger 1977).

—— (1980). Gray partridge winter movements and habitat use in north-central North Dakota. In *Proceedings of Perdix II* (ed. S. P. Peterson and L. Nelson Jr.) pp. 147–55. Contribution No. 211 to the Gray Partridge Workshop, 18–20 March, Moscow, Idaho. Forest, Wildlife, and Range Experiment Station, Moscow, Idaho.

Schwartz, C. W. and Schwartz, E. R. (1950). The California quail in Hawaii. *Auk* **67**, 1–38.

Scott, T. G. (1985). *Bobwhite thesaurus*. International Quail Foundation, Edgefield, S.C. (c. 3000 references.)

Sharp, P. J. *et al.* (1986). Photoperiodic and endocrine control of seasonal breeding in grey partridge. *J. Zool.* **209**, 187–200.

Sibley, C. G. and Ahlquist, J. E. (1972). A comparative study of the egg white proteins of non-passerine birds. *Bull. Peabody Mus. Nat. Hist.* (New Haven) **39**, 1–276.

—— and —— (1985). The relationships of some groups of African birds, based on comparisons of the genetic material, DNA. In Schuchmann, K.-L. (ed.), *Proceedings of the International Symposium on African Vertebrates*. Zoolische Forschung and Museum Alexander Koenig, Bonn.

—— and —— (1986). Phylogeny of non-passerine birds, based on DNA comparisons. Abstract of paper presented at the XIXth International Ornithological Congress, 22–29 June, Ottawa.

Skutch, A. F. (1947). Life history of the marbled wood-quail. *Condor* **49**, 217–32.

Slud, P. (1964). The birds of Costa Rica, distribution and ecology. *Bull. Am. Mus. Nat. Hist.* **128**, 1–430.

Smith, D. S. and Cain, J. R. (1984). Criteria for age classification of juvenile scaled quail. *J. Wildl. Manage.* **48**, 187–91.

Smith, R. H. (1961). Age classification of the chukar partridge. *J. Wildl. Manage.* **25**, 84–6.

Smythies, B. E. (1968). *The birds of Borneo*. Oliver & Boyd, Edinburgh.

Snow, D. W., ed. (1978). *An atlas of speciation in African non-passerine birds*. British Museum (Natural History), London.

Southwood, T. R. E. and Cross, D. J. (1967). The ecology of the partridge. II. The role of pre-hatching influences. *J. Anim. Ecol.* **36**, 557–62.

—— and —— (1969). The ecology of the partridge. III. Breeding success and the abundance of insects in natural habitats. *J. Anim. Ecol.* **38**, 497–509.

Sowls, L. K. (1960). Results of a banding study of Gambel's quail in southern Arizona. *J. Wildl. Manage.* **24**, 185–90.

Spana, S. (1978). Nuovi ibridi naturali *Alectoris rufa rufa* × *Alectoris graeca saxatilis* (Bechstein) sulle Alpi marittime e relative considerazoini tassonomische. *Estr. Daglo Annali del Museo Civico de Storia Nat. di Genova* **82**, 154–64. (English summary.)

—— and Csermely, D. (1980). (Parental care by the male *Alectoris rufa* in captivity.) *Avocetta* **4**, 31–4. (In Italian.)

Stanford, J. A. (1972a). Second broods in bobwhite quail. In *Proceedings of the First National Bobwhite Quail Symposium* (ed. J. A. Morrison and J. C. Lewis) pp. 21–7. Oklahoma State University, Stillwater.

—— (1972b). Bobwhite quail population dynamics: relationship of weather, nesting, production patterns, fall population characteristics and harvest in Missouri quail. In *Proceedings of the First National Bobwhite Quail Symposium* (ed. J. A. Morrison and J. C. Lewis) pp. 115–39. Oklahoma State University, Stillwater.

Stettner, L. J., Garreffa, L. F., and Missakian, E. (1971). Factors affecting monogamous behaviour in the bobwhite quail (*Colinus virginianus*). *Commun. Behav. Biol.* **6**, 137–64.

Stiver, S. J. (1984). Himalayan snowcocks—Nevada's newest upland game. Transactions of the 1984 meeting of the California–Nevada section of the Wildlife Society, pp. 55–8.

Stock, A. D. and Bunch, T. D. (1982). The evolutionary implications of chromosome banding pattern homologies in the bird order Galliformes. *Cytogenet. Cell Genet.* **34**, 136–48.

Stoddard, H. L. (1931). *The bobwhite quail: its habits, preservation and increase*. Charles Scribner's Sons, New York.

Stokes, A. W. (1961). Voice and social behavior of the chukar partridge. *Condor* **63**, 111–27.

—— (1963). Agonistic and sexual behavior in the chukar partridge (*Alectoris graeca*). *Anim. Behav.* **11**, 121–34.

—— (1967). Behavior of the bobwhite, *Colinus virginianus*. *Auk* **84**, 1–33.

—— and Williams, H. W. (1968). Antiphonal calling in quail. *Auk* **85**, 83–9.

—— and —— (1971). Courtship feeding in gallinaceous birds. *Auk* **88**, 543–59.

—— and —— (1972). Courtship feeding calls in gallinaceous birds. *Auk* **89**, 177–80.

Stronach, B. (1966). The feeding habits of the yellow-necked francolin (*Pternistis leuscoscepus*) in northern Tanzania. *E. Afr. Wildl. J.* **4**, 76–81.

Swarth, H. S. (1909). Distribution and molt of the Mearns quail. *Condor* **11**, 39–43.

Summers, D. D. B. (1972). Pterylography, plumage development and moult of Japanese quail in captivity. *Ibis* **114**, 79–88.

Tait, H. D. (1968). Index to Federal Aid publications in sport, fish and wildlife restoration, and selected Cooperative Research Project Reports, March 1968. US Department of the Interior, Washington D.C.

Taka-Tsukasa, N. (1967). *The birds of Nippon*. Maruzen, Tokyo.

Tanaka, K., Mather, F. B., Wilson, W. O., and McFarland, L. Z. (1965). Effect of photoperiods on early growth of gonads and on potency of gonadotrophins of the anterior pituitary in *Coturnix* quail. *Poult. Sci.* **44**, 662–5.

Thompson, D. R. and Kabat, C. (1950). The wing molt of the bobwhite. *Wilson Bull.* **62**, 20–31.

Thornhill, J. W. (1981). Captive breeding of jungle bush quail *Perdicula asiatica*. *Wld Pheasant Ass. J.* **6**, 53–9.

Tobler, S. L. and Lewis, J. C. (1972). A partial bibliography of the bobwhite quail. In *Proceedings of the First National Bobwhite Quail Symposium* (ed. J. A. Morrison and J. C. Lewis) pp. 377–90. Oklahoma State Unviersity, Stillwater. (*c*. 900 references.)

Todd, W. E. C. (1920). A revision of the genus *Eusychortyx*. *Auk* **37**, 189–220.

Toschi, A. (1959). *La Quaglia: Vita, allevamento*. Cassia, Bologna (not seen).

Traylor, M. A. (1960). *Francolinus schlegelii* Hueglin in Cameroon. *Bull. Br. Ornithol. Club* **80**, 86–8.

Urban, E. K., Fry, C. H., and Keith, S. (1986). *The birds of Africa* Vol. II. Academic Press, London.

Vaurie, C. (1965). *The birds of the Palearctic fauna. Non-passeriformes.* Witherby, London.

Verheyen, R. (1956). Contribution à l'anatomie et à la systématique des Galliformes. *Inst. R. Sci. Nat. Belg.* **32**(42), 1–24.

Waggerman, G. L. (1968). The critical period for development of primary social habits among bobwhites. M.S. thesis, Oklahoma State University, Stillwater.

Wallmo, O. (1954). Nesting of Mearns quail in southeastern Arizona. *Condor* **56**, 125–8.

—— (1956). *Ecology of scaled quail in west Texas.* Texas Game and Fish Commission, Texas.

Warner, D. W. (1959). The song, nest, eggs, and young of the long-tailed partridge. *Wilson Bull.* **71**, 307–12.

—— and Harrell, B. E. (1957). The systematics and biology of the singing quail, *Dactylortyx thoracicus*. *Wilson Bull.* **69**, 123–48.

Watson, G. E. (1962a). Sympatry in Palearctic *Alectoris* partridges. *Evol.* **16**, 11–19.

—— (1962b). Three sibling species of *Alectoris* partridges. *Ibis* **104** 353–67.

—— (1962c). Molt, age determination and annual cycle in the Cuban bobwhite. *Wilson Bull.* **74**, 28–42.

Weaver, H. R. and Haskell, W. L. (1968). Age and sex determination of the chukar partridge. *J. Wildl. Manage.* **32**, 46–50.

Weigand, J. P. (1977). A technique for sexing live Hungarian partridge in the field. In *Proceedings of Perdix I, Hungarian partridge workshop*, Minot, N. Dakota, pp. 150–4. N. Dakota Chapter Wildlife Society and N. Dakota Game and Fish Department, Bismarck.

—— (1980). Ecology of the Hungarian partridge in north-central Montana. Wildl. Monogr. No. 74.

Welch, G. and Welch, H. (1984a). The rare Tadjoura francolin found in East Africa. *Wld Pheasant Ass. News* **6**, 8–12.

—— and —— (1984b). Djibouti expedition 1984: a preliminary survey of *Francolinus ochropectus* and the birdlife of the country. Published by the authors, Knottingley.

Wennrich, G. (1983). Hattung und Zucht der Frankolinwachtel (*Perdicula asiatica*). *Gefiederte Welt* **106**, 321–4.

Westerskov, K. (1965). Winter ecology of the partridge (*Perdix perdix*) in the Canadian prairie. *Proc. N.Z. Ecol. Soc.* **12**, 23–30.

—— (1966). Winter food and feeding habits of the partridge (*Perdix perdix*) in the Canadian prairie. *Can. J. Zool.* **44**, 303–22.

Wetherbee, D. K. (1961). Investigations into the life history of the common coturnix. *Am. Midl. Nat.* **65**, 168–86.

Wetmore, A. (1965). The birds of the Republic of Panama. Pt. I. Tinamidae (tinamous) to Rhynchopidae (skimmers). *Smithsonian Misc. Coll.* **150**, 1–483.

Whitley, H. (1927). The breeding of the crowned wood partridge (*Rollulus roulroul*). *Avic. Mag.* Ser. **4**(5), 253–6.

Wikramanayake, E. B. (1969). Some rare and vanishing birds of Ceylon. *Loris* **11**, 374–6.

Willard, L. E. (1973). Plumage changes and growth rates of Hungarian partridge (*Perdix perdix*) chicks. M.S. thesis, Western Illinois University, Macomb.

Williams H. E. and Stokes, A. W. (1965). Factors affecting the incidence of rally calling in the Chukar partridge. *Condor* **67**, 31–43.

Williams, H. W. (1969). Vocal behavior of adult California quail. *Auk* **84**, 631–59.

Wilson, K. A. and Vaughn, E. A. (1944). *The bobwhite quail in eastern Maryland.* Maryland Game and Inland Fish Commission, Baltimore.

Wolff, S. W. and Milstein, P. le S. (1976). *A guide to the terrestrial gamebirds of the Transvaal.* Transvaal Provincial Administration, Pretoria.

Wolters, H. E. (1975–82). *Die Vogelarten der Erde.* Paul Parey, Hamburg.

Yocom, C. F. (1943). The Hungarian partridge, *Perdix perdix* (L.), in the Palouse region, Washington. *Ecol. Mongr.* **13**, 167–201.

—— and Harris, S. W. (1952). Food habits of mountain quail (*Oreortyx picta*) in eastern Washington. *J. Wildl. Manage.* **17**, 204–7.

Yoho, N. S. and Dimminck, R. W. (1972). Habitat utilization by bobwhite quail in winter. In *Proceedings of the First National Bobwhite Quail Symposium* (ed. J. A. Morrison and J. C. Lewis) pp. 90–9. Oklahoma State University, Stillwater.

Zorig, G. and Bold, A. (1983). (Breeding biology of the Altai snowcock.) *Ornitologiya* **18**, 109–11. (In Russian, not seen.)

Index

This index includes all generic and specific names considered as valid taxa in this book, and also all English vernacular names as used herein. Commonly used alternative English vernacular names for species and subspecies are provided for cross-reference, but all non-English vernacular names and taxa considered as synonyms are excluded. Complete page citations are shown only under the English names used in the text. Principal accounts of each genus and species are indicated by *italics*; maps or drawings are indicated by **bold face**.

acacia francolin 127, 128, **163**, **168**, *167–9*, Plate 76
adansonii, Coturnix 192, 198, 204
adspersus, Francolinus 127, 129, 160
afer, Francolinus 126, 128, 134, 138, 139
African blue quail 22, 192, 198, *202–6*, **203**, Plate 98
African harlequin quail 192, 193, **197**, *197–9*, Plate 93
African painted quail, *see* African blue quail
africanus, Francolinus 127, 128, 162, 165, 166
ahanta francolin 127, 129, *147–9*, **148**, Plate 59
ahantensis, Francolinus 127, 129, 147
albogularis, Francolinus 127
Alectoris 4, 6, **8**, 10, 11, 18, 29, 31, 32, **33**, 38, 96, 197, *111–21*, 124, 128, 173, 183, 195, 241, 245
Altai snowcock 103, **104**, **105**, *109–10*, Plate 38
altaicus, Tetraogallus 109
Ammoperdix 3, 6, 8, 96, *122–5*, 241
Anderson's bamboo-partridge, *see* mountain bamboo-partridge
Annamese hill-partridge, *see* scaly-breasted hill-partridge
Anurophasis 5, 6, **8**, 96, *206–7*
Arabian red-legged partridge *111–14*, **112**, **113**, Plate 39
Arabian see-see partridge, *see* sand partridge
Arborophila 6, 7, **8**, 15, 19, 96, 97, 191, *215–34*, 235–8
Archer's grey-winged francolin, *see* acacia francolin
ardens, Arborophila 215, 216, 224, 225
argoonda, Perdicula 208, 210
Asian blue quail 10, 11, 12, **35–7**, *202*, 203, Plate 97
Asian migratory quail 13, 21, 22, 26, 27, 31, 192, **193**, *195–7*, Plate 92
asiatica, Perdicula 208, 210, 211
atrifrons, Odontophorus 71, 79, 83
atrogularis, Arborophila 215, 216, 223, 225

balliviani, Odontophorus 71, 83, 85
Ballivian's partridge, *see* stripe-faced wood quail
Bambusicola 3, 6, **8**, 97, 243, *244–6*
banded quail, *see* barred quail, tawny-faced quail

barbara, Alectoris 30, 111, 118, 119, 120, 121
Barbary partridge 11, 12, 111, *119–20*, **119**, Plate 44
barbatus, Dendrortyx 42, 43
bare-throated francolin, *see* red-necked francolin
bar-backed hill-partridge, *see* brown-breasted hill-partridge
bare-throated hill-partridge 215, **226**, *226–8*, Plate 114
barred quail 17, 18, **48**, *47–8*, Plate 4
bearded partridge, *see* Daurian partridge
bearded tree-quail (or wood-partridge) 42, *43–5*, Plate 2
Benson quail, *see* elegant quail
bicalcarata, Galloperdix 247, 249
bicalcaratus, Francolinus 127, 129, 150, 152, 153, 158
black-breasted bobwhite, *see* northern bobwhite
black-breasted quail 10, 22, 192, 193, **197**, *199–200*, Plate 94
black-breasted wood-quail 71, *78–9*, **80**, 81 Plate 20
black-eared wood-quail, *see* rufous-fronted wood-quail
black francolin (or partridge) 15, 19, 126, 128, *129–32*, Plate 48
black-fronted wood-quail 71, **80**, **81**, 83, Plate 22
black-headed bobwhite, *see* northern bobwhite
black-throated bobwhite (or quail) 18, 60, *64–6*, **65**, 69, Plate 11
black wood-partridge **186**, *190–1*, Plate 90
blue-breasted quail, *see* Asian blue quail
blue quail, *see* African, Asian blue quails, and scaled quail
bobwhite, *see* northern bobwhite
Boehm's francolin, *see* red-necked francolin
Bolivian partridge, *see* rufous-breasted wood-quail
Borneo hill-partridge 215, **229**, *231*, Plate 116
Boulton's hill-partridge, *see* Sichuan hill-partridge
brown-breasted hill-partridge 215, **226**, *228–9*, Plate 115
brown quail 192, **197**, *201–2*, Plate 96
brunneopectus, Arborophila 215, 225, 226, 228, 231
Buckley's partridge, *see* marbled wood-quail

buffy-crowned tree-quail **42**, *45–6*, Plate 3
buffy-fronted wood-partridge, *see* buffy-crowned tree-quail
Burmese francolin, *see* Chinese francolin
Bütterkofer's francolin, *see* acacia francolin

Cabanis' francolin, *see* yellow-necked francolin
California quail 11–15, 17–27, 29, 30, 31, 49, 51, 54, *55–6*, Plate **57**, Plate 9
californica, Callipepla 51, 56
Callipepla 6, **7**, 11, 12, 19, 42, 50, *51–9*
Caloperdix 5, 6, **8**, 9, 97, *235–6*
cambodiana, Arborophila 215, 230, 231
Cameroon Mountain francolin **145**, *146–7*, Plate 58
camerunensis, Francolinus 127, 129, 146
Campbell's hill-partridge, *see* bare-throated hill-partridge
Cape Francolin 15, 127, 129, **151**, *159–60*, Plate 70
capensis, Francolinus 127, 129, 159
capuiera, Odontophorus 71, 73
capuiera wood-quail, *see* spot-winged wood-quail
Caspian snowcock 104, **103**, **105**, *106–7*, Plate 35
caspius, Tetraogallus 103, 106, 109
castaneicollis, Francolinus 126, 129, 143, 144
Caucasian snowcock *103–6*, **104**, **105**, Plate 34
caucasicus, Tetraogallus 103, 109
Ceara partridge, *see* spot-winged wood-quail
Ceylon spurfowl 15, **242**, 247, *249–50*, Plate 127
charltonii, Arborophila 215, 232
Chestnut-bellied hill-partridge, *see* Javan hill-partridge
chestnut-breasted hill-partridge, *see* scaly-breasted hill-partridge
chestnut-eared partridge, *see* rufous-fronted wood-quail
chestnut-headed hill-partridge 215, **229**, *230–1*, Plate 106
chestnut-naped francolin 126, 129, *143–4*, **145**, Plate 56
chestnut-throated partridge, *see* chestnut wood-quail
chestnut wood-quail 75, **76**, *76–7*, Plate 17
chinensis, Coturnix 192, 202, 204, 205

Index

Chinese bamboo-partridge **242**, 244, *245–6*, Plate 124
Chinese francolin 126, 128, **130**, *133–4*, Plate 50
Chinese painted quail, *see* Asian blue quail
Chiriqui partridge, *see* marbled wood-quail
chukar, *Alectoris* 29, **33**, 111, 114–16, 120, 121
Chukar partridge 11, 12, 14, 18, 20, 31, **90**, *114–16*, Plate 42
cinctus, Rhynchortyx 94
clappertoni, Francolinus 127, 129, 153, 155, 157, 158
Clapperton's francolin 127, 139, **151**, *153–4*, Plate 65
close-barred francolin, *see* red-billed francolin
Colinus 5, 6, **7**, 11, 12, 41, 48, 50, *60–70*
collared hill-partridge 215, **216**, **220**, *221*, Plate 108
columbianus, Odontophorus 71, 79, 82
common bobwhite, *see* northern bobwhite
common quail, *see* European migratory quail
common hill-partridge, *see* necklaced hill-partridge
coqui francolin 127, 128, 161, *171–2*, **171**, 179, Plate 79
coqui, Francolinus 127, 128, 171, 172, 173, 174
coromandelica, Coturnix 192, 193, 197, 199, 201, 204
coturnix, Coturnix 192, 193, 196, 197, 202, 206, 207
Coturnix 3, 6, **8**, 9, 11, 12, 19, 27, 38, 96, 189, 195, 198, 200, 210
cottontop quail, *see* scaled quail
Coyolcos bobwhite, *see* northern bobwhite
Cranch's francolin, *see* red-throated francolin
crested bobwhite (or colin) 60, **65**, **67**, *68–70*, Plate 13
crested francolin 127, 129, **160**, *160–1*
crested quail, *see* crested bobwhite, elegant quail
crested wood-partridge **237**, *239–40*, Plate 121
crimson-headed wood-partridge **237**, *237–8*, Plate 120
cristatus, Colinus 60, 64, 66, 68
crowned wood-partridge, *see* crested wood-partridge
crudigularis, Arborophila 215, 216, 224
Cuban bobwhite, *see* northern bobwhite
Cyrtonyx 5, 6, **7**, 11, 41, 88, *90–3*, 95

Dactylortyx 5, 6, **7**, 11, 41, 95, *87–9*
dark-backed wood-quail **75**, **76**, *77–8*, Plate 18
Daurian partridge 180, **181**, *183–4*, Plate 86
dauuricae, Perdix 180, 183, 185
davidi, Arborophila 215, 229
David's hill-partridge, *see* orange-necked hill-partridge

delagorguei, Corturnix 192, 193, 197, 200, 204
Dendrortyx 5, 6, **7**, 11, 41, *42–6*, 48
desert quail, *see* Gambel's quail
dialeucos, Odontophorus 71, 79, 80, 83
Dickey's quail, *see* spot-bellied bobwhite
Djibouti francolin, *see* pale-bellied francolin
Dorst's francolin, *see* pale-bellied francolin
double-spurred francolin 127, 129, **151**, *150–2*, Plate 62
Douglas quail, *see* elegant quail
douglasii, Callipepla 51, 53
Duida partridge, *see* marbled wood-quail

eastern bobwhite, *see* northern bobwhite
Egyptian quail, *see* European migratory quail
elegant quail 17, 18, 51, **52**, *53–4*, Plate 7
erckelii, Francolinus 126, 129, 141
Erckel's francolin, 126, 129, *141–2*, **145**, Plate 55
erythrops, Odontophorus 71, 74, 77, 81, 85, 86
erythrorhyncha, Perdicula 208, 211
European migratory quail 10–12, 14, 15, 22, 25, 26, 192, **193**, *193–5*, 198, Plate 91
Eyton's hill-partridge, *see* scaly-breasted hill-partridge

fasciatus, Philortyx 47
ferruginous wood-partridge **232**, *235–6*, Plate 119
finschi, Francolinus 127, 128, 162, 165, 170
Finsch's francolin 127, 128, **163**, *170–1*, Plate 78
Fischer's francolin, *see* Hildebrandt's francolin
Florida bobwhite, *see* northern bobwhite
Francolinus 6, **8**, 9, 19, 97, 124, *126–79*, 187, 195, 236, 246
francolinus, Francolinus 126, 129, 133, 134
French partridge, *see* red-legged partridge
fulvous-breasted quail, *see* Gambel's quail
Fytche's bamboo-partridge, *see* mountain bamboo-partridge
fytchii, Bambusicola 244

Galloperdix 3, 6, **8**, 97, 243, 245, *247–50*
gambelii, Callipepla 51, 54, 57
Gambel's quail 12, 14, 17–24, 30, 31, 51, *54–5*, 57, Plate 8
gariep francolin, *see* acacia francolin
Gedge's francolin, *see* Clapperton's francolin
gingica, Arborophila 215, 216, 221, 223
Godman's bobwhite, *see* northern bobwhite
gorgeted wood-quail 71, *79–80*, **80**, **81**, Plate 21
graeca, Alectoris 111, 114, 117, 118, 120, 121

Grant's francolin, *see* crested francolin
Grayson's bobwhite, *see* northern bobwhite
green-legged hill-partridge, *see* scaly-breasted hill-partridge
green wood-partridge, *see* crested wood-partridge
grey-breasted francolin (or spurfowl) 15, 126, 128, **135**, *139–9*, Plate 43
grey francolin 10, 11, 13–16, 18–26, 32, *180–3*, 184, Plate 83
grey partridge (*see also* grey francolin) 10, 11, 13–16, 18–24, 25, 26, 32, **34**, **80**, **181**, *180–3*, Plate 85
grey quail, *see* stubble quail, European migratory quail
grey-striped francolin 127, 129, **148**, 149, *150*, Plate 61
grey-winged (or greywing) francolin 11, 19, 20, 127, 128, *166–7*, 169, Plate 75
griseogularis, Ammoperdix 3, 122, 125
griseostriatus, Francolinus 127, 129
Guatemalan bobwhite, *see* northern bobwhite
Guatemalan partridge, *see* buffy-crowned tree-quail
Guianan partridge, *see* marbled wood-quail
gujanensis, Odontophorus 71, 72, 81, 83, 85
gularis, Francolinus 128, 129, 178
guttatus, Odontophorus 71, 85

Haematortyx 5, 6, **8**, 9, 96, *237–8*
Hainan hill-partridge, *see* white-eared hill-partridge
handsome francolin 126, 129, **145**, *145–6*, Plate 63
harlequin quail, *see* African harlequin quail, Montezuma quail
hartlaubi, Francolinus 127, 129, 154–6
Hartlaub's francolin, 127, 129, **151**, *156–7*, Plate 63
harwoodi, Francolinus 127, 129, 152, 153, 154, 157, 159, 160
Harwood's francolin 127, 129, **151**, *157–8*, Plate 68
heyi, Ammoperdix 122, 124
highland partridge, *see* buffy-crowned tree-quail
hildebrandti, Francolinus 127, 129, 144, **151**, *154–5*
Hildebrandt's francolin, 127, 129, 144, **151**, *154–5*, Plate 66
Himalayan snowcock **1**, 103, **104**, **105**, *107–8*, 109, Plate 36
Himalayensis, Tetraogallus 103, 106, 107, 109
hodgsoniae, Perdix 180, 184
Honduran partridge, *see* rufous-fronted wood-quail
Horvaths's quail, *see* crested bobwhite
Hubbard's francolin, *see* coqui francolin
Humboldt's francolin, *see* red-necked francolin
Hungarian partridge, *see* grey partridge
hyperythra, Arborophila 215, 231
hyperythrus, Odontophorus 71, 76, 78, 79

Index

icterorhynchus, Francolinus 127, 129, 152, 154, 158
Indian grey partridge, *see* grey francolin
Indian hill-partridge, *see* necklaced hill-partridge
Indian mountain-quail 214–15, Plate 104

jacksoni, Francolinus 126, 129, 144
Jackson's francolin 126, 129, *144–5*, **145**, 147, Plate 57
Japanese grey quail, *see* Asian migratory quail
Japanese rice quail, *see* Asian migratory quail
japonica, Coturnix 192, 193, 195, 200
Jaumave bobwhite, *see* northern bobwhite
Javan hill-partridge 215, **220**, *225–6*, 226, Plate 113
javanica, Arborophila 215, 216, 221, 225, 228
Johnston's francolin, *see* Hildebrandt's francolin
jungle bush-quail *208–10*, **209**, Plate 100

king quail, *see* Asian blue quail
Kirk's francolin, *see* crested francolin

lathami, Francolinus 127, 129, 174, 175
Latham's forest francolin 15, 127, 128, 161, **174**, *174–5*, Plate 82
least bobwhite, *see* northern bobwhite
Leland's quail, *see* spot-bellied bobwhite
Lerwa 5, 6, **8**, 97, *98–9*, 207
lerwa, Lerwa 98
leucolaemus, Odontophorus 71, 78, 83
leucophrys, Dendrortyx 42, 45
leucopogon, Colinus 60, 64, 66, 68, 69
leucoscepus, Francolinus 126, 129, 138, 139
levaillantii, Francolinus 127, 128, 162, 165, 167, 169
levaillantoides, Francolinus 127, 128, 165, 167
Levaillant's (red-winged) francolin 11, 19, 20, 127, **163**, 166, **168**, *169–70*, Plate 77
long-billed wood-partridge (or partridge) 15, **186**, *186–7*, 228, Plate 88
longirostris, Rhizothera 186
long-legged colin, *see* tawny-faced quail
long-tailed tree-quail (or partridge) **42**, *42–3*, 44, Plate 1
long-toed quail (or partridge), *see* singing quail
lunulata, Galloperdix 247, 248, 250

macroura, Dendrortyx 42
madagarensis, Margaroperdix 188
Madagascar partridge *188–9*, **203**, Plate 89
magna, Alectoris 11, 114, 116
mandellii, Arborophila 215, 216, 219, 221, 223
Mandell's hill-partridge, *see* red-breasted hill-partridge

Manipur bush-quail 208, **212**, *212–13*, Plate 103
manipurensis, perdicula 218, 212, 214
marbled wood-quail 71, **72**, *72–3*, Plate 14
Margaroperdix 3, 5, 6, **8**, 96, *188–9*
masked bobwhite, *see* northern bobwhite
Massena quail, *see* Montezuma quail
Mearns' quail, *see* Montezuma quail
melanocephala, Alectoris 111, 119, 120
melanonotus, Odontophorus 71, 77, 78
Melanoperdix 5, 6, **8**, 9, 96, *190–1*
Messina quail, *see* European migratory quail
Mexican harlequin quail, *see* Montezuma quail
Mocquery's quail, *see* crested bobwhite
monorthonyx, Anurophasis 206
montane red-winged francolin 127, 128, *163–4*, **163**, **164**, Plate 73
Montezuma quail 14, 17, 18, **90**, *90–2*, Plate 28
montezumae, Cyrtonyx 90, 92, 93
moorland francolin, *see* montane red-winged francolin
mountain bamboo-partridge **242**, *244–5*, Plate 123
mountain quail (*see also* Indian mountain-quail) 14, 17–20, 24, **47**, *49–50*, 58, Plate 5

nahani, Francolinus 128, 175
Nahan's (forest) francolin 128, **174**, *175–6*, Plate 63
Natal francolin 127, 129, 138, **151**, *155–6*, 160, 161, Plate 67
natalensis, Francolinus 127, 129, 154, 155, 159
necklaced hill-partridge 215, **216**, *216–18*, 218, Plate 105
Nelson's bobwhite, *see* northern bobwhite
nigra, Melanoperdix 190
nigrogularis, Colinus 60, 61, 64, 66, 68
nobilis, Francolinus 126, 129, 145
northern bobwhite *12–30*, 60, *61–4*, 66, 69, Plate 10
novaezeelandiae, Coturnix 192, 200

obscurus, Tetraophasis 100, 102
ocellated quail **90**, *92–3*
ocellatus, Cyrtonyx 90, 92, Plate 29
ochropectus, Francolinus 126, 129, 142
oculea, Caloperdix 235
Odontophorus 5, 6, **7**, 11, 41, 43, *71–86*, 88, 95
Ophrysia 3, 5, 6, **8**, 196, *214–15*
orange-necked hill-partridge 215, **229**, *229–30*, Plate 106
Orange River francolin, *see* acacia francolin
Oreortyx 5, 6, **7**, 11, 41, 48, *49–50*
orientalis, Arborophila 215, 216, *224–6*, 231

painted bush quail 208, *211–12*, **212**, Plate 102

painted francolin (or partridge) 126, 128, **130**, *132–3*, Plate 49
painted quail, *see* mountain quail
painted spurfowl 15, 247, *248–9*, Plate 126
pale-bellied francolin 126, 129, *142–3*, **145**, Plate 63
paramba partridge, *see* rufous-fronted wood-partridge
partridge, *see* grey partridge
pectoral quail, *see* stubble quail
Perdicula 3, 6, **8**, 11, 19, 96, *208–13*, 214
perdix, Perdix 3, 180, 184, 185
Perdix 3, 6, **8**, 96, 97, *180–5*
petrosus, Ptilopachus 241
pheasant-grouse, *see* Szecheny's monal-partridge
philbyi, Alextoris 111, 118
Philby's rock-partridge 111, **112**, **113**, *118–19*, Plate 43
Philortyx 5, 6, **7**, 11, 41, 43, *47–8*
pictus, Francolinus 126, 128, 131, 132, 134
pictus, Oreortyx 49, 129
pintadeanus, Francolinus 126, 129, 133
plains bobwhite, *see* northern bobwhite
plumed quail, *see* mountain quail
pondicerianus, Francolinus 128, 129, 176
Przevalski's rock partridge 111, **112**, **113**, **114**, Plate 40
psilolaemus, Francolinus 127, 128, 162, 165, 166, 169, 170
Ptilopachus 3, 5, 6, **8**, 9, 98, 197, *241–3*
puebla bobwhite, *see* northern bobwhite

rain quail, *see* black-breasted quail
red-billed francolin 15, 127, 129, **151**, *158–9*, Plate 69
red-billed hill-partridge 215, **229**, *231–2*, Plate 117
red-breasted hill-partridge 215, **216**, *219–21*, **220**, Plate 107
red-breasted tree-partridge, *see* Borneo hill-partridge
red-crested wood-partridge, *see* crested wood-partridge
red-legged partridge *10–13*, 25, 26, 29, 32, 111, **112**, **113**, *120–1*, Plate 45
red-necked francolin (or spurfowl) 15, 126, 128, *134–7*, **135**, **136**, 161, Plate 51
red spurfowl 15, **242**, *247–7*, Plate 125
red-throated francolin, *see* red-necked francolin
redwing francolin, *see* Levaillant's red-winged francolin
Rhizothera 5, 61, **8**, 9, 96, 97, *186–7*, 191
Rhynchortyx 5, 6, 41, *94–5*
Rickett's hill-partridge, *see* collared hill-partridge
ring-necked francolin 127, 129, **160**, *161–2*, Plate 71
rock bush-quail 208, **209**, *210–11*, Plate 101
rock partridge *10–12*, 17, 19, 25, 26, 29, 32, **112**, **113**, *114–15*, Plate 41
Rollulus 5, 6, **8**, 9, 39, 191, *239–40*
rouloul, Rollulus 239
roulroul, *see* crested wood-partridge
rubirostris, Arborophila 215, 231

rufa, Alextoris 30, 32, **33**, 111, 116, 118, 120
rufipectus, Arborophila 215, 216, 218, 225
rufogularis, Arborophila 215, 216, 221, 223
rufopictus, Francolinus 126, 129, 137, 138
rufous-breasted wood-quail **75**, **76**, 78, Plate 19
rufous-fronted wood-quail 71, *74–5*, **75**, **76**, Plate 16
rufous-throated hill-partridge 215, **218**, *221–3*, **222**, Plate 109
Rüppell's francolin, *see* acacia francolin
Rüppell's partridge, *see* Arabian red-legged partridge

Salle's quail, *see* Montezuma quail
Salvin's bobwhite, *see* northern bobwhite
sand partridge 122, **123**, *124–5*, Plate 47
San Lucas quail, *see* California quail
sanguiniceps, Haematortyx 237
Santa Catalina quail, *see* California quail
scaled quail 14, 17, 18–21, 26, 28, 48, 50, *51–3*, **52**, 53, 62, Plate 6
scaly-breasted hill-partridge 215, **232**, **233**, *232–5*, Plate 118
scaly francolin 127, 129, 144, *148–50*, Plate 60
schlegelii, Francolinus 127, 128, 172, 173, 175
Schlegel's banded francolin 15, 127, 128, **171**, *172–3*, Plate 80
Schuett's francolin, *see* scaly francolin
Sclater's quail, *see* spot-bellied bobwhite
see-see partridge 12, *122–25*, **123**, Plate 46
sephaena, Francolinus 127, 129, 160, 162
shelleyi, Francolinus 127, 128, 163, 164, 166, 167, 169
Shelley's francolin 16, 127, 128, **163**, **164**, *164–6*, 167, Plate 74
short-crested quail, *see* crested bobwhite
Sichuan hill-partridge 215, **216**, **218**, *218–19*, Plate 106
silver quail, *see* brown quail
Simon's partridge, *see* marbled wood-quail
singing quail 17, **80**, *87–9*, Plate 27
Snow Mountain quail 15, **200**, *206–7*, Plate 99
snow partridge *98–9*, **99**, Plate 33

Soderstrom's partridge, *see* rufous-breasted wood-quail
Sonnini's bobwhite, *see* crested bobwhite
spadicea, Galloperdix 247, 249
speciosus, Odontophorus 71, 77, 78, 85
spot-bellied bobwhite 60, **65**, **67**, *66–8*, Plate 12
spotted francolin, *see* crested francolin
spotted wood-quail 18, 71, **84**, *85–6*, Plate 25
spot-winged wood-quail 71, **72**, *73–4*, Plate 15
squamata, Callipepla 51
squamatus, Francolinus 127, 129, 148
starred wood-quail 71, **84**, *84–5*, Plate 25
stellatus, Odontophorus 71, 84
stone partridge **242**, *241–3*, Plate 122
streptophorus, Francolinus 127, 129, 161
stripe-faced wood-quail 71, **84**, *82–4*, Plate 24
strophium, Odontophorus 71, 79, 82
stubble quail 10, 12, 26, 192, **200**, *200–1*, Plate 95
Sumatran hill-partridge, *see* bare-throated hill-partridge
superciliosa, Ophrysia 214
swainsoni, Francolinus 126, 129, 137
Swainson's francolin (or spurfowl), 126, 128, **135**, *137–8*, 161, Plate 52
swamp francolin 15, 128, **176**, *178–80*, Plate 84
swamp partridge, *see* brown quail, swamp francolin
swamp quail, *see* brown quail
swierstrai, Francolinus 127, 129, 146
Swierstra's francolin 127, 129, **145**, *147*, Plate 63
szechenyi, Tetraophasis 100, 101
Szecheny's monal-partridge (or partridge-pheasant) 100 **101**, *101–2*, Plate 32
Szechwan hill-partridge, *see* Sichuan hill-partridge

Tarcarcuna wood-quail 71, **80**, *80–2*, Plate 22
Tadjoura francolin, *see* pale-bellied francolin
Taiwan hill-partridge, *see* white-throated hill-partridge
Tasmanian quail, *see* brown quail
tawny-faced quail 90, *94–5*, Plate 30
Tetraogallus 6, **8**, 12, 96, 98, 99, *103–10*
Tetraophasis 6, **8**, 96, *100–2*, **102**

Thayer's bobwhite, *see* northern bobwhite
thick-billed partridge, *see* marbled wood-quail
thoracica, Bambusicola 244, 245
thoracicus, Dactylortyx 87
Tibetan partridge 180, **181**, *184–5*, Plate 87
Tibetan snowcock 103, **104**, 105, *108–9*, Plate 37
tibetanus, Tetraogallus 103, 107, 108, 110
Tiburon Island quail, *see* Gambel's quail
torqueola, Arborophila 215, 216, 220, 223, 224

valley quail, *see* California quail
variegated partridge, *see* black-fronted wood-quail
Venezuelan wood-quail 71, **80**, *82–3*, Plate 23
Veraguan partridge, *see* rufous-fronted wood-quail
Verraux's monal-partridge (or pheasant-partridge) **100**, *100–1*, Plate 31
Virginian colin, *see* northern bobwhite
virginianus, Colinus 60, 61, 65

white-breasted bobwhite, *see* spot-bellied bobwhite
white-cheeked hill-partridge 215, **220**, **222**, *223–4*, Plate 110
white-eared hill-partridge 215, **220**, **222**, 225, Plate 112
white-eared quail, *see* crested bobwhite
white-faced bobwhite (or colin), *see* spot-bellied bobwhite
white-throated francolin 15, 127, 128, **171**, *173–4*, Plate 81
white-throated hill-partridge 215, **220**, **222**, *224–5*, Plate 111
white-throated wood-quail, *see* black-breasted wood-quail

yaqui quail, *see* elegant quail
yellow-billed francolin 127, 129, **151**, *152–3*, Plate 64
yellow-necked francolin (or spurfowl) 15, 126, 128, **135**, *139–41*, 254
ypsilophorus, Coturnix 192, 201, 206, 207